Power Apps Tips, Tricks, and Best Practices

A step-by-step practical guide to developing robust
Power Apps solutions

Andrea Pinillos

Tim Weinzapfel

‹packt›

Power Apps Tips, Tricks, and Best Practices

Associate Group Product Manager: Aaron Tanna

Publishing Product Manager: Uzma Sheerin

Book Project Manager: Prajakta Naik

Senior Content Development Editor: Rosal Colaco

Technical Editor: Vidhisha Patidar

Copy Editor: Safis Editing

Indexer: Rekha Nair

Production Designer: Nilesh Mohite

Business Development Executive: Saloni Garg

First published: November 2024

Production reference: 1200924

Published by Packt Publishing Ltd.

Grosvenor House

11 St Paul's Square

Birmingham

B3 1RB, UK

ISBN 978-1-83508-007-8

www.packtpub.com

To my family—Mom, Dad, Raf, and Angelo—thank you for your unwavering support, love, and belief in me. You have been my constant source of strength, and I couldn't have done this without you.

To Uday, my soon-to-be husband, your love, encouragement, and belief in my potential inspire me every day. Thank you for always pushing me to grow, dream bigger, and reach new heights.

- Andrea Pinillos

To my amazing family, Alli, Kyle, and Kelsey, who may not have shared my enthusiasm for the technical details of this book but supported me every step of the way. Your love, patience, and encouragement mean the world to me.

To Mac and Juna, my trusty canine companions, who spent endless nights and weekends at my side.

- Tim Weinzapfel

Contributors

About the authors

Andrea Pinillos is a dedicated Power Platform Software Engineer at Microsoft. She began her career as a full-stack developer before discovering the transformative potential of Power Apps in 2018. Her passion for low-code solutions quickly took root, leading her to become a fervent advocate and educator within the Power Platform community.

Throughout her journey, Andrea has facilitated App in a Day workshops, participated in hackathons, and spoken at numerous conferences. She also shares her expertise through her YouTube channel, where she educates users of all skill levels about the powerful capabilities of Power Apps. Currently, Andrea is transitioning into the technical program management space, where she writes technical specifications for Power Platform solutions, continuing to inspire and guide others in harnessing the full potential of this innovative technology.

Tim Weinzapfel is passionate about all things data, analytics, and process improvement. Although he spent over 20 years in a non-IT related career, he started dabbling early on in various technical areas including Microsoft Access, SQL, web development, and later IBM Cognos. In 2017 he was introduced to Microsoft Power BI and the Power Platform, was immediately hooked, and saw the vast potential these tools offered. From that moment on, his focus has been on all things related to the Power Platform and specializing in how these applications can integrate to build more comprehensive solutions. He has a passion for teaching others and has done numerous trainings including live, through YouTube, and presenting at various user groups.

About the reviewer

Alexandru Badiu, who works as the Head of Business Analytics, is a data expert with over 12 years of experience in BI and Project Management. He has held key roles in top multinational companies, specializing in BI strategy, data governance, visualization, and digital transformation using Microsoft solutions. Passionate about agile methodologies, he has successfully led multicultural teams on complex projects. As a recognized Super User in the Power BI/Fabric community, he is also a speaker and the author of the online course: Data Storytelling and User-Centered Design in Power BI. He is known for his contributions to innovation and education in the field of Business Intelligence.

Table of Contents

Part 1: Overall Project Planning

1

Understanding Requirements and Project Planning   3

2

Working with Solutions   17

3

Power Platform Environments   43

4

Choosing the Right Tool – Navigating Canvas Apps, Power Pages, and Model-Driven Apps   65

5

Data Connections   95

Part 2: Advanced Power App Techniques

6

Variables, Collections, and Data Filtering   133

7

Canvas App Formulas   161

8

Conditional Formatting and URL Deep Linking   181

Part 3: Power Platform and Other Integrations

9

Integration with Power Automate/Teams/Outlook   207

10

Integration with Power BI   249

11

Integrating Power Apps with SharePoint   281

12

Part 4: Governance, Security, and Deployment

13

14

15

Registering a Power App in Azure   381

Preface

The Microsoft Power Apps platform has become a key tool for modern businesses, offering a highly flexible and efficient way to develop custom applications. By simplifying the development process with its drag-and-drop interface and low-code/no-code capabilities, Power Apps allows users—regardless of technical expertise—to quickly build apps tailored to their business needs. Pre-built templates and seamless integration with Microsoft services and external data sources further enhance its utility, promoting a cohesive and optimized business environment. For companies seeking to innovate, streamline operations, and drive digital transformation, Power Apps serves as an essential tool that enables agility, cost-efficiency, and scalability.

Although Power Apps excels in providing an accessible low-code/no-code environment, fully unlocking its potential requires looking beyond these fundamentals. This guide will walk you through every phase of the app development lifecycle. From identifying initial concepts to applying advanced development techniques and maintaining your applications, we'll explore how to maximize the capabilities of Power Apps. No matter the project—whether building apps, websites, or new products—the first critical step is understanding the underlying business need, ensuring a strong foundation for success.

We have designed this book to focus on key areas of overall Power App development. We begin with key foundational aspects including project planning, using Power Platform environments and solutions, understanding canvas apps versus model-driven apps, as well as data sources.

We then move into more advanced development areas to help you build more complex apps. In addition, we cover integrating Power Apps with other applications such as Power Automate, Teams, Outlook, SharePoint, and Power BI. Finally, we touch down on governance and development strategies.

Power Apps provides you the ability to develop robust low code/no code applications. However, as you move beyond basic apps, there are many areas to need to be considered. Our goal is to strengthen your expertise to go beyond just the basics.

Who this book is for

This book is designed for developers who have some familiarity with building out applications within Power Apps and want to expand their knowledge across a variety of areas. If you are brand new to Power Apps, this book is not the best area to start with and there are other books out there that provide a more introductory use of Power Apps.

However, if you have begun using Power Apps and want to expand your knowledge then this book is for you. Building Power Apps goes well beyond the actual development. From initial project planning to using Power Platform environments and solutions all the way to overall governance, there is much to consider.

What this book covers

Chapter 1, Understanding Requirements and Project Planning, introduces the overall concept of planning your overall Power App development. Proper planning is an integral part of ensuring a successful outcome.

Chapter 2, Working with Solutions, discusses the concept of creating and using solutions within the Power Platform. Solutions are crucial for managing, packaging, and deploying changes across different environments.

Chapter 3, Power Platform Environments, explores Power Platform environments, how they are created, and how they can be effectively deployed. Understanding how environments are used is also important in overall app governance and security which are covered in *Chapter 13*.

Chapter 4, Choosing the Right Tool – Navigating Canvas Apps, Power Pages, and Model-Driven Apps, discusses the different types of apps within Power Apps including canvas apps, model-driven apps, and Power Pages. In this chapter, we explore the different uses of each type.

Chapter 5, Data Connections, explores different data connections that be used as you build out your Power Apps. We cover three of the most popular data connections including Excel, SharePoint, and Dataverse.

Chapter 6, Variables, Collections, and Data Filtering, discusses three fundamental areas within Power Apps. Variables and collections are used to temporarily store data and are essential in building more complex apps. Furthermore, we'll discuss data filtering to refine the data.

Chapter 7, Canvas App Formulas, continues discussing key functions within Power Apps. In this chapter, we cover important functions around form submission and app navigation. In addition, we cover how to connect a form to a Dataverse table.

Chapter 8, Conditional Formatting and URL Deep Linking, discusses two common capabilities that enrich user interaction. First, we cover how to use and apply conditional formatting within your app. We then cover URL deep linking which allows you to create hyperlinks taking users directly to your app. Both concepts can enhance the overall user experience and the user interface.

Chapter 9, Integration with Power Automate/Teams/Outlook, shows how Power Apps can be integrated with Power Automate, Teams, and Outlook to add additional functionality. We cover sending Outlook emails and calendar invites using Power Automate. In addition, we show how you can also use Power Automate to send Teams notifications using adaptive cards.

Chapter 10, Integration with Power BI, shows how you can integrate Power BI dashboards and reports into your Power Apps. In addition, we show how you can integrate a Power App into a Power BI report.

Chapter 11, Integrating Power Apps with SharePoint, shows how you can integrate Power BI with SharePoint. This includes embedding Power Apps into a SharePoint site, creating custom SharePoint list forms, and also covering the `SharePointIntegration` component.

Chapter 12, Integration with Power Virtual Agents/CoPilot, introduces Microsoft Copilot Studio and Copilot which allows you to integrate AI. Microsoft Copilot Studio allows you to create virtual chatbots while Copilot leverages AI to assist in app creation.

Chapter 13, Governance, Security, and Application Life Cycle Management, covers key areas important to the Power Apps area. We discuss how to establish a Power Platform environment strategy, discuss the use of service accounts, and how to set up data policies to support Data Loss Prevention (DLP). Finally, we will introduce the concept of Application Lifecycle Management (ALM).

Chapter 14, Error Handling, discusses approaches to handling errors within your app. This includes understanding an overview of error handling, using built-in functions in your app, using the `OnError` app property, and creating custom error messages.

Chapter 15, Registering a Power App in Azure, closes out the book by showing how you register an app in Azure. This includes understanding the importance of registering an app in Azure, how to register your app, and finally when to add the Azure app ID, client ID, and secret ID within Power Apps.

To get the most out of this book

The following software are needed for this book:

- Power Apps
- Power Automate
- SharePoint
- PowerBI
- Azure

Download the example code files

You can download the example code files for this book from GitHub at `https://github.com/PacktPublishing/Power-Apps-Tips-Tricks-and-Best-Practices`. If there's an update to the code, it will be updated in the GitHub repository.

We also have other code bundles from our rich catalog of books and videos available at `https://github.com/PacktPublishing/`. Check them out!

Conventions used

There are a number of text conventions used throughout this book.

`Code in text`: Indicates code words in text, database table names, folder names, filenames, file extensions, pathnames, dummy URLs, user input, and Twitter handles. Here is an example: "Locate the formula that defines the `Items` property of the copied `Gallery1`."

A block of code is set as follows:

```
If(IsBlank(CurrentVendor), BrowseGallery1.Selected, LookUp('Preferred
Vendor List', ID = CurrentVendor) )
```

When we wish to draw your attention to a particular part of a code block, the relevant lines or items are set in bold:

```
Param("VendorID")
Param("Screen")
```

Bold: Indicates a new term, an important word, or words that you see onscreen. For instance, words in menus or dialog boxes appear in **bold**. Here is an example: "Click on **Solutions** on the left navigation."

> **Tips or important notes**
> Appear like this.

Get in touch

Feedback from our readers is always welcome.

General feedback: If you have questions about any aspect of this book, email us at customercare@packtpub.com and mention the book title in the subject of your message.

Errata: Although we have taken every care to ensure the accuracy of our content, mistakes do happen. If you have found a mistake in this book, we would be grateful if you would report this to us. Please visit www.packtpub.com/support/errata and fill in the form.

Piracy: If you come across any illegal copies of our works in any form on the internet, we would be grateful if you would provide us with the location address or website name. Please contact us at copyright@packt.com with a link to the material.

If you are interested in becoming an author: If there is a topic that you have expertise in and you are interested in either writing or contributing to a book, please visit authors.packtpub.com.

Share Your Thoughts

Once you've read *Power Apps Tips, Tricks, and Best Practices*, we'd love to hear your thoughts! Scan the QR code below to go straight to the Amazon review page for this book and share your feedback.

https://packt.link/r/1-835-08007-3

Your review is important to us and the tech community and will help us make sure we're delivering excellent quality content.

Download a free PDF copy of this book

Thanks for purchasing this book!

Do you like to read on the go but are unable to carry your print books everywhere?

Is your eBook purchase not compatible with the device of your choice?

Don't worry, now with every Packt book you get a DRM-free PDF version of that book at no cost.

Read anywhere, any place, on any device. Search, copy, and paste code from your favorite technical books directly into your application.

The perks don't stop there, you can get exclusive access to discounts, newsletters, and great free content in your inbox daily

Follow these simple steps to get the benefits:

1. Scan the QR code or visit the link below

https://packt.link/free-ebook/B21327

2. Submit your proof of purchase
3. That's it! We'll send your free PDF and other benefits to your email directly

Part 1:
Overall Project Planning

In this part, you will get a comprehensive overview of the key topics essential for successful Power Apps development. You'll begin by understanding the critical importance of thorough planning and requirement gathering to ensure project success. You'll then learn about solutions within the Power Platform, which are important for managing, packaging, and deploying changes efficiently across different environments. This part also explores Power Platform environments, including how they are created and deployed, with a focus on their role in app governance and security. Additionally, you'll gain insights into the different types of Power Apps—canvas apps, model-driven apps, and Power Pages—and their specific uses. Finally, you'll delve into data connections, covering popular options such as Excel, SharePoint, and Dataverse, and how they can be effectively integrated into your Power Apps. These topics collectively provide a solid foundation for planning, building, and managing your Power Apps projects.

This part has the following chapters:

- *Chapter 1, Understanding Requirements and Project Planning*

- *Chapter 2, Working with Solutions*

- *Chapter 3, Power Platform Environments*

- *Chapter 4, Choosing the Right Tool – Navigating Canvas Apps, Power Pages, and Model-Driven Apps*

- *Chapter 5, Data Connections*

1

Understanding Requirements and Project Planning

In this chapter, we are going to begin covering critical aspects of developing a Power App. Understanding the client's requirements, defining the overall scope, and preparing a well-thought-out project plan are fundamental components of any project. Not only will it potentially save you time, and most likely rework, but it can also help set overall expectations with the end users.

In this chapter, we are going to cover the following topics:

- Understanding client requirements
- Importance of defining a project plan
- Building the foundation before painting the walls

By the end of this chapter, you will have a solid understanding of the importance of defining a project's scope and how it is integral to the success of the project.

Technical requirements

In order to follow all the subsequent chapters in this book, you will need to have access to Microsoft Power Apps. If you currently do not have access to Power Apps, such as access through your employer, you are still able to sign up for a Power Apps Developer plan. This will give you full access to the Power Platform, with some limited exceptions, without having to pay for your own tenancy. You can sign up for your free Developer Plan. In addition, please note that some areas covered in upcoming chapters may be a "premium" feature. We will identify them as they arise.

Obtaining a Developer Plan

If you currently do not have access to Power Apps, Microsoft offers a free Developer Plan that can be used for development and testing purposes. This also includes access to other applications, such as Power Automate, which we will use later in the book.

To obtain a free plan, simply go to `https://powerapps.microsoft.com/en-us/developerplan/` to sign up, as shown in *Figure 1.1*. Please be aware that personal email addresses (Gmail, Yahoo, Hotmail, etc.) are not allowed. A valid work or school email address is required.

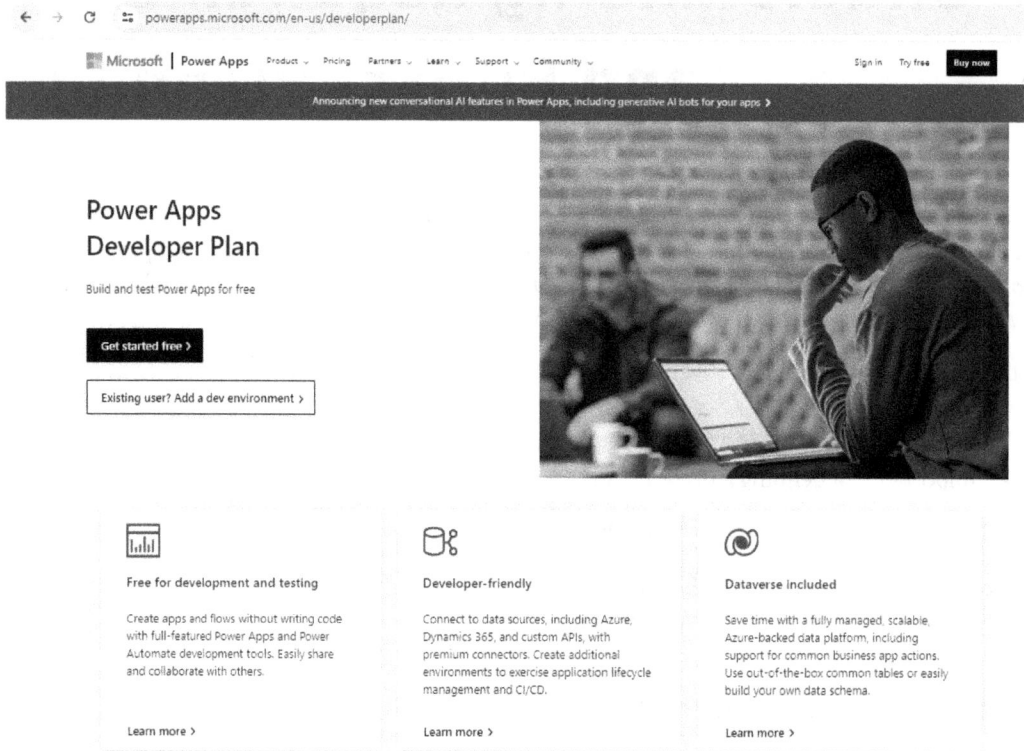

Figure 1.1 – Obtaining a Power Apps Developer Plan

As we explore the process of acquiring the Power Apps Developer Plan, it's essential to understand the capabilities of Microsoft Power Apps with respect to business application development. To begin with, let's first get a broad overview of Power Apps.

Power Apps overview

Microsoft Power Apps is a leading solution in the world of business application development, offering unparalleled flexibility and efficiency. This innovative platform empowers users, regardless of their technical expertise, to create custom applications tailored to their specific business needs. By using its intuitive drag-and-drop interface, its low code/no code concept, and robust pre-built templates, Power Apps simplifies the app development process, enabling rapid deployment and significant cost savings. Its seamless integration with other Microsoft services and a wide range of external data sources further enhances its functionality, ensuring a cohesive and interconnected business environment. The platform's emphasis on democratizing app development aligns with the growing need for agile, responsive, and customizable solutions in the dynamic business landscape. Power Apps thus emerges as an invaluable tool for businesses seeking to innovate, optimize workflows, and drive digital transformation efficiently and effectively.

Power Apps provides a low code/no code solution in which apps can be quickly developed with little to no coding. In fact, a great way to start learning how to use Power Apps is to build one using this approach and then reverse-engineer the app to understand the functionality. However, to fully tap into the power of Power Apps, you will need to go beyond the low code/no code concept, and throughout this book, we are aiming to do just that. We will cover different topics that arise throughout the overall development process, starting at the very beginning, when an overall concept is being identified, then looking at advanced techniques that will help you expand the app's capabilities, and ending with how to maintain your app. With many projects, whether you are building an app, creating a website, or even developing a new product, the first step is almost always understanding what the overall need is.

Understanding client requirements

In most cases, the underlying motivation to build a Power App is to satisfy some sort of need. Whether this is developing an app for a small team to update a simple list of data to an enterprise-wide application, every project is going to have one or more end users. As a result, clearly understanding the intended purpose, the audience, available data sources, and other factors will contribute to either a successful project or one that ends up sitting on a shelf collecting dust.

Building Power Apps can be a fun and exciting endeavor. It is not uncommon for anxious developers to jump in and begin building out the application prior to fully understanding client needs. It may also be that the developer just presumes they know what the client wants. Another situation may be that work begins on development, yet the developer does not fully understand the overall processes, requirements, or data. This section will cover all those key elements for understanding the client's requirements, including engaging with the client and defining the overall project scope. Engaging with the client is all about fully understanding the overall need. And, as that need is identified, the next step is to develop a clearly defined project scope. We cover both next.

Engaging with the client

One of the initial steps should be an engagement with the end user as early as possible. This discussion should go beyond just "tell me what you want" and should be a deep dive covering all the bases. Here are some recommended areas to consider.:

- **Business objectives**: What are the overarching goals the client aims to achieve with the Power App? This could include improving efficiency, enhancing customer engagement, or automating specific processes.

- **User needs and expectations**: Identify the needs and expectations of the end users who will interact with the Power App and what success looks like for them. Consider their technical skills, workflow, and how the app will solve their problems or improve their work processes. It is also good to inquire about how their needs relate to any company/organizational goals and vision so that future enhancements can be anticipated. Lastly, understand who the user population will be. Is this limited to a specific set of users, a department, or the entire organization?

- **Functional requirements**: Inquire about what functional capabilities the Power App must have. This may include data input and output, integrations with other systems, user interface features, and any specific functionalities unique to the client's operations.

- **Data management**: Understand how data will be handled within the app. This covers data collection, storage, retrieval, and security. Discuss the sources of data, data formats, and any compliance requirements for data handling.

- **Integration requirements**: Determine if the Power App needs to integrate with existing systems or platforms used by the client, such as CRM systems, databases, or other Microsoft 365/third-party applications.

- **Compliance and security requirements**: Discuss any legal, regulatory, or company policy compliance requirements. Also, understand the security needs of the app, including whether there is a need for user authentication or to secure any specific data. For example, will certain data only be viewable to certain users?

- **Scalability and future-proofing**: Consider how the app may be able to scale to accommodate future growth or changes in the client's business. Discuss plans for updates, expansions, or additional features that might be required down the line. Scope creep is a common occurrence. One example is if you are building an app for a specific department in your organization and additional organizations will want to utilize the app in the future, having this information up front may allow you to plan appropriately.

- **User interface and experience**: Discuss the design and usability aspects of the Power App. This includes layout, navigation, accessibility, and overall user experience. Make sure to consider any accessibility concerns that users with vision, hearing, or other impairments may have. Also, discuss how users will access the app itself. Will it be embedded within a Microsoft Teams or a SharePoint site? Will users be provided a hyperlink for access?

- **Overall app deployment and maintenance**: Discuss how the app will be deployed as well as maintained. Will the app be part of an overall solution? It will also be necessary to determine what environment(s) will be used. We cover solutions and environments in *Chapter 2* and *Chapter 3* respectively. Will other components be integrated, such as Power Automate flows? Understanding these technical details will help in the overall planning phase.

- **Reporting and analytics**: Identify any reporting capabilities or analytics the client may want. This could involve tracking user activity, generating reports, or providing insights based on app usage. We will cover embedding Power BI reports in *Chapter 10*.

- **Budget and time constraints**: Clarify if there are any budgetary limits and the timeline for the development and deployment of the Power App. This helps in aligning the project scope with the client's financial and scheduling expectations.

- **Training and support needs**: Determine if there is a requirement for user training and ongoing support after the app is deployed. This includes preparing user manuals, help guides, or providing technical support.

- **Feedback and iteration process**: Establish a process for ongoing client feedback and iterations. Power Apps development often involves agile, iterative development, so setting expectations for feedback loops and updates is important.

> **Important Note**
>
> An adage in buying real estate is *"Location, Location, Location"*. In developing Power Apps, this should be *"Document, Document, Document"*. In other words, maintaining good documentation throughout the entire Power App lifecycle is critically important. And there is no better place to start than the earliest engagements with end users. We also recommend a common location for documentation that other team members/collaborators will have access to.

Defining project scope

After gathering client requirements, defining the project scope is perhaps one of the most important aspects of Power Apps development. It lays the foundation for the intended purpose and key requirements of the project. It also sets initial expectations for the client. Some aspects of the project scope include the following:

- **Objective**: What is the overall intended purpose of the Power App? The *Understanding client requirements* section, and specifically knowing the overall business objectives, helps drive this.

- **Deliverables**: Detail what the project will deliver upon completion. In addition to the Power App, the project will have other components as well, such as the underlying data sources, connected Power Automate Flows, technical documentation, and so on.

- **Project boundaries**: Define what is included and, just as importantly, what is excluded from the project. This helps manage client expectations and prevent or minimize scope creep.

- **Milestones and timeline**: Outline the major milestones and the timeline for the project. This includes start and end dates and key checkpoints along the way. Large projects may require executive/steering committee meetings to track overall progress and identify any risks and uncertainties that may arise.

- **Requirements**: Specify the project requirements, including technical, user, and functional requirements. This should be based on a thorough understanding of client needs.

- **Data sources and other technical requirements**: Clearly define where the data will reside. Do you have access to the data, or will you need to collaborate with others? Are there any other technical requirements (such as critical data security requirements that you may need to address)?

- **Constraints and assumptions**: Identify any limitations (such as budget, resources, and regulations) and assumptions (such as the availability of technology or resources) that may impact project execution.

- **Risk management**: Highlight potential risks and their mitigation strategies. This includes risks related to scope, resources, timeline, and external factors.

- **Stakeholder analysis**: Identify all stakeholders involved in the project, their roles, and how they will be engaged. This includes clients, team members, suppliers, and others who have an interest in the project. A **RACI chart**, also known as a **responsibility matrix**, is a common way to track roles and responsibilities for all team members. RACI is an acronym for four key areas:

  - **Responsible**: Who is responsible for the project and/or task? This role is the decision maker.

  - **Accountable**: Who has accountability for the project or task?

  - **Consulted**: Who are the subject matter experts that will be consulted with for information?

  - **Informed**: Who needs to receive communication and updates on the project or task?

- **Resource allocation**: Detail the resources (human, financial, and technological) required for the project and how they will be allocated.

- **Quality management**: Define the quality standards and metrics that the project deliverables must meet.

- **Change management process**: Outline a process for handling changes in scope, including who has the authority to approve changes and how they will be documented and communicated.

- **Acceptance criteria**: Establish the criteria for accepting deliverables, including who will sign off on them and how they will be tested or reviewed.

- **Documentation and reporting**: Determine what documentation will be needed (project plans, progress reports) and how the information will be communicated to stakeholders.

You have met the end user and have a much better understanding of their overall objective. In addition, you have clearly articulated the overall scope of work. Finally, you have even started to organize and document your notes. It's time to jump in and start developing the Power App, correct? Not so fast. Just like heading off on a long-awaited road trip, an important first step is to plan your route. This also holds true for Power App development. Sure, it is possible to jump into the car and start driving, and you may still end up at your destination. However, if your goal is to arrive as efficiently as possible and without any unnecessary stops, a plan is a must. The next section looks at this area.

Importance of project planning

As the old saying goes, *"If you fail to plan, you plan to fail."* Proper planning is critical to overall success as it helps define the overall project. This includes areas such as the overall scope with clearly defined requirements, identifying the necessary resources and defining roles and responsibilities, determining a timeline, as well as identifying any potential risks or challenges.

This book is not intended to cover in-depth aspects of project management. However, there are several common areas of project management that we believe are important for developers to be aware of. With many projects, there are different roles, project management methodologies, and tools. We will touch on different aspects of each of these.

Project management roles

In a project management process, there are many different roles that might be called upon. Depending on the size and scope of your effort, it is important to consider which of these roles might be beneficial:

- **Project manager**: The primary leader of the project, responsible for planning, executing, and closing projects. They oversee all aspects of a project, manage the team, and ensure that project goals are met.

- **Team members**: Individuals who work on various tasks within the project. Their expertise and efforts are critical to the completion of project tasks. This may include a Power App developer, **database administrator** (**DBA**), systems administrator, and others who will be directly involved with the overall development.

- **Stakeholders**: These are people or groups who are interested in the project's outcome. They can include clients, customers, sponsors, and others who are impacted by the project's results.

- **Quality assurance**: This role ensures that the project meets the required quality standards. They are involved in testing and ensuring that the app performs as it should.

- **Subject Matter Experts** (**SMEs**): Individuals who provide specialized knowledge or expertise in a specific area relevant to the project.

- **Business analysts**: Professionals who analyze the business needs and ensure that the project plan meets these needs.

- **Sponsors**: Typically, these are the people or the organization that provides the financial resources for the project. They play a critical role in project initiation and can also influence key decisions.

- **Project steering committee**: A group of high-level stakeholders who provide strategic direction, ensure the project is aligned with the organization's goals, and make decisions on key project issues.

- **End users**: The final recipients of the project's outcome. Their feedback can be crucial in the project's planning and development stages.

- **Project coordinators/administrators**: They assist the project manager in administrative tasks such as scheduling meetings, updating project documents, and communication.

Project management methodologies

Along with differing roles that might be involved, there are also different project management methodologies. While there are many different approaches to project management, some examples of methodologies in software development are Waterfall, Agile, Scrum, and Lean. The following table provides a broad overview:

Waterfall	Agile	Scrum	Lean
1. Project done in a linear approach 2. Clearly defined sequence 3. Each phase is completed before the next phase begins 4. Tasks may have dependencies on each other	1. Project work is adaptive 2. Focus is on iteration, collaboration, continuous testing, and feedback 3. Open to change as development progresses	1. Follows the Agile framework 2. Short "sprints"(usually 2–4 weeks) are used to manage projects 3. Usually involve short regular/daily meetings	1. Uses cross-functional teams to work in short iterations 2. Focuses on eliminating non-value-added steps from processes

Table 1.1 – Sample Project Management Methodologies

This is only a short summary of different approaches and is not meant to be all-inclusive. Experienced project managers are a key asset in determining how best to approach your overall project.

Project management tools

In addition to different roles and methodologies, there are a number of applications that can manage the overall project. And, similar to the section on project management methodologies, our intention here is to provide awareness of some commonly used project management tools. These tools can be very helpful for tracking tasks, timelines, responsible individuals, and so on.

Microsoft Project

Microsoft Project is Microsoft's overall suite of project management tools. There are both cloud-based and on-premises solutions. Microsoft Project (desktop) is an on-premises solution that is installed directly on your computer. In addition, there are two cloud-based solutions, Project for the web and Project Online.

Project for the web

Project for the web is built on the Microsoft Power Platform. It is easy to begin using and simple to use and can also integrate with other Power Platform applications, including Power Apps, Power BI, and Power Automate. It is also good for simple and small-scale projects.

Project Online

Project Online is built on Microsoft SharePoint. It is better for more complex projects.

Azure DevOps

Separate from Microsoft Project, **Azure DevOps** is another application that can be used for both project management and overall application life cycle management. DevOps allows you to address various project management approaches (Agile, Scrum, etc.), create sprints, and create, track, and assign work items.

Other project management tools

While Microsoft Project and Azure DevOps offer project management tracking, there are other commonly used applications. These include the following:

- Jira
- Trello
- Smartsheet
- Asana
- monday.com
- GitHub Projects

And, although we are providing a list of commonly used applications, the intent is not to provide an overall recommendation but rather awareness. Each tool will have its own strengths and weaknesses depending on the size of the project, the team, the overall approach, and other factors.

> **Interested in learning more?**
>
> There are many books available on Packt's website that cover project management and some of the applications listed previously. Visit `https://www.packtpub.com/` for more details.

So far, we have covered important steps to consider when planning the overall project. First is understanding overall client requirements. Next is developing an overall project plan and considering utilizing a management tool to track tasks. Finally, there are various areas to plan for that may impact your overall development. These are covered next.

Establishing a Power Apps Foundation

The importance of this concept is probably best described by the fable, "The Three Little Pigs." This story features three pigs that each build a house to protect themselves from the big bad wolf. The first pig uses straw, the second uses wood, and the final pig builds the house out of bricks. Although the wolf can blow down the first two houses, it is unable to destroy the brick house. The moral of the story is that proper planning prevents catastrophe.

This also applies when developing your Power App. As noted previously, it can be very tempting to immediately jump in and start building out screens, galleries, and forms. Doing so, however, can greatly increase the risk of significant rework down the road.

We use the concept of building a foundation as a way to think about the importance of planning ahead. When developing a Power App, there are areas to consider during the process beyond just adding screens, galleries, and forms. Accounting for these areas early in the process may not only save you rework but also enable you to build out a much more robust and scalable app. In this section, we are going to cover key areas that we believe are the foundation for a successful project. These include data sources, planning the overall layout, understanding key security requirements, integrating with other applications, and finally where the Power App and any associated components will reside. We can't overemphasize that proper planning will not only greatly reduce the chance for rework, but will provide a greater chance for building a more efficient and most likely scalable project.

Data sources

Understanding where the data will come from is critical. A simple app may just be connecting to a single data source, such as a SharePoint list. However, more complex apps may include multiple connections, different data sources, and so on. There are some questions to consider in relation to data sources.

Will end users have access to the data sources? It is important to remember that when developing your app, you as the developer may have access but if an end user does not, this could prevent the app from working for them. A simple example of this is when using a SharePoint list as your data source. If someone accessing the published app doesn't have the necessary permissions to access the SharePoint list, they would be unable to access it through the app. Thus, it is highly recommended to evaluate where your data is coming from.

Do you understand or have a clear picture of the overall data model? It is also possible that the model does not exist and will be created as part of building out the app (or perhaps just some new tables). Planning this in advance will greatly help as you consider the layout of your app.

For the purposes of this book, we are going to be using an example of creating an event planning Power App. This is going to require a new dataset to be created and will include tables for areas such as events, venues, attendees, and vendors. It is best to draw a diagram of the tables we anticipate needing and how they all relate to each other.

Here is an example of a conceptual data model. This image is not meant to be all-inclusive but to serve as an example.

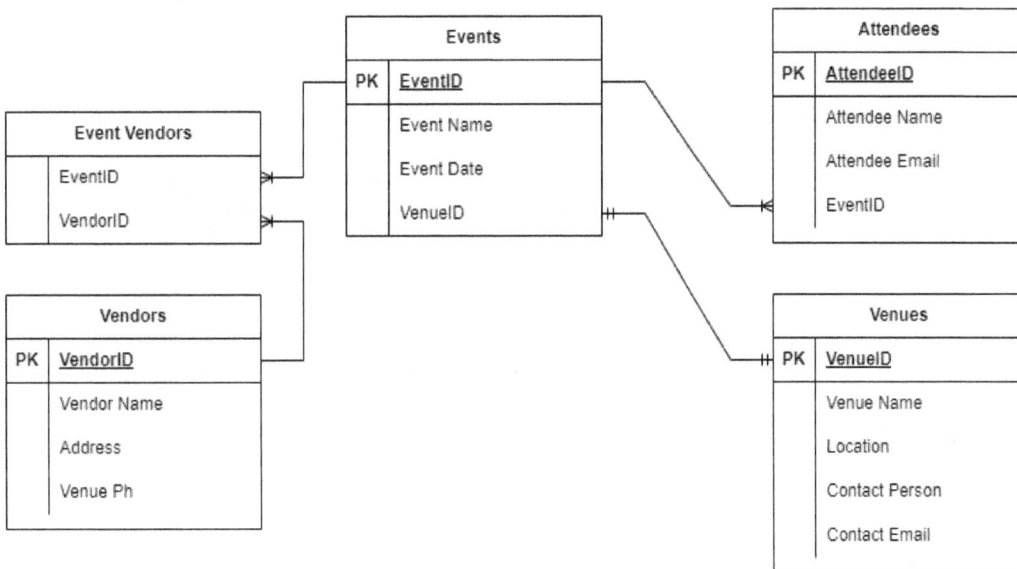

Figure 1.2 – Example data model

Overall Power App layout

Effective app development requires prior planning, particularly in the initial stages. A common pitfall is diving into development with a focus on individual screens, only to later realize the need for additional features such as a uniform background, a cohesive header, and standard icons across pages. This approach, while common, leads to unnecessary work or even revisions. For example, developing a single, well-planned template that incorporates these elements can be replicated across different screens, streamlining the development process. By investing a little more time in planning at the outset, significant time and effort can be saved later. Taking just a bit of extra planning time at the beginning can save significant time later. Trust us on this one!

Security requirements

Will your Power App have any security requirements? This might include restricting what data users are able to see, displaying only certain fields to specific users, and/or displaying certain areas of the Power App (such as an Admin screen).

A good example of this is having a section of the Power App devoted to administrators only. This can be achieved simply by showing or hiding an administration section based on the user's login. What is important is being aware of this upfront so that as you are developing the Power App, this functionality can be built in from the beginning rather than having to update or perform rework on already built sections.

We will cover additional security requirements in *Chapter 13*.

Integration with other applications

This is another big area to consider. Power Apps, as a standalone application, provides great functionality. However, you may find situations where integrating your app with other applications will expand your overall solution.

Here are some examples. First and foremost is integration with Power Automate. Power Apps and Power Automate go together like chocolate and vanilla ice cream. In fact, one of the most common integrations is between these two applications as Power Automate facilitates automation.

However, there are other applications that Power Apps integrates nicely with. SharePoint and Power BI are two examples. Thus, it is important to plan for them. We will cover integration with other Power Platform and other applications in *Part 3*.

Where will the Power App reside?

Lastly, it is important to understand where the Power App will reside, both in terms of the development process but also when published. What Power Platform environment will be used? Do you intend to have separate environments for developing, testing, and production? Do you anticipate having other components, such as Power Automate flows and/or environment variables? Related to this, will the Power App be included in a solution?

We will be covering solutions in *Chapter 2* and environments in *Chapter 3*.

Summary

In this chapter, we covered several critical steps that should be considered when developing your Power App. First, we described the importance of understanding client requirements. This included engaging with the client to discuss the overall objective, key requirements, and other areas to consider. In addition, we provided a number of areas to consider when developing the overall project scope.

Next, we explored key steps in developing a project plan. We provided various roles that can be useful through the project lifecycle, such as project manager, team members, and end users, as well as other useful roles. We looked at different project management methodologies to be considered, as well as applications that can help support the process.

Finally, we covered important foundational topics that should be considered for the development process. Areas such as understanding the data sources, the overall layout, any security requirements, integrating with other applications, and where the Power App will reside.

In the next chapter, we are going to start understanding solutions and how they play a crucial role in the overall process.

2

Working with Solutions

Establishing solid foundations through adherence to best practices is paramount right from the start. In this chapter, we delve into the essential steps of creating a solution and a publisher, providing a comprehensive guide on incorporating various components into your solution. Emphasizing the critical role of creating a solution in every project, we underscore its significance as a key factor during project deployment to test our production environments. Solutions are crucial for managing, packaging, and deploying changes across different environments. They serve as containers that hold various components such as apps, flows, and customizations.

In this chapter, we will learn about the following topics:

- Strategic decision-making for solutions and environments

- The crucial role of solutions in Power Apps

- Step-by-step guide – solutions, publishers, and versions

In this journey, you will gain a deep understanding of how solutions serve as the cornerstone of efficient Power Apps development, ensuring seamless deployment, version control, and management of applications across different environments. By the end of this chapter, you will be equipped with the knowledge and skills needed to effectively leverage solutions to streamline your Power Apps development workflow, enhancing productivity and project success.

Technical requirements

- Microsoft Power Apps environment

- Access to Power Apps Maker Portal

- Basic understanding of Power Apps components and environments

Strategic decision-making for solutions and environments

Before diving into the technical details, it is crucial to make strategic decisions regarding environments and security. These decisions lay the groundwork for effective use of solutions.

Environments

Environments in Power Apps are containers that isolate data, apps, and flows, crucial for **Application Lifecycle Management** (**ALM**). They ensure controlled development, testing, and deployment processes.

Here are the types of environments:

- **Default environment**: This is the initial environment created for each tenant. It's primarily intended for personal productivity and should not be used for production apps due to its broad access permissions.

- **Sandbox environment**: These are non-production environments, ideal for development and testing. They allow for experimentation without affecting live data or applications.

- **Production environment**: These environments are used for live apps and data. They have higher stability and security measures to ensure reliability and integrity.

- **Trial environment**: Trial environments are temporary environments for exploring features and capabilities. They are suitable for short-term evaluations and typically expire after a set period.

- **Developer environment**: Developer environments are personal environments for individual developers. They are ideal for personal development, learning, and prototyping.

Example scenario – sandbox environment for event planning

Let's start with the context. You are tasked with developing an event planning solution for a corporate events management team using Power Apps. Before diving into development, you need to set up a sandbox environment to ensure isolated testing and development without affecting production data or applications.

> **Why choose a sandbox environment?**
>
> A sandbox environment is chosen for developing this solution to first create a **Proof of Concept** (**PoC**). This approach allows for testing various components and functionalities in a controlled setting before deploying them to a live environment. Using a sandbox environment helps identify and resolve potential issues early, ensuring a smoother transition to production.

Here are the steps to set up a sandbox environment:

1. **Navigate to the Power Apps admin center**: Log in to the Power Apps admin center using your administrator credential using this link: `https://admin.powerplatform.microsoft.com/`.

2. **Access environments**: In the left-hand menu, click on **Environments** to view existing environments or create a new one.

3. **Create a new sandbox environment**:

    I.   Click on **New environment** to start creating a new environment.

    II.   Enter the following details:

    - **Environment name**: `Event Planning Sandbox`

    - **Environment type**: Select **Sandbox**

    - **Description**: Briefly describe the purpose of this environment, such as `Development and testing for event planning solution`

    - **Region**: Choose the appropriate geographic region based on your organization's location

    - **Environment URL**: Automatically generated based on the environment name

    - **Capacity**: Configure capacity based on expected usage and resource requirements

4. **Validate and create**:

    I.   Review all entered details to ensure accuracy.

    II.   Click **Create** to initialize the sandbox environment.

5. **Environment setup complete**:

    I.   Once created, the new sandbox environment will appear in the list of environments in the top-right corner of the Maker Portal.

    II.   Click on the environment name to access its details and further configure settings if necessary.

Setting up a sandbox environment for the event planning solution provides a controlled environment for development and testing purposes. This ensures that live data and applications remain unaffected during the process. By following these steps, you establish a foundation for building and refining the solution without impacting production data or applications, ensuring a smooth transition to deployment when ready.

Each environment type serves a specific purpose, and understanding when and why to use each type is critical for effective application management.

Environment types matrix

The following table shows the environment types matrix:

Environment Type	Purpose	Usage	Access Control	Expiration
Default	Initial environment for each tenant	Personal productivity	Broad access	No
Sandbox	Non-production for development/testing	Experimentation	Controlled, dev/testing teams	No
Production	Live apps and data	Operational reliability	Restricted	No
Trial	Temporary for exploring features	Short-term evaluations	Controlled	Yes, typically 30–90 days
Developer	Personal for individual developers	Development, learning, and prototyping	Individual developer	No

Table 2.1 – Environment types matrix

Security and governance

In the context of Power Apps, ensuring robust security and governance practices is crucial for maintaining the integrity and confidentiality of your applications and data. This section addresses key elements that contribute to a secure and well-governed environment.

- **Security roles and permissions**: Define security roles and ensure that proper permissions are set to control access and actions within Power Apps.

- **Data Loss Prevention (DLP) policies**: DLP policies are critical for securing data. Define and enforce DLP policies across environments to prevent unauthorized data access and sharing.

- **Service principals and accounts:** Use service principals or service accounts for better security and management, ensuring that automated processes have the necessary permissions without exposing user credentials.

For more details on security solutions and managing permission, refer to *Chapter 13* on governance and security, where we explore the necessary measures to protect data and ensure process access control throughout the development and deployment lifecycle.

The crucial role of solutions in Power Apps

In Power Apps, solutions are vital for organizing, managing, and deploying your apps, flows, and customizations. They act as containers that streamline the development process and ensure that changes can be efficiently moved across different environments, such as development, testing, and production. Understanding how to use solutions effectively is key to maintaining a well-organized and scalable Power Apps environment.

Creating publishers

To effectively manage and track your solutions, it's important to define publishers. Publishers help identify the origin of the components within a solution and maintain consistency across different environments. Here is a closer look at the key aspects:

- **Publishers**: Before creating a solution, define a publisher. This helps in identifying the origin of the components and maintaining consistency. A well-defined publisher enhances organization and clarity within your development and deployment processes.

- **Naming conventions**: Establish clear naming conventions for solutions to ensure consistency and clarity. Proper naming helps in identifying and managing solutions effectively, reducing confusion and ensuring that components are easily recognizable and manageable.

Example scenario – creating an Event Planning Engineers publisher

Let's walk through how to create a publisher and define naming conventions specifically tailored for event planning solutions. This example will illustrate the practical steps involved in setting up a publisher and applying naming conventions to organize and manage your solutions effectively.

Creating a publisher

To start, follow these steps to create a publisher in the Power Apps Maker Portal:

1. **Access the Power Apps Maker Portal**: Go to Power Apps Maker Portal and log in with your credentials.

2. **Select the environment**: Choose the environment where you want to manage your event planning solutions, such as Development or Production. In this case, choose the **Event Planning Sandbox** environment.

3. **Navigate to Publishers**: In the left-hand menu, click on **Solutions** and then select **Publishers**.

4. **Create a new publisher**: Click on **New publisher** to begin creating a new publisher.

5. Enter the following details:

   - **Display name**: `Event Planning Engineers`
   - **Name**: `EventPlanningEngineers`
   - **Description**: Optionally, describe the purpose, such as `Engineers dedicated to building the event planning application`.
   - **Prefix**: `epe`
   - **Choice value prefix**: `11111`

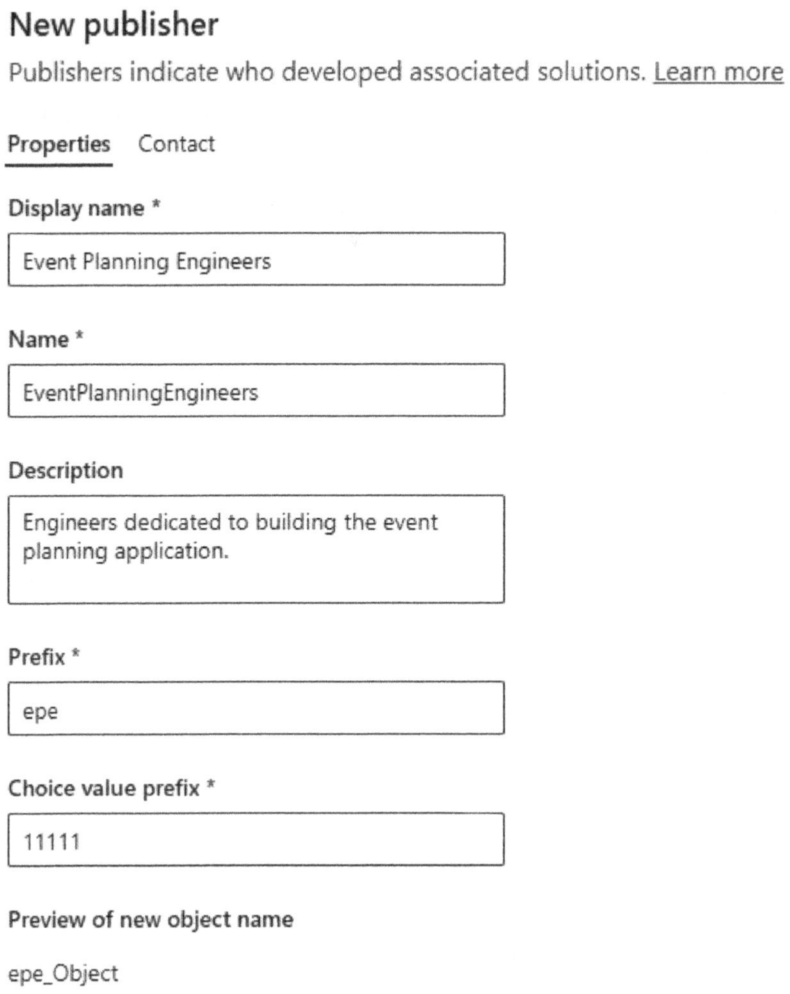

New publisher

Publishers indicate who developed associated solutions. Learn more

Properties   Contact

Display name *

Event Planning Engineers

Name *

EventPlanningEngineers

Description

Engineers dedicated to building the event
planning application.

Prefix *

epe

Choice value prefix *

11111

Preview of new object name

epe_Object

Figure 2.1 – Publisher form example

6. **Save the publisher**: Click **Save** to finalize and create the publisher.

7. **Verify the publisher**: Ensure that the new publisher, **Event Planning Engineers**, is successfully created and listed under publishers for the chosen environment.

Naming conventions

Establishing consistent naming conventions is crucial for maintaining clarity and organization across your event planning solutions. By following a standardized approach to naming, you ensure that all components are easily identifiable, manageable, and aligned with your organizational practices. Here's how you can define naming conventions for **Event Planning Engineers** (**EPE**):

Example scenario – naming conventions

1. **Prefix**: Use `EPE-` to identify solutions developed by Event Planning Engineers.

2. **Descriptive names**: Choose descriptive names indicating the purpose of each solution, such as `EPE-EventManagement-v1.0` for the initial version of an event management solution.

3. **Versioning**: Incorporate version numbers (e.g., v1.0, v2.0, etc.) to distinguish different iterations of solutions.

4. **Documentation**: Document these conventions to maintain consistency and facilitate future management.

By following these steps, EPE ensures that its event planning solutions are well-organized and easily identifiable within Power Apps. Clear naming conventions and a proper publisher setup contribute to streamlined management and improved clarity throughout the development and deployment processes.

Defining and creating solutions

A solution is a container for apps, flows, and customizations in Power Apps. It helps in managing, packaging, and deploying changes efficiently.

Managed versus unmanaged solutions

Understanding the difference between managed and unmanaged solutions is essential for effective solution management. Here's a breakdown of each type:

- **Managed solutions**: These are intended for deployment to production environments. They are locked to prevent modifications, ensuring consistency and stability.

- **Unmanaged solutions**: These are used in development environments where changes are still being made. They allow for full customization and modifications.

Example scenario – unmanaged event planning solution

By default, when you create a solution in a development environment, it will be unmanaged. The choice between managed and unmanaged solutions is made during the export process.

1.   Navigate to the Power Apps Maker Portal.

2.   Click on **Solutions** in the left-hand menu.

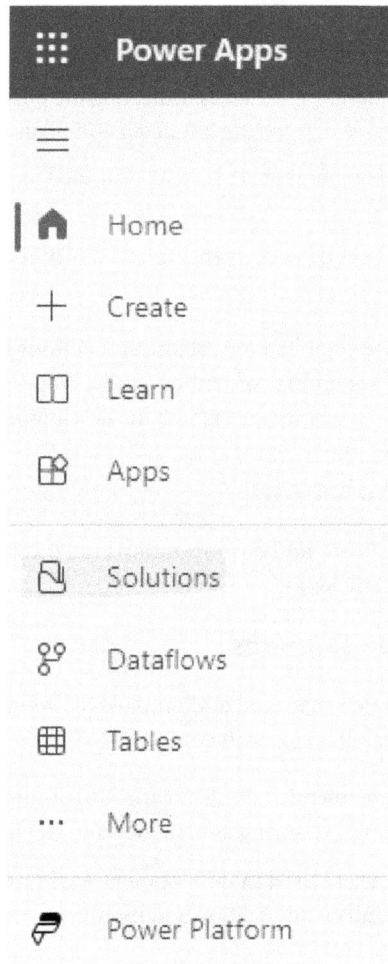

Figure 2.2 – Left navigation with Solutions highlighted

3.   Click on **New Solution**.

4.   Enter Event Planning Project as the name of the solution.

5.   **Unmanaged** will be selected as the default package type.

6. Choose **Event Planning Engineers** as the publisher.

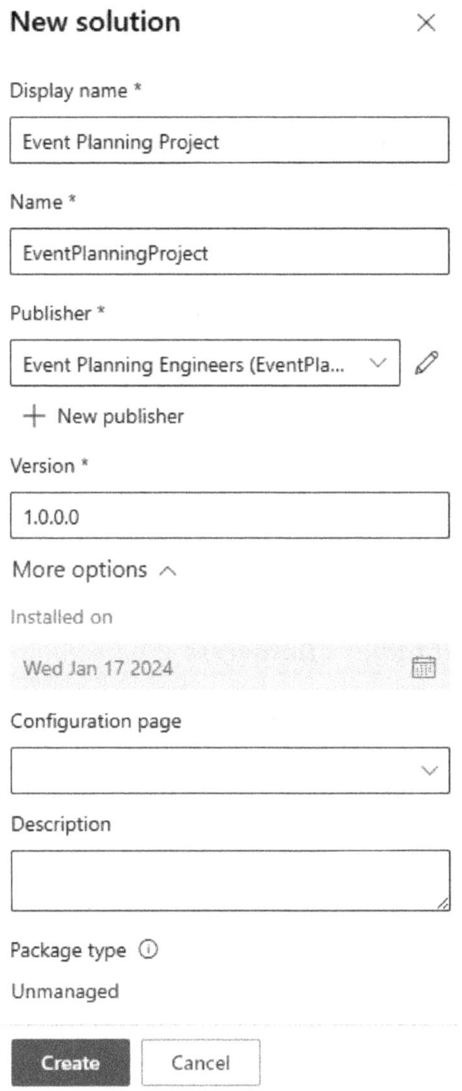

New solution   ✕

Display name *

Event Planning Project

Name *

EventPlanningProject

Publisher *

Event Planning Engineers (EventPla...   ⌄   ✎

+ New publisher

Version *

1.0.0.0

More options ⌃

Installed on

Wed Jan 17 2024   📅

Configuration page

⌄

Description

Package type ⓘ

Unmanaged

[Create]   [Cancel]

Figure 2.3 – Example of the New solution form with the fields filled in

7. Click **Create** to initialize the solution.

So far, we have created a solution and a publisher, establishing the structure, which is currently empty. Next, we will focus on building our apps and showcase how we can add data. This will allow us to develop the event planning solution further and illustrate how various components are integrated.

Dataverse and ALM

To fully leverage Power Apps, it's crucial to understand how data is managed and how applications are developed and maintained. This includes grasping the role of Dataverse in data storage and the principles of ALM for managing app development and deployment:

- **Dataverse overview**: Dataverse is the backbone for data storage and management in Power Apps. It provides a scalable, secure, and versatile platform for storing application data. Dataverse and ALM are interconnected because Dataverse's robust data management capabilities provide a reliable foundation for managing the entire lifecycle of solutions. Dataverse ensures consistent data handling, security, and integrity, which is essential for the smooth deployment and operation of solutions as they transition through various environments, from development to testing and production. This integration allows for a seamless data flow and helps maintain the quality and stability of applications throughout their lifecycle.

- **Application Lifecyle Management (ALM)**: ALM is important in managing the lifecycle of apps from development to production. Effective ALM practices ensure that changes are systematically tested, documented, and deployed. ALM is closely tied to environments such as DEV, TEST, and PROD, as well as to solutions, which are packaged artifacts that move between these environments. This alignment ensures that applications are developed, tested, and released in a controlled manner, maintaining their quality and performance across different stages of the lifecycle.

Example scenario – event table in Dataverse

In our event planning scenario, Dataverse serves as the centralized data storage and management platform within Power Apps. It provides a structured and scalable environment to store, organize, and manage all data related to our event planning application.

1. **Create the event table**: Navigate to the solution named **Event Planning Project** within Power Apps.

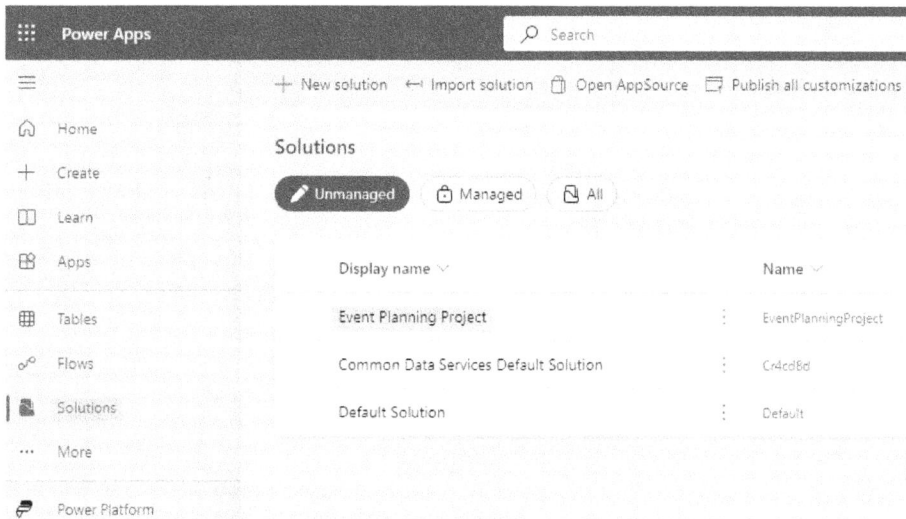

Figure 2.4 – Displaying the Event Planning Project solution

2. Create a new table called Events to store information about various events managed by our application.

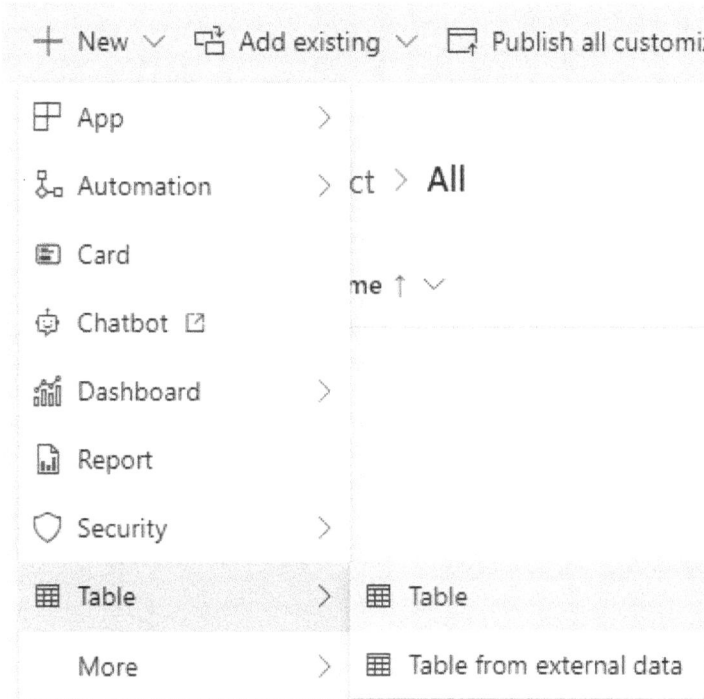

Figure 2.5 – Table button within the New tab

3. **Define table columns**: Add relevant columns to the Events table to capture essential details such as the following:

   - **Event name**: Text field to store the name of the event
   - **Event start date**: Date field to record the start date of the event
   - **Event end date:** Date field to record the end date of the event
   - **Description**: Memo field to include a detailed description of the event
   - **Organizer:** Text field to identify the person or team responsible for organizing the event
   - **Attendee count**: Number field to track the number of attendees registered for the event

- **Event status**: Choice field (e.g., **Draft**, **Confirmed**, or **Canceled**) to indicate the current status of the event

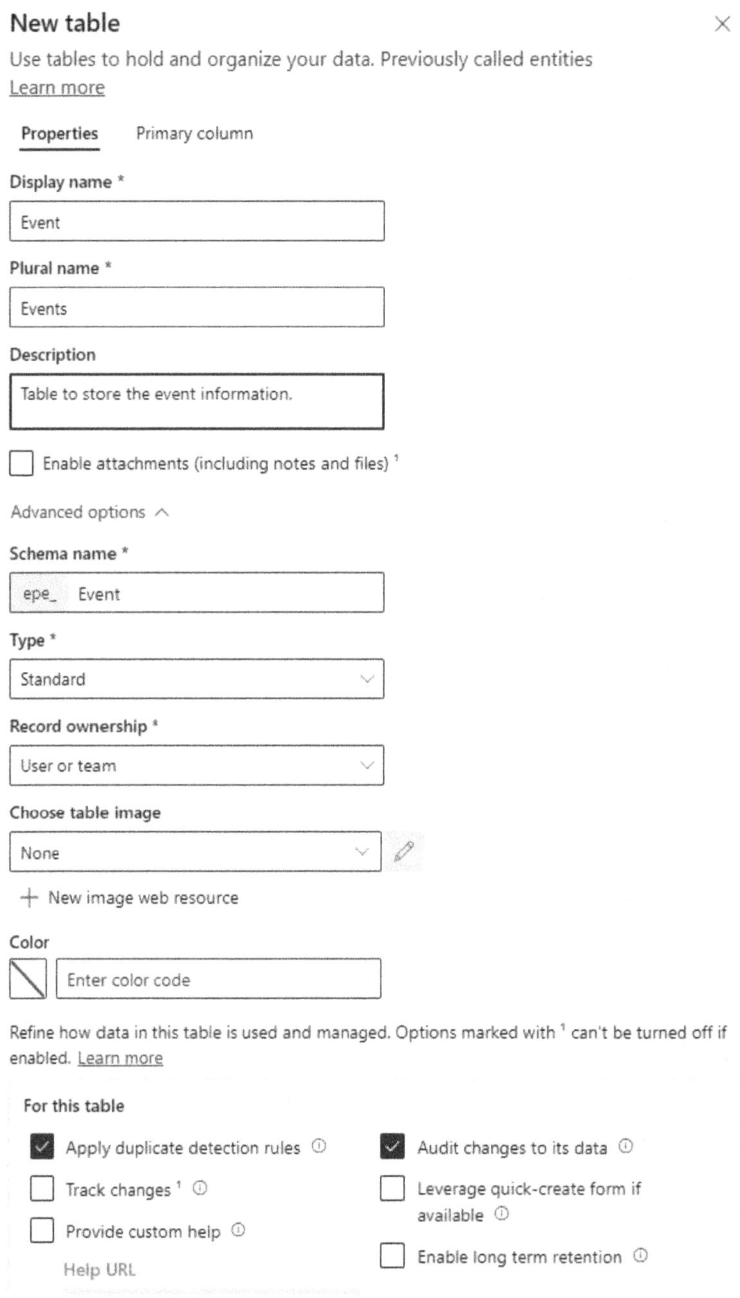

New table   ×

Use tables to hold and organize your data. Previously called entities
Learn more

Properties   Primary column

Display name *

Event

Plural name *

Events

Description

Table to store the event information.

☐ Enable attachments (including notes and files) [1]

Advanced options ∧

Schema name *

| epe_ | Event |

Type *

Standard   ∨

Record ownership *

User or team   ∨

Choose table image

None   ∨   🖉

\+ New image web resource

Color

⬜ Enter color code

Refine how data in this table is used and managed. Options marked with [1] can't be turned off if enabled. Learn more

For this table

☑ Apply duplicate detection rules ⓘ   ☑ Audit changes to its data ⓘ

☐ Track changes [1] ⓘ   ☐ Leverage quick-create form if available ⓘ

☐ Provide custom help ⓘ

Help URL   ☐ Enable long term retention ⓘ

Figure 2.6 – Event table form

Benefits of using Dataverse for event planning

When managing event data, choosing the right data storage and management solution is crucial for ensuring that your application remains efficient, secure, and adaptable. Dataverse offers several advantages that make it an ideal choice for event planning solutions:

- **Scalability**: As your event planning needs grow, Dataverse accommodates the growth and expansion of event data as the application scales to manage larger events and increased volumes of data. Whether you're handling a single small event or managing large-scale events, Dataverse scales with your needs, ensuring that performance remains optimal and data management remains efficient.

- **Security**: Dataverse provides robust security features to protect your event data. With granular role-based permissions, you can ensure that sensitive information is accessible only to authorized users. This built-in security of Dataverse helps safeguard against unauthorized access and ensures that data integrity is maintained throughout your event management process.

- **Versatility**: Dataverse supports a wide range of data types and complex relationships, making it highly versatile for comprehensive event management. Beyond basic event information, you can manage intricate details such as attendee interactions, scheduling conflicts, and more, all within a unified platform. This versatility allows for a more holistic approach to event planning and management.

Experience tips

Here are some tips for your better understanding:

- **Why not use SharePoint Lists?** SharePoint Lists is often considered for data management due to its integration with other Microsoft products. However, there are significant differences when compared to Dataverse. With Dataverse, your data is automatically synchronized across environments (DEV, TEST, and PROD) without additional effort. In contrast, using SharePoint Lists requires manually recreating and duplicating your data multiple times. This process involves managing variables inside your solutions to adjust addresses, which increases the risk of manipulation errors and complicates data management.

- **Consistency and automation**: Dataverse ensures that data handling is consistent and automated, which reduces the risk of errors and streamlines the development process. This automation is especially beneficial in complex scenarios where manual data handling could lead to inconsistencies and inefficiencies.

By leveraging Dataverse, you streamline your event planning processes, ensure data integrity, and support scalable and secure management of event data, making it a superior choice for managing comprehensive event planning solutions.

Setting relationships in Dataverse

In the **Event Planning Project** solution developed using Power Apps and Dataverse, setting relationships between tables is crucial for organizing and managing event data effectively. Let's explore how to establish relationships between the Events table and the Attendees table to track attendees for specific events.

Example scenario – setting relationships in the event planning solution

Let's start with the context. As part of the event planning application, we need to manage various aspects of events, including attendee information. Establishing relationships allows us to link attendee data with specific events, facilitating comprehensive event management. Imagine a scenario where an event planner needs to view attendee lists for upcoming events within **Event Planning Project**. By navigating to the Events table and accessing the Attendees lookup field, they can quickly retrieve and manage attendee details associated with each event. This streamlined approach ensures that event data remains organized and accessible throughout the planning and execution phases:

1. **Open Event Planning Project**: Select **Event Planning Project** where the Events table is defined.

2. **Access Data and Tables**: Navigate to the **Data** or **Tables** section within Power Apps.

3. **Add an Attendees table**: Create a new table named Attendees within the same solution.

Figure 2.7 – Attendee table

4. **Open the Events table**: Locate and open the `Events` table that was previously created for storing event details.

5. **Create a lookup field**:

   - Within the `Events` table, create a new column to establish a relationship with the `Attendees` table

   - Name this field appropriately, such as `Attendees` or `Event Attendees`

6. **Define the lookup relationship**:

   - Specify that the field type is **Lookup**

   - Configure the lookup to reference the `Attendees` table

   - Select the relevant field in the `Attendees` table that will be linked to events (e.g., `Attendee ID` or `Name`)

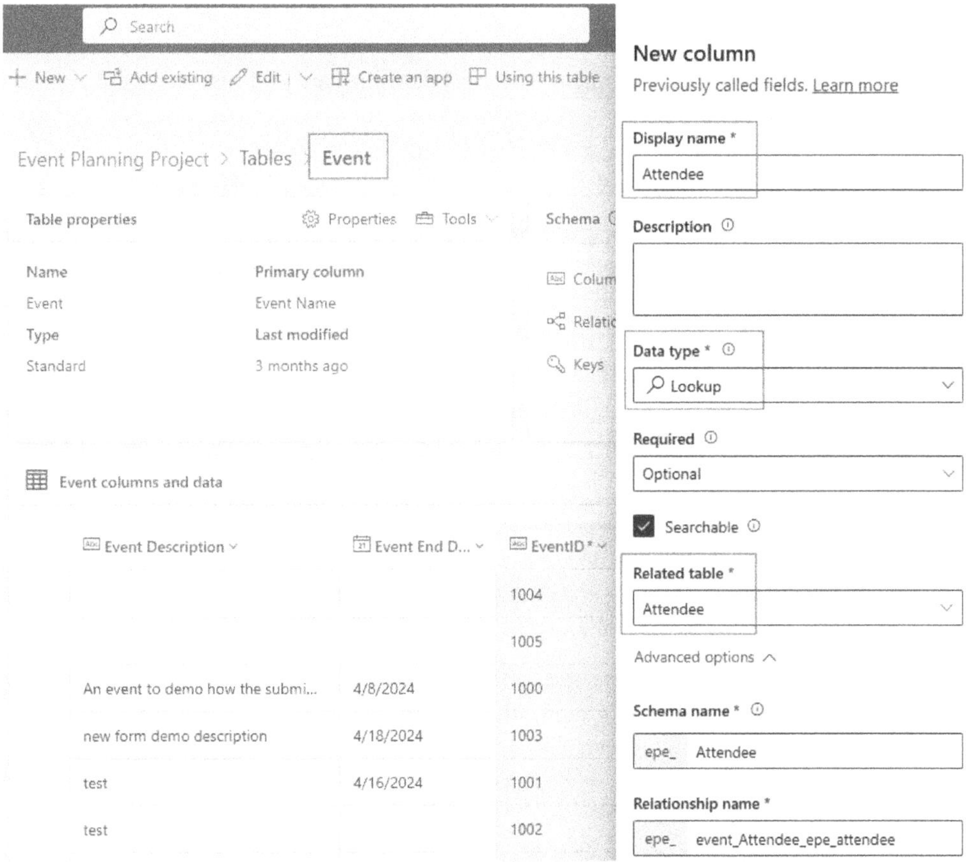

Figure 2.8 – Attendee lookup column within the Event table

7. **Save and validate**: Save the relationship configuration to establish the lookup relationship between the `Events` and `Attendees` tables.

8. Let's create a Venue table with the same relationship as Attendee. Follow the same steps to create the Venue table that you just did for Attendee.

Benefits of setting relationships

In event management, establishing relationships between different data entities is essential for maintaining an organized and efficient system. By linking related data, such as attendees with specific events, you can enhance data accuracy, accessibility, and reporting capabilities. Here's how setting these relationships benefits your event planning process:

- **Data integrity**: Ensures that attendee information is accurately associated with specific events, avoiding duplication or inconsistency. Properly set relationships prevent data anomalies and ensure that each piece of information is correctly linked to its corresponding event, maintaining the reliability of your data.

- **Efficient data access**: Facilitates easy access to attendee details directly from the event record, enhancing usability for event planners. When relationships are well-defined, event planners can quickly retrieve relevant attendee information without having to search through multiple data sources, streamlining their workflow and improving efficiency.

- **Enhanced reporting**: Simplifies reporting and analytics by providing structured data relationships, allowing for insights into attendee participation across events. Well-established relationships enable more accurate and comprehensive reports, helping you analyze trends, measure engagement, and make informed decisions based on structured data.

Defining views and forms in Dataverse

To effectively manage and interact with event data, it is crucial to create and customize views and forms within Dataverse. These elements allow you to tailor how data is presented and edited, ensuring that the interface aligns with specific user workflows and requirements:

- Create custom views and forms within Dataverse to display and interact with event data effectively. Custom views allow you to define which fields are visible and how they are presented, while custom forms enable users to enter and edit information according to their needs.

- Customize the layout and fields displayed on forms to match the workflow and user requirements for managing events. This customization enhances usability by ensuring that users have access to relevant information and tools for their tasks.

Example scenario – views and forms in the Events table

Here is the context. As part of the event planning application, custom views and forms allow event planners to visualize and interact with event data efficiently. This customization ensures that the user interface aligns with specific workflow requirements and enhances usability. Consider an event planner using **Event Planning Project** to manage a series of corporate conferences. By accessing the custom **Event Details** view, they can instantly see scheduled events, including details such as event names, dates, locations, and status. When editing event details, the customized **Event Details** form allows them to update attendee lists, venue information, and event statuses seamlessly. This personalized approach ensures that event data is organized and accessible, supporting efficient planning and execution of corporate events.

1. **Open the Event table**: Locate and open the `Event` table that was previously created for storing event details.

2. **Create custom views**: Click on **Views** within the `Event` table settings to manage views and forms.

3. **Define custom views**:

    I.   Click on **New View** to create a custom view tailored for event management.

    II.   Name the view appropriately, such as `Event Details`.

    III. Select which fields from the `Event` table should be displayed in this view, such as `Event Name`, `Date`, `Location`, `Organizer`, `Attendee Count`, and `Status`.

    IV.   Arrange the fields in a logical order that suits the user's needs, ensuring easy readability and access to critical information.

    V.   Click **Save and publish** then go back to the `Event` table.

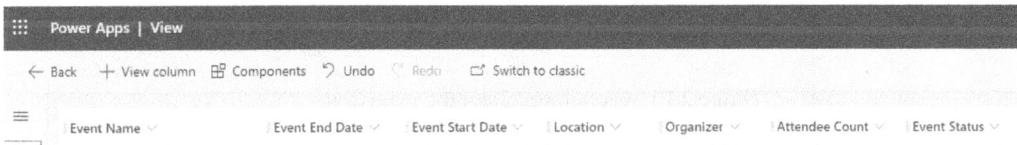

Figure 2.9 – Event details view

4. **Configure the form layout**: Click on **Forms** to define the layout and fields displayed when viewing or editing event records. Create a new form, `Event Details`, or modify an existing form (main form), if available.

5. **Customize form fields**:

    • Drag and drop fields from the `Event` table onto the **Form** canvas to include them in the Form layout.

    • Adjust field properties such as label names, data types, and validation rules to match specific user requirements.

    • Use sections and tabs to organize related fields and improve form usability.

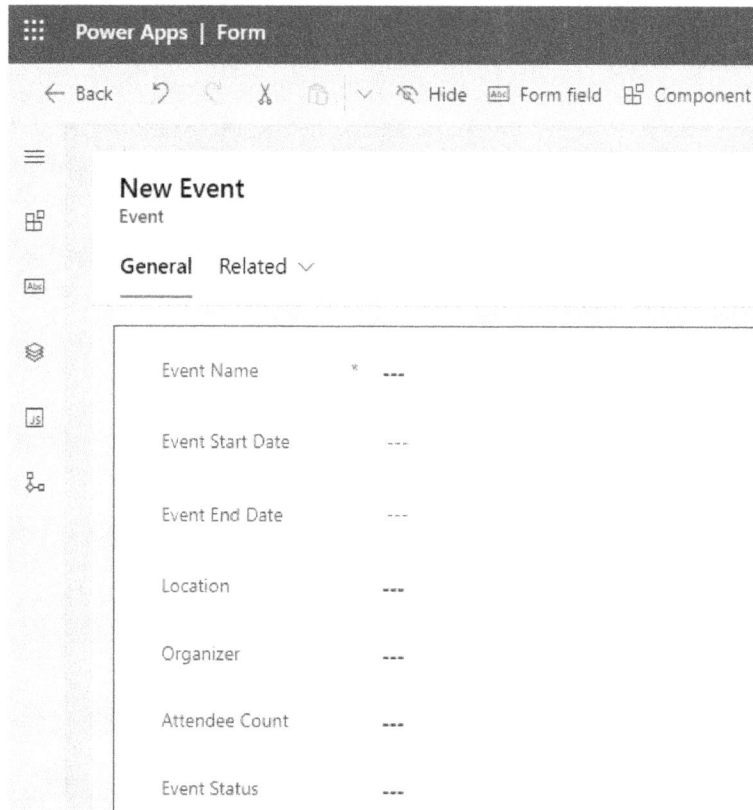

Figure 2.10 – Event form under the main form

6. **Save and publish**:

    - Save the custom views and forms to apply the configurations to **Event Planning Project**
    - Publish the changes to make the updated views and forms available for use by event planners

Benefits of custom views and forms

Custom views and forms in Dataverse offer several advantages that significantly enhance the management and interaction with event data. By tailoring these elements to fit specific user needs and workflows, you can achieve the following benefits:

- **User-centric design**: Tailoring views and forms ensures that event planners can quickly access and interact with relevant event data, enhancing productivity. For example, a custom view might highlight upcoming events and key deadlines, allowing planners to focus on their most pressing tasks. This targeted approach enhances productivity by reducing the time spent searching for necessary information.

- **Improved data visibility**: Custom views provide focused insights into critical information such as upcoming events, attendee lists, and event details. By filtering and organizing data according to specific criteria, these views facilitate informed decision-making. For instance, a view showing only upcoming events helps planners prioritize their efforts and track progress more effectively.

- **Enhanced user experience**: Optimized form layouts and intuitive interfaces streamline data entry and management tasks. Custom forms designed with user needs in mind can reduce errors and increase efficiency. For example, a form with clearly labeled fields and logical flow can simplify the process of updating event details, making it easier for users to input accurate information quickly.

Implementing business rules in Dataverse

Business rules in Dataverse automate processes and enforce data integrity, ensuring consistent behavior across applications. Let's walk through the steps to apply a business rule in the **Event Planning Project** scenario.

Example scenario – setting the event status to Confirmed

In our event planning application, we want to automatically update the status of an event to **Confirmed** when the number of registered attendees reaches a predefined threshold. Imagine a scenario where an event planner updates the attendee list for a conference in **Event Planning Project**. As the attendee count reaches or exceeds the predefined threshold (e.g., 100 attendees), the system automatically updates the event status from **Pending** to **Confirmed**. This automated process ensures that event statuses accurately reflect registration progress, allowing planners to focus on other aspects of event management with confidence in data accuracy and system reliability.

1. **Open Event Planning Project**: Select **Event Planning Project** where the Event table and related components are defined.

2. **Access Business Rules**: Navigate to the Event table within the **Data** or **Tables** section.

3. **Define the business rule**: Click on **Business Rules** to manage rules that apply to the Event table.

4. **Create a new business rule**: Click on **New Business Rule** to create a new rule for setting the event status.

5. **Name the business rule**: Provide a descriptive name for the business rule, such as Set Event Status to Confirmed.

6. **Set conditions**: Define the condition that triggers the business rule, as in this example:

   - **Condition**: *Number of Attendees >= Predefined Threshold (100)* – Set the threshold number of attendees that triggers the status change.

   - Click **Apply**

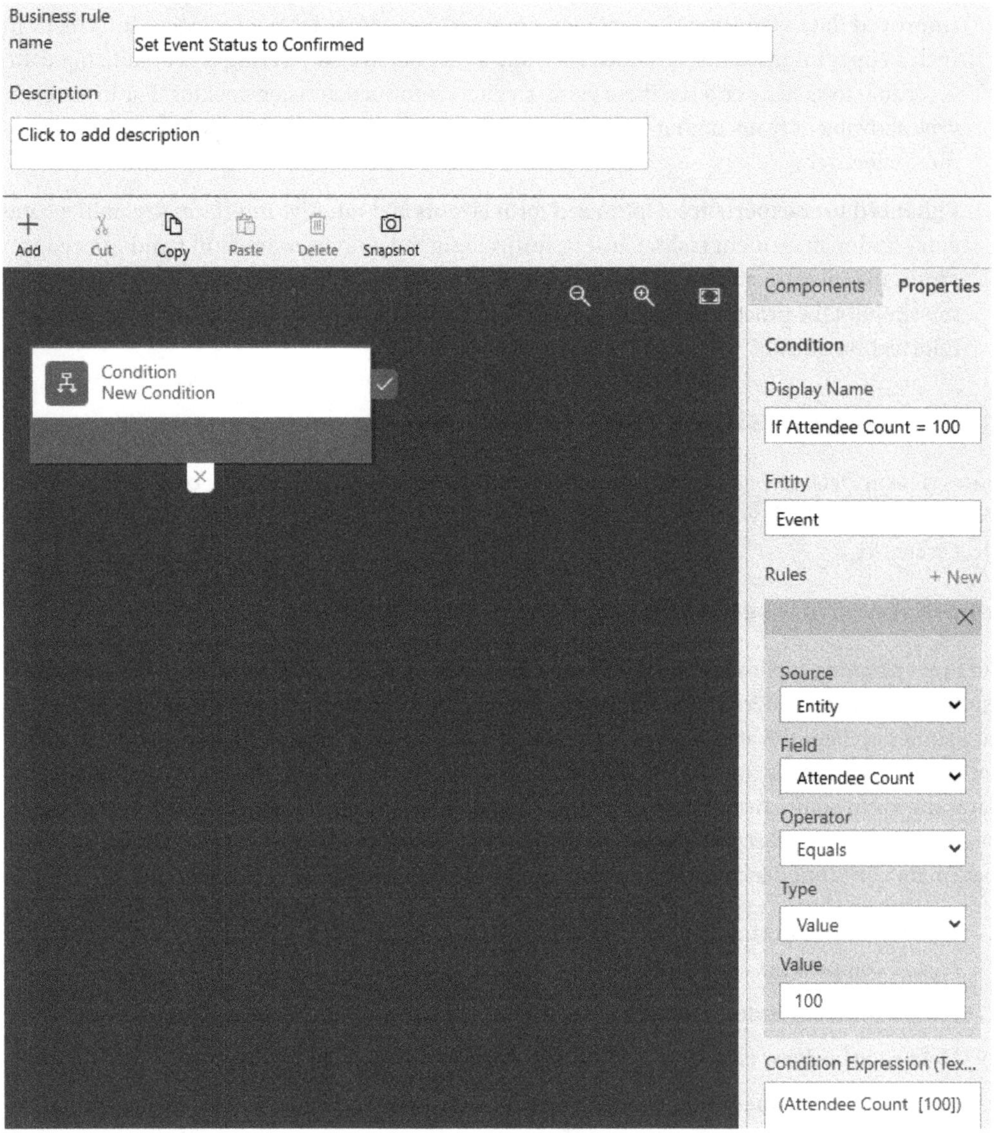

Business rule name

Set Event Status to Confirmed

Description

Click to add description

Add   Cut   Copy   Paste   Delete   Snapshot

Condition
New Condition

Components   Properties

Condition

Display Name

If Attendee Count = 100

Entity

Event

Rules   + New

Source

Entity

Field

Attendee Count

Operator

Equals

Type

Value

Value

100

Condition Expression (Tex...

(Attendee Count [100])

Figure 2.11 – Condition for business rule

7. **Define actions**: Specify the actions to be performed when the condition is met:

- **Action – Set Field Value**:

    - Choose the **Status** field of the Event table
    - Set the value to **Confirmed** when the condition is satisfied

- Click **Apply**

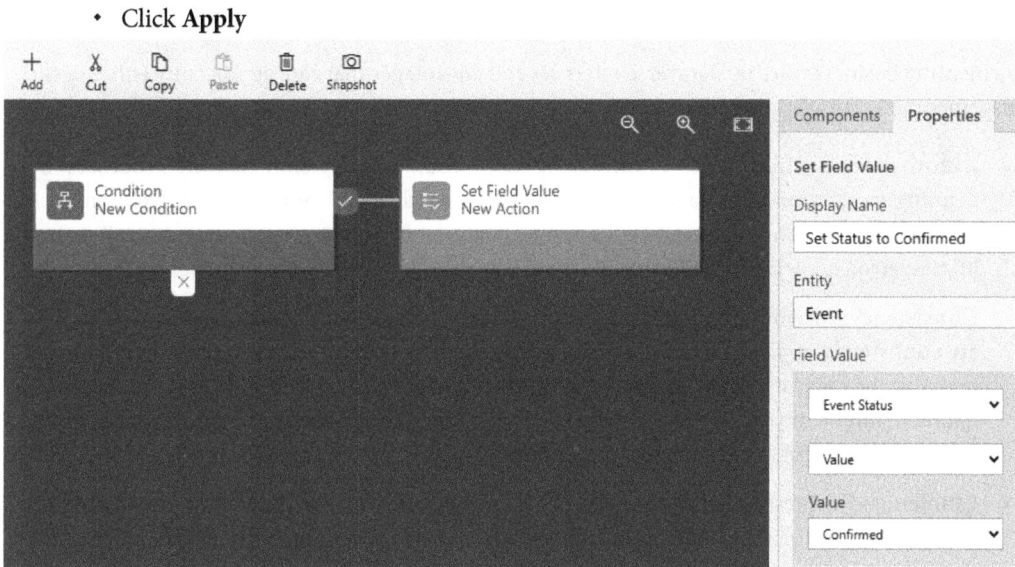

Figure 2.12 – Setting the status to Confirmed in the business rule

8. **Save and publish**:

   - Save the business rule configuration
   - Publish the changes to apply the business rule to **Event Planning Project**

Testing and validation

To ensure that business rules are functioning correctly and that your event management system operates as intended, it is crucial to perform thorough testing and validation. Here's how to effectively test and validate business rules in Dataverse:

1. **Test the business rule**: Create or update event records to check that the status updates automatically when the attendee count reaches the defined threshold (e.g., 100 attendees). Ensure that the rule triggers as expected without any manual intervention.

2. **Verify data integrity**: Ensure that the business rule correctly enforces the status update, maintaining data accuracy and integrity. Check that the event status changes to **Confirmed** precisely when the attendee count crosses the threshold, and verify that no erroneous updates occur.

Benefits of business rules

Implementing business rules in Dataverse offers several advantages that can significantly enhance the management of event planning processes:

- **Automation**: Business rules automate repetitive tasks and decision-making processes. By defining specific conditions and actions, you can reduce the need for manual intervention, such as manually updating event statuses. This automation not only saves time but also minimizes human error, allowing you to focus on more strategic aspects of event management.

- **Consistency**: By enforcing standardized processes, business rules ensure that certain actions are consistently applied across all event management operations. For instance, automatically updating the event status to **Confirmed** when the attendee count reaches a specific number helps maintain uniformity in how event statuses are managed. This consistency helps in maintaining reliable and predictable data across the application.

- **Efficiency**: Business rules streamline workflows by automating critical updates based on predefined conditions. This efficiency improves the overall operational performance by reducing the time and effort required to manage event data. For example, automatically changing event statuses reduces the manual effort needed to track and update event progress, leading to more efficient event planning and execution.

ALM considerations

Throughout the development and deployment phases of our event planning application, we'll consider the following points for better understanding:

- **Version control**: Use versioning in Dataverse to track changes made to event data and configurations over time, ensuring transparency and facilitating rollback if needed.

- **Testing and deployment**: Utilize DEV (development) and TEST (testing) environments within Dataverse for testing new features and updates before deploying them to PROD (production).

- **DEV environment**: Primarily used for initial development and experimentation, where new features are built and tested in isolation.

- **TEST environment**: Used for rigorous testing, including **User Acceptance Testing** (**UAT**) and performance testing, to ensure that updates function as expected and do not introduce issues.

- **PROD environment**: This is the live environment where the final, stable version of the application is deployed. It is crucial to minimize risks and disruptions to ongoing event management activities by thoroughly testing in DEV and TEST environments first.

- **Continuous improvement**: Implement feedback loops and iterative development cycles to enhance the event planning application based on user feedback and evolving requirements.

Dataverse provides a robust foundation for developing and managing the event planning application. By leveraging its capabilities for data storage, relationships, and ALM practices, we ensure that our application meets the needs of event organizers effectively while maintaining scalability, security, and flexibility.

Versions and deployment pipelines

Managing versions and deployment pipelines is essential for maintaining the stability and evolution of your Power Apps solutions. Proper version control and a structured deployment process help ensure that updates are applied smoothly and efficiently, minimizing disruptions to your users.

Solution versions

Effective version management is crucial for tracking the development and progression of your solutions. Implementing a robust versioning strategy helps in identifying and managing different stages of your solution's lifecycle:

- **Versioning**: Keeping track of changes and updates using version numbers is essential for maintaining a clear and organized development process. Versioning ensures that each iteration of your solution is distinctly identified, facilitating smooth upgrades and maintenance. For example, assigning a unique version number such as v1.0 for the initial release, v1.1 for minor updates, and v2.0 for major changes helps in clearly delineating each stage of your solution's development.

Managing drafts and releases

In addition to versioning, differentiating between drafts and released versions is critical for effective solution management:

- **Drafts and releases**: Drafts are versions of your solution that are still in development and may not be stable or fully tested. These are typically used for ongoing work and internal testing. Released versions, on the other hand, are stable and ready for deployment to production environments. By clearly distinguishing between drafts and releases, you ensure that only fully tested and stable versions are deployed, reducing the risk of issues in production.

Example scenario – event management solution

To illustrate how versioning and draft management work in practice, consider the following example:

- **Initial version**: Create `EventManagement-v1.0` as the initial version of your event management solution

- **Updates**: Increment the version number for subsequent updates, such as `EventManagement-v1.1`, `EventManagement-v2.0`, and so on

- **Documentation**: Document changes made in each version to facilitate understanding and future updates

By following these guidelines, you can effectively implement versioning in your Power Apps solutions, ensuring smooth upgrades, maintenance, and transparency in change management processes.

Deployment pipelines

Effective management of solution deployments is crucial for maintaining the stability and reliability of applications as they move from development to production environments. Deployment pipelines play a key role in this process by automating and streamlining the transition of solutions, which helps to minimize errors and ensure that updates are delivered efficiently. We'll use the following points to achieve this:

- **Concept and importance**: Deployment pipelines streamline the process of moving solutions from development to production, ensuring a smooth and error-free deployment. By automating the steps involved in deploying updates, pipelines help reduce manual effort and the potential for human error. This is essential for maintaining the integrity of your solutions and providing a seamless experience for end users.

- **Evolution of pipelines**: Deployment pipelines have evolved significantly over time, introducing advanced features such as **Continuous Integration/Continuous Deployment (CI/CD)**, automated testing, and version control. These enhancements have transformed deployment practices by making them faster, more reliable, and more secure. Understanding these advancements is crucial for leveraging the full potential of deployment pipelines, which ultimately leads to more efficient and effective solution management.

Benefits of deployment pipelines for event planning solutions

Here are the benefits of deployment pipelines for even planning solutions:

- **Faster deployment**: Automated deployment processes reduce deployment time and ensure timely delivery of updates

- **Improved reliability**: Automated testing and validation mitigate risks associated with manual errors, ensuring reliable deployments

- **Enhanced security**: Controlled deployment stages and version control enhance security by preventing unauthorized changes

By leveraging deployment pipelines with CI/CD practices and automated testing, EPE can streamline the deployment of their Power Apps solutions. This approach not only accelerates time-to-market for new features but also improves overall application reliability and security.

Summary

In this chapter, we explored essential practices for managing solutions and environments within Power Apps. We began by emphasizing strategic planning, highlighting the significance of environment selection and security considerations. Understanding the roles of default, sandbox, and production environments ensures that applications are developed and tested effectively.

Next, we explored the concept of solutions, distinguishing between managed and unmanaged solutions. Practical guides were provided for creating publishers, establishing naming conventions, and implementing versioning strategies. These skills are vital for maintaining consistency and control throughout the application lifecycle.

We also covered Dataverse as a foundational data management platform and explored advanced techniques such as business rule automation and deployment pipelines. Refining these skills equips developers to build scalable applications aligned with organizational needs.

In the upcoming chapter, we will focus on Power Platform environments. We will explore how to set up and manage environments across the Power Platform ecosystem, including Power Apps, Power Automate, and Power BI. Understanding these environments is crucial for deploying and integrating solutions seamlessly across different Microsoft cloud services.

3

Power Platform Environments

Welcome to the dynamic world of **Power Platform environments**, where every facet is carefully curated to elevate your development experience. In this chapter, we embark on a comprehensive exploration, guiding you through crucial aspects that define your workspace. We will explore strategic considerations for creating new environments, navigating the intricate process of environment creation, exporting and importing solutions, and the choice between unmanaged and managed solutions. This chapter aims to empower you with the skills to craft, manage, and deploy solutions efficiently within the Power Platform environment.

In this chapter, we're going to cover the following topics:

- What is an environment?
- When to create a new environment
- Create a new environment
- Unmanaged versus managed solutions
- Exporting and importing a solution

By the end of this chapter, you will have gained the practical skills and knowledge to craft, manage, and deploy solutions with precision and efficiency within the Power Platform environment.

Technical requirements

Before diving into the chapter, ensure you have the following technical prerequisites, including Power Apps admin access, to optimize your learning experience:

- **Power Apps license and admin access**: Obtain a valid Power Apps license, and ensure you have administrative access to Power Apps. Admin permissions are crucial for creating and managing environments, a key focus in the upcoming chapters.

- **Power Platform environment**: Familiarize yourself with the basics of Power Platform environments as the chapters will extensively cover the creation and management of environments.

- **Permissions**: Verify that you have the necessary permissions to create, import, and manage solutions in Power Apps, especially with administrative privileges.

By ensuring these technical prerequisites, including Power Apps admin access, you'll be well prepared to navigate the chapters, fostering a seamless and productive learning journey with Power Apps.

What is an environment?

An **environment** in Power Apps refers to a container that holds apps, data, and other related resources. It provides a segregated space where you can develop, test, and deploy your Power Apps solutions without impacting other environments. Environments play a crucial role in organizing and managing the lifecycle of your applications.

There are seven types of environments in Power Platform:

- **Default**: This environment is automatically created in every Power Platform licensed tenant. It can be used for evaluating, proof of concepts, and so on, but should not be used for complex solution development or production, due to lack of access control.

- **Trial**: This is a temporary environment, best suited for testing specific product features, third-party solutions, demonstration purposes, and so on.

- **Developer**: This specific environment is provisioned with the Power Apps Developer Plan license. This environment can have only the owner as a single user. The provisioning of this environment type needs to be performed using a specific URL: Power Apps Community Signup. Do not consider using the developer environment type for team development in large projects. It is only suitable for individual developers.

- **Sandbox environment**: This environment can be used for pre-production purposes such as development, testing, training, support, and so on. However, it is not intended for production purposes.

- **Production environment**: This environment is typically used for running a deployed solution in production. It is important to clarify that **development**, **testing**, and **quality assurance (QA)** environments are all considered types of production environments. This distinction should be made explicit to prevent any potential misunderstanding. While having a QA environment is considered best practice, it is not mandatory. Most companies use three environments. In some scenarios, companies try to restrict the number of workspaces to a bare minimum, with even two environments sufficing.

- **Microsoft Teams**: This new type of environment is used for creating and running Power Platform solutions directly within Microsoft Teams. This environment type is created from Microsoft Teams directly in the background when a Teams user starts building a Power Apps or Power Automate flow.

- **Support**: This specific environment cannot be created by the customer, only by Microsoft support personnel, to resolve service case issues. It is usually created as a copy of the existing troublesome environment and deleted after the issue is resolved.

In addition to these environments, it is essential to understand the concept of **Managed Environments**, a focus area increasingly highlighted by Microsoft. Managed Environments offer several benefits, which include the following:

- Deeper environment-level visibility

- Welcome message

- Sharing controls

- Native deployment pipelines

- Solution checker enforcement

- Weekly digest emails

- No license reporting

- Extended backup retention period

- Default environment routing

- Other advantages

Understanding the different types of environments in Power Apps is essential for effective application development, testing, and deployment. Each environment serves a distinct purpose in the software development lifecycle, providing developers and organizations with the flexibility and control needed to manage their Power Apps solutions efficiently.

When to create a new environment

If you are just starting with a Power Apps tenant, you will have the default environment to start with.

Best practice: Rename this to `Personal Productivity`. This is a guaranteed feature for all tenants that eliminates ambiguity. *Figure 3.1* provides an example and highlights where the current environment is located. In this example, the default environment is active. This environment can be used but there will be limitations.

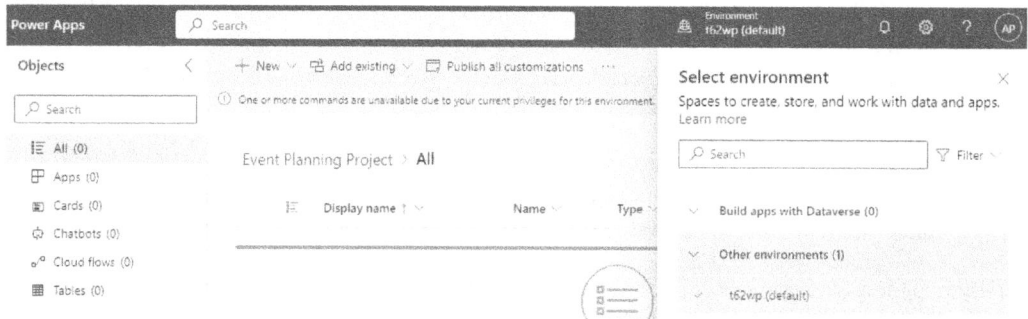

Figure 3.1 – The location of the environment

You'll notice there is a warning banner saying there are limitations to this environment. Since this is the default environment, there is no Dataverse database created, which means we are unable to create tables to store data. This is shared by all users. Restrictions related to DLP are usually more strict in this environment as well as increased risk of orphaned apps and resources, lack of deployment flexibility, and challenges in collaboration and coordination: `https://learn.microsoft.com/en-us/power-platform/admin/database-security`.

For our purpose, we will need to create an environment to be able to create tables with columns and start to add data to our app. It's important to consider when and why to create a new environment instead of using the default environment that Power Apps provides. The decision should be based on the type of work being conducted: whether it is for personal productivity or enterprise- and team-based production work.

Considerations for creating a new environment

Creating environments in Power Apps is a critical aspect of managing your applications' lifecycle. Here are key considerations for when to create a new environment:

1. **Project lifecycle stages**: Different environments are essential for different stages of the project lifecycle—development, testing, and production. This separation ensures that changes can be made without affecting the production environment and allows for thorough testing before deployment.

2. **Organizational structure**: Environments can be created to reflect the organizational structure. For instance, separate environments for different departments or teams can help in managing access control and ensuring that only relevant users have access to specific resources.

3. **Geographical locations**: If your organization operates in multiple geographical locations, creating environments for each location can help comply with regional data residency requirements.

4. **Storage capacity limits**: Storage capacity limits apply to all Power Platform environments created in one tenant together. There are separate capacity limits for the file, log, and database storage types, and the capacity depends on the various Power Platform license types and the number of user licenses a customer has purchased. Currently, the basic storage capacity every new customer gets with their first subscription is as follows:

   - **Database**: 10 GB

   - **Log**: 2 GB

   - **File**: 20 GB

   The following environments are excluded from the storage capacity limits:

   - Trial

   - Preview

   - Support

   - Developer

   The database storage capacity, the file storage capacity, and the log storage capacity from all the environments in a tenant are added together and compared against the capacity limits. If the database capacity limits are reached, no new environment can be created. It's worth noting that each new environment provisioned with Dataverse incurs a cost of 1 GB of storage

5. **Overall integration**: Considering the overall integration of Power Apps within your organization, it's essential to provide a clear picture of how the platform interacts across different teams. This approach not only engages your interest but also deepens your understanding of how Power Apps can be strategically implemented.

   Clients often ask the following questions:

   - What are the necessary interactions with security, architecture, and procurement teams?

   - What are the best practices for maintaining control and governance?

   - How can we ensure our environment remains clean and optimized?

   - What steps should be taken to manage orphaned apps, unused resources, and other control-related issues?

- What are the anticipated costs, and how can we ensure disaster recovery and versioning?

- How does Power Apps differentiate itself from tools such as Power BI, and what unique capabilities does it offer?

Addressing these questions comprehensively will help you understand how Power Apps can seamlessly integrate into and enhance your organizational workflows, improving productivity and efficiency.

Licensing considerations

When planning to create and manage environments in Power Apps, it is crucial to understand the licensing requirements and limitations. Here are some key points to consider:

- **Environment licenses**: Each environment may require specific licenses depending on its purpose (e.g., development, production). Ensure you have the appropriate licenses for your environments.

- **User licenses**: Users accessing the environments need valid Power Apps licenses. Consider the number of users and their roles when planning licenses.

- **Capacity add-ons**: If you anticipate high usage or large data volumes, consider purchasing additional capacity add-ons to meet your needs.

Effective license management ensures compliance and prevents unexpected costs, enabling smooth and uninterrupted access to Power Apps environments.

Deployment pipelines

Managing deployment pipelines is essential for efficient **application lifecycle management** (**ALM**) in Power Apps. Deployment pipelines allow you to automate the process of moving applications and solutions through different environments (e.g., development, test, production). Key practices include the following:

- **Continuous integration/continuous deployment (CI/CD)**: Implement CI/CD practices to automate testing and deployment, ensuring quick and reliable delivery of updates

- **Version control**: Use version control systems to track changes and manage different versions of your applications

- **Rollback strategies**: Prepare rollback strategies to quickly revert changes if issues arise during deployment

By establishing robust deployment pipelines, you can enhance the quality and speed of your application releases, reducing downtime and improving user satisfaction.

Co-development and code review

Collaborative development and code review are crucial for maintaining high-quality applications in Power Apps. Effective practices include the following:

- **Code review processes**: Establish code review processes to ensure code quality, security, and compliance with standards

- **Collaboration tools**: Use collaboration tools to facilitate communication and coordination among development teams

- **Peer reviews**: Encourage peer reviews to identify issues early and promote knowledge sharing

By fostering a collaborative development environment, you can improve the quality and reliability of your Power Apps solutions, ensuring they meet user needs and expectations.

Governance for Power BI and Power Platform

Governance is essential for managing Power BI and Power Platform environments effectively. Key governance practices include the following:

- **Policy enforcement**: Implement policies to control access, data usage, and compliance with organizational standards

- **Monitoring and auditing**: Continuously monitor and audit environments to detect and address issues promptly

- **Role-based access control (RBAC)**: Use RBAC to manage user permissions and ensure only authorized users have access to sensitive data and functions

Effective governance ensures that Power BI and Power Platform environments are secure, compliant, and aligned with organizational goals.

To complement these considerations, let's now dive into a step-by-step process for creating a new environment in Power Apps, ensuring you can effectively implement these best practices.

Creating a new environment

Let's get started. If you are the admin of your tenant, you should have access to create a new environment. However, it's crucial to exercise caution when handling requests to convert sandbox environments directly to production, as this practice can lead to complications and is generally considered an anti-pattern. If you're not the admin, please reach out to them and request the creation of a development environment. This ensures you have the proper setup to continue building your project smoothly. Follow these steps:

1. Click on the gear icon on the top right corner of the Power Apps maker portal.
2. Click on **Admin Center**.

3. If this is the first time you are visiting the admin center, you will get a popup welcoming you; feel free to exit out of the popup or click on **Get Started** to take a tour.

4. Navigate to **Environments** in the left navigation and click **New**.

5. Provide a name for your environment, including `Dev` or `Development` in the **Name** field.

6. Choose a value for **Sandbox Type**.

7. Add a **Purpose** option for your environment.

8. Select **YES** for **Add a Dataverse data store**:

    A. This step is crucial as it enables the environment to leverage Dataverse capabilities for storing and managing data effectively.

9. Keep **Pay-as-you-go with Azure** as **NO**.

10. Click **Save**.

11. Before clicking **Save**, ensure the **New Environment** pane resembles the following figure. Pay special attention to selecting a **Region** value for legal compliance and optimal performance (choose a region close to your users).

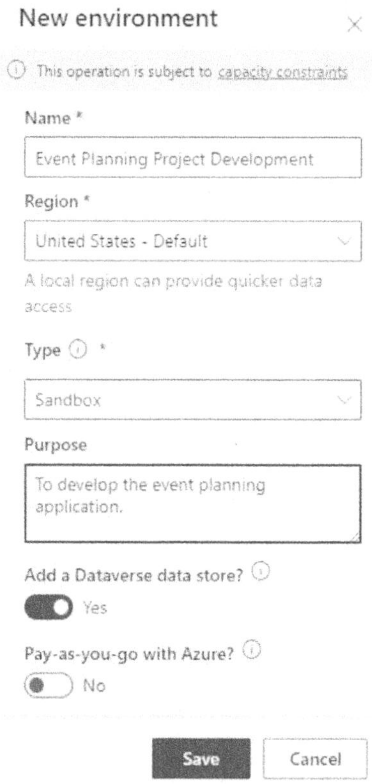

Figure 3.2 – New environment form filled out

12. The resultant **Environments** screen should be like the following figure.

Figure 3.3 – A new environment populated in the admin center

> **Important note**
>
> If you are using the developer subscription, you may be limited to using the **Trial** or **Developer** type since there are capacity constraints for the other types. You can read more about capacity constraints here: `https://learn.microsoft.com/en-us/power-platform/admin/capacity-storage#changes-for-exceeding-storage-capacity-entitlements`.

13. Now that your environment has been created, navigate back to the **Non-admin** tab in Power Apps.

14. Click on **Environment** on the top purple bar and you should see your newly created environment:

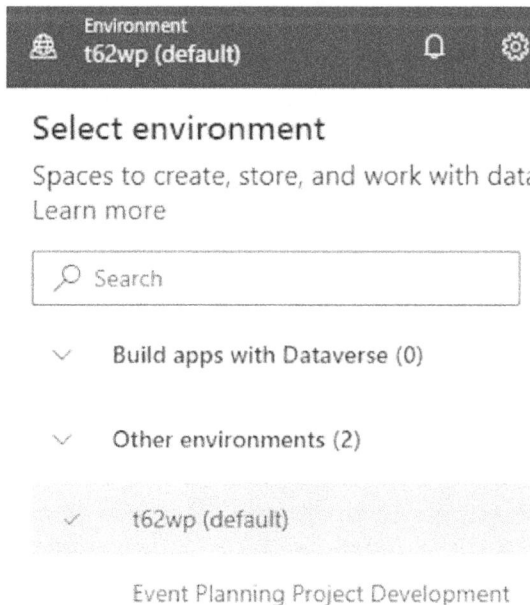

Figure 3.4 – Location of where to find recently created environment

15. Choose your newly created environment. Ensure that the use case you intend to develop aligns with the capabilities and objectives of Power Platform. Consider the following guidelines to evaluate suitability:

    - Assess whether the use case involves business processes that can benefit from automation or data management.

    - Determine whether the use case requires integration with other systems or data sources that Power Platform can facilitate.

    - Ensure the use case aligns with your organization's security and compliance policies.

    This step ensures that the environment is used effectively for suitable projects that can leverage Power Platform's capabilities.

16. Click on **Solutions** on the right navigation pane. Depending on your organization's governance model, you may have structured guidelines regarding solution management:

    - **Central governance**: In environments with centralized governance, administrators typically oversee and manage solutions to ensure consistency, compliance, and optimal resource utilization. Users may require permissions or approvals to create and deploy solutions.

    - **Self-service**: Alternatively, in environments supporting self-service capabilities, users have the autonomy to create and manage their solutions within defined boundaries. This approach promotes agility but requires clear guidelines to maintain security and compliance.

    Consider your organization's governance approach when proceeding with solution management in your newly created environment.

17. You'll notice a message that no database has been found; click the **Create a Database** button. This is only if you intend to use Dataverse as your database.

18. Select your preferences for **Currency** and **Language** (e.g., **USD** and **English (United States)**).

19. In the slide-out pane for **New database**, deselect **Include sample apps and data** to maintain clarity in your solution unless you plan to explore sample apps.

20. Click on **Create my database**.

21. You should see the following message and it will take a few minutes for your database to configure.

> **Best practice**
> Stay on the page and avoid refreshing for a few minutes to allow for background configuration.

After a few minutes, refresh your screen.

No solutions yet. Building your database...

Solutions are built on a Dataverse database. As soon as we're done
building your database, you'll be ready to start creating solutions. Learn
more

Figure 3.5 – Example of the No solutions yet screen

22. You should see a similar screen to the following figure, indicating the ability to create a new
    solution or import an existing one.

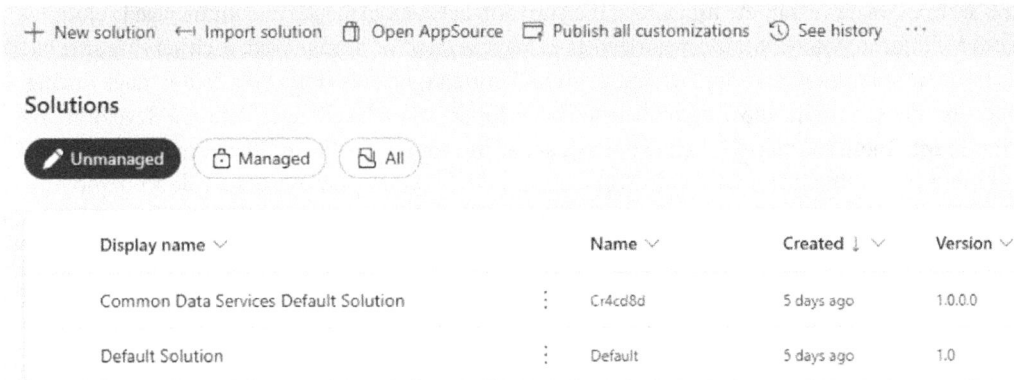

+ New solution   ←⊣ Import solution   📋 Open AppSource   🖼 Publish all customizations   🕐 See history   ···

Solutions

✏ Unmanaged   🔒 Managed   🗗 All

Display name ⌄		Name ⌄	Created ↓ ⌄	Version ⌄
Common Data Services Default Solution	⋮	Cr4cd8d	5 days ago	1.0.0.0
Default Solution	⋮	Default	5 days ago	1.0

Figure 3.6 – Example of the Solutions grid

Additional security considerations

When considering environments, it's crucial to address security considerations, including the following:

- **Security roles**: Define roles and permissions within environments to control access
- **Sharing methods**: Utilize security groups (e.g., Azure AD groups) for efficient sharing
  and collaboration
- **Use case suitability**: Establish criteria to evaluate if a use case is appropriate for Power Platform
- **Governance model**: Decide between a centralized governance model with controlled access
  and self-service capabilities for users to create and manage solutions

These considerations ensure that your Power Apps environments are secure, compliant, and aligned with organizational objectives.

As we wrap up this section, you've taken the first strides toward creating a new environment in Power Apps. If you're the admin, you now have a dedicated space ready for your projects. However, if you're not the admin, reaching out to them to create a development environment ensures you have the optimal playground for building your project with ease. This chapter is not just about creating an environment; it's about crafting a space where your ideas come to life. Whether you're the architect or the requester, you've set the stage for something special. Now, as we transition to the next phase of our exploration, let's harness this newfound environment to unleash the full potential of your ideas.

Now, let's delve into another crucial aspect of solution management: understanding the difference between unmanaged and managed solutions. By grasping this concept, you gain greater control over the lifecycle of your applications, optimizing their efficiency and maintainability. Explore the import-export dynamics and witness your innovations in action!

Unmanaged versus managed solutions

In this section, we delve into the fundamental distinction between managed and unmanaged solutions in Power Platform. A **solution** is categorized as either managed or unmanaged, each serving specific roles in the development and deployment process. **Unmanaged solutions** take center stage during application development, allowing makers and developers to make changes within a development environment. These solutions, when exported, act as the source for Power Platform assets in your development pipeline. **Managed solutions**, on the other hand, play a crucial role in deploying applications to various environments beyond development. Now, let's explore each side individually, uncovering the unique characteristics, best practices, and considerations associated with managed and unmanaged solutions.

In the dynamic landscape of Power Apps and Power Automate ALM, understanding key solution concepts is fundamental for effective development and deployment. Here, we explore the nuanced interplay between managed and unmanaged solutions, delve into the realm of solution components, navigate the solution lifecycle, define best practices for solution publishers, and highlight tips for managing dependencies.

Managed and unmanaged solutions

A solution in the Power Platform ecosystem is classified as either managed or unmanaged, with each type serving distinct purposes in the ALM process.

Unmanaged solutions

Unmanaged solutions are utilized in development environments, acting as a dynamic workspace for changes to the application. Exported unmanaged versions serve as the source for Power Platform assets. Key characteristics include the following:

- **Flexibility**: They allow dynamic customization and rapid iteration during application development

- Source of Truth: They act as the foundational source for exporting and versioning Power Platform assets

Managed solutions

Managed solutions take center stage during deployments to various environments beyond development. These solutions are packaged and deployed as controlled artifacts, ensuring consistency and security across deployments. Important considerations include the following:

- **Deployment readiness**: They are designed for deployment to production and other controlled environments.

- **Restricted editing**: Components within managed solutions are generally managed solution are generally restricted from direct editing to maintain stability and adherence to governance. This restriction ensures that changes made in production environments do not inadvertently impact solution integrity during updates.

- **Automatic dependency management**: Dependencies are automatically managed when components are added to an unmanaged solution. Managed solutions handle dependencies automatically, ensuring consistency and reliability across deployments.

Best practices for solution management

Here are the best practices and tips for solution management:

- **Best practices**: Utilize unmanaged solutions in development environments for flexible customization. Embrace managed solutions for deployment to non-development environments.

- **Tips**: Establish a process to automate the regular export of unmanaged solutions as versioned backups. This ensures robust version history and facilitates integration with source control systems such as Git or Azure DevOps.

Solution components

Components represent customizable elements, encompassing anything that can be included in a solution. The solution lifecycle involves critical actions such as creating, updating, upgrading, and patching solutions. Solutions support ALM processes, allowing for seamless upgrades, patching, and the creation of solutions with distinct layers.

Before we explore best practices for solution management, let's establish some foundational principles of efficiency:

- **Best practices**: Maintain an organized and modular structure within solutions for clarity and ease of management. Adhere to solution size limitations for efficient import and export processes.

- **Tip**: Leverage solution components to create reusable elements, fostering consistency across different projects.

- **Example**: In a project management application built with Power Apps, create reusable components such as custom controls for task lists, project timelines, and resource allocation views. By encapsulating these components as reusable elements within your solution, you can ensure consistency across different projects. Developers can easily incorporate these components into new projects, promoting efficiency and maintaining a uniform user experience across applications.

Solution lifecycle

The lifecycle of a solution in Power Apps involves key stages for efficient development, maintenance, and deployment:

- **Create**: Creating solutions is the foundation of building applications within the Power Platform. When creating solutions, it's essential to approach the process strategically and adhere to best practices. Author and export unmanaged solutions during development for flexibility and customization. Implement a clear naming convention for enhanced organization and clarity. By following these steps, you can establish a solid foundation for your applications, allowing for seamless development and management throughout the lifecycle of your projects:

  - **Best practice**: When authoring solutions during development, prioritize agility by opting for unmanaged solutions.

  - **Tip**: Enhance the organization by implementing a clear naming convention for solution components.

- **Update**: Updating solutions is integral to maintaining the functionality and relevance of your applications. When updating solutions, it's important to proceed with caution and follow best practices. Craft updates within managed solutions to facilitate changes deployed to the parent solution effectively. Exercise caution, as components with updates cannot be deleted. Prioritize thorough testing to ensure that updates do not introduce any unexpected issues or errors. By following these steps, you can ensure that your solutions remain efficient and up to date, providing optimal performance for your users:

  - **Best practice**: For changes deployed to the parent solution, craft updates within a managed solution.

  - **Tip**: Exercise caution, as components with updates cannot be deleted.

- **Upgrade**: Upgrading solutions is a critical aspect of solution management, ensuring that your applications remain up-to-date and efficient. When upgrading solutions, it's essential to plan and execute the process systematically. Consider importing a solution as an upgrade to manage versions effectively and maintain consistency across your environment. Test upgrades in a controlled environment to identify and resolve any conflicts or issues before deployment. Choose between immediate or staged upgrades based on your specific needs and requirements. By following these steps, you can ensure a smooth and successful upgrade process while minimizing disruptions to your applications:

  - **Best practice**: To manage versions effectively, import a solution as an upgrade.

  - **Tip**: For seamless conflict resolution, test upgrades in a controlled environment.

- **Patch**: Patch management is essential for maintaining the stability and security of your solutions. When applying patches, assess their impact on existing configurations to minimize risks. Consider implementing patches for small updates, resembling hotfixes, layered on the parent solution. Exercise caution during the patching process, testing thoroughly to ensure compatibility and functionality. By prioritizing thorough testing and careful evaluation, you can effectively manage patches and safeguard the reliability of your solutions:

  - **Best practice**: Use patches for small updates, resembling hotfixes, layered on the parent solution.

  - **Tip**: Assess the impact of patches on existing configurations.

- **Removing dependencies**: Removing dependencies within your solutions is crucial for maintaining a streamlined and efficient application. Before removal, thoroughly document and understand dependencies to identify which components rely on others. When uninstalling, prioritize removing dependent solutions before the base solution to prevent disruptions. Ensure thorough testing to avoid unexpected issues, and communicate changes to relevant stakeholders. This systematic approach ensures a smooth transition and preserves the integrity of your solutions:

  - **Best practice**: Ensure a clean removal process by thoroughly documenting and understanding dependencies.

  - **Tip**: For smooth removal, uninstall dependent solutions before the base solution.

- **Solution publisher**: In line with discussions from *Chapter 2* on solution management, it's emphasized that every app and solution component is a part of a solution with an associated publisher. Creating a custom publisher is recommended, and it specifies the owner of a component. A solution publisher includes a prefix to avoid naming collisions, ensuring smooth integration of solutions from different publishers:

  - **Best practices**: Create a custom solution publisher for ownership clarity and efficient management. Use solution publisher prefixes wisely to prevent naming collisions and streamline solution installation.

  - **Tip**: Choose a meaningful solution publisher name to enhance understanding and collaboration within development teams.

- **Solution dependencies**: Managed solutions can have dependencies on components in other managed solutions, enabling modular and streamlined customization. The system tracks dependencies, preventing the installation or uninstallation of solutions when dependent components are missing:

  - **Best practice**: Leverage modular solutions with dependencies for a scalable and maintainable application structure.

  - **Tip**: Document and track solution dependencies to ensure smooth installations and uninstallations.

In the intricate landscape of Power Apps and Power Automate, mastering the nuances of managed and unmanaged solutions, understanding solution components, navigating the solution lifecycle, defining solution publishers, and managing dependencies are key elements for a robust and efficient application development and deployment process.

Join us as we explore the next vital step: exporting and importing a solution, where your creativity flourishes in your new Power Apps environment!

Exporting and importing a solution

Now that we have two different environments, let's export the solution that we created in the previous chapter since it was built in the default solution and import it into our new environment so we can build tables and add data. This process allows us to begin building tables and adding data effectively:

1. Navigate to the default environment using the top **Environment** button.

2. Click on **Solutions** on the left navigation.

3. Click on the **Event Planning Project** solution.

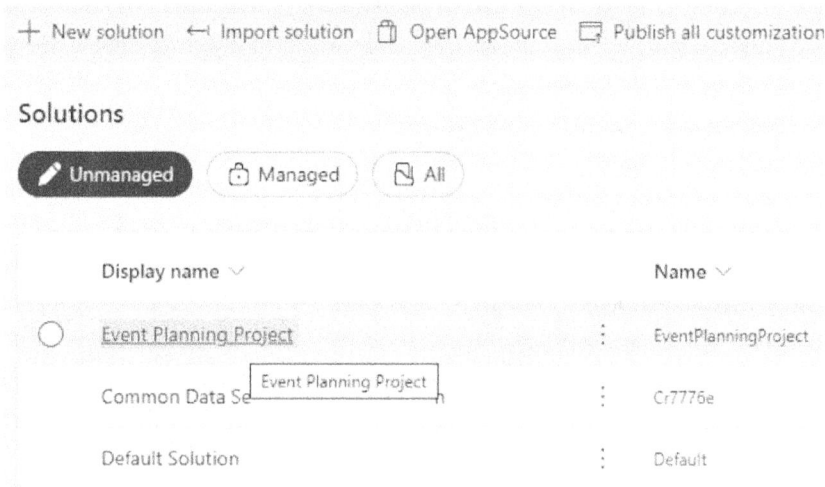

Figure 3.7 – Location of the Event Planning Project solution

4. Click on the **Overview** option on the left navigation panel.

5. Click **Export** on the top navigation panel.

6. It is best practice to click **Publish** before you continue to export to ensure all your changes have been saved and published to the solution.

7. You'll notice that the version number has increased.

8. Select the **Unmanaged** option:

   - In the context of a development environment, selecting the **Unmanaged** option ensures flexibility and agility in making adjustments to the solution as needed. It supports the iterative nature of software development and allows for continuous improvement without imposing restrictions on further modifications. This approach is ideal during the development phase when the solution is still evolving and undergoing testing and refinement.

   - Therefore, opting for the **Unmanaged** option in a development environment helps maintain a streamlined development process and facilitates efficient collaboration among team members working on the solution.

9. Click **Export**. A banner confirms your solution is being exported.

10. Do not refresh your screen.

11. After a few moments, you will see the successful banner that the solution has been successfully exported!

12. Click on the **Download** button.

13. It will appear in your Downloads folder as a .zip file.

14. Leave it as is or place it in a folder of your choice without extracting the .zip file.

15. Navigate up to **Environment** in the purple banner.

16. Click on **Environment** and select the environment you previously created in the admin center, such as **Event Planning Project**.

17. Click on **Solutions** on the left navigation panel.

18. Click on **Import solution** on the top navigation panel.

Consider your security role; in some environments, only admins can import into production to enforce governance. To avoid bottlenecks, you can do the following:

- **Custom user roles**: Define roles granting import permissions based on responsibilities, decentralizing the process while ensuring security.

- **Deployment pipelines**: Automate deployments across environments to reduce dependency on individual user profiles. This ensures controlled, auditable deployments and enhances continuity amid organizational changes.

By implementing these strategies, organizations balance governance with operational flexibility, creating a secure environment for solution management.

19. In the slide-out panel for **Import a solution**, click on the **Browse** button.

20. Go to the folder where you placed the .zip file or go to Downloads if you did not move the .zip file.

21. Select it and click **Open**.

22. You will see your file added next to the **Browse** button.

23. Click **Next**.

24. Click **Import**.

A loading banner appears during the import process.

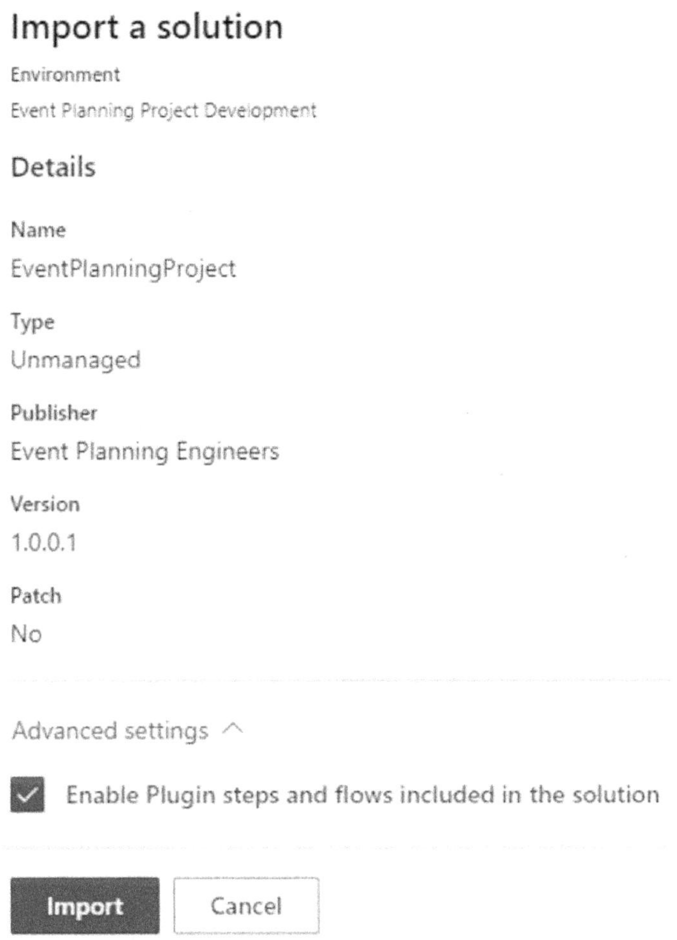

Import a solution

Environment

Event Planning Project Development

Details

Name

EventPlanningProject

Type

Unmanaged

Publisher

Event Planning Engineers

Version

1.0.0.1

Patch

No

Advanced settings ∧

☑ Enable Plugin steps and flows included in the solution

Import   Cancel

Figure 3.8 – Example of what the Import a solution panel will look like

Important note

Do not refresh your screen; it will automatically change to the successful banner once it has finished importing. Once imported successfully, a banner confirms and the solution appears on your list.

Deployment pipelines and implications

In addition to exporting and importing solutions, consider the following implications when deploying solutions using deployment pipelines:

- **Managed environments**: Every target environment for deployment must be a managed environment. This requirement necessitates a premium license for every user accessing this environment.

- **Pipeline configuration**: While the pipeline host does not need a managed environment, other users involved in the deployment process require access to managed environments to increase visibility and control.

- **Service accounts**: Deploy solutions using either service accounts or specific user accounts. For solutions utilizing premium features such as RPA, ensure the corresponding licenses are allocated appropriately, as service accounts cannot be assigned premium licenses.

- **Power Automate requirements**: Users running Power Automate cloud flows in managed environments must possess a standalone Power Automate per user license. Consider using **service principal names** (**SPNs**) for flow ownership and establishing connections, though limitations may apply, such as with SharePoint.

- **Flow licensing**: Premium actions within flows necessitate per-flow licensing, dependent on who executes the flow. Service principals require per-flow licenses if premium actions are utilized.

- **API limits**: If flows managed by SPNs do not utilize premium features, API limits may apply. Be mindful of cumulative API limits across all SPN-owned flows, capped at 25,000.

- **Security roles**: Installing the Power Platform pipelines application introduces two essential security roles:

  - **Deployment Pipeline User**: Capable of executing shared pipelines.
  - **Deployment Pipeline Administrator**: Empowered to oversee all pipeline configurations without system admin role membership.

These considerations ensure effective deployment and management of Power Apps solutions across environments, aligning with organizational objectives and enhancing operational efficiency. As we conclude, congratulations are in order for seamlessly navigating the importing and exporting of Power Apps solutions between environments. Your adept handling of this process ensures consistency and scalability, effortlessly transitioning your creations. This section opens a realm of possibilities, empowering you to shape the trajectory of your Power Apps projects.

Summary

In this comprehensive exploration of Power Apps environments, we covered key aspects crucial for effective application development and management. The chapter began by understanding what constitutes an environment and the strategic considerations for creating a new one. Subsequently, a step-by-step guide unfolded, detailing the process of creating a new environment, ensuring readers are equipped with the necessary tools to seamlessly set up their development spaces.

The discussion then explored the nuanced distinctions between unmanaged and managed solutions, offering practical insights into their purposes, constraints, and optimal usage scenarios. Understanding these concepts provides greater control over the lifecycle of your applications, optimizing their efficiency and maintainability.

Moreover, it's essential to recognize the necessity of aligning governance practices between Power BI and Power Platform. As these tools are often managed by different teams, the risk of operational silos can impede collaboration and efficiency. By implementing shared environments, robust deployment pipelines, and leveraging Entra ID groups, organizations can mitigate these risks and optimize collaborative efforts across integrated platforms. Exploring the distinctions between service principals and service accounts further enhances our understanding of secure and efficient management practices.

In the next chapter, we will explore the different types of applications available in Power Apps, including canvas apps, power pages, and model-driven apps, and understand their respective strengths and use cases.

Choosing the Right Tool – Navigating Canvas Apps, Power Pages, and Model-Driven Apps

In this chapter, we'll delve into the nuanced decision-making process behind selecting the appropriate app type within Power Apps – whether it's a canvas app, a Power Pages app, or a model-driven app. We'll explore the distinctive strengths and best use cases for each type, equipping you with the knowledge to make informed choices based on project requirements.

In this chapter, we're going to cover the following main topics:

- Understanding when to use a canvas app

- Exploring the scenarios where Power Pages shine

- Determining the ideal circumstances for leveraging a model-driven app

By the end of this chapter, you'll have a comprehensive understanding of when and why you should opt for a canvas app over a model-driven app or vice versa, empowering you to create tailored solutions that align perfectly with your business needs. You'll also be equipped with the tools and insights needed to harness the full potential of canvas apps and unlock a new realm of possibilities in app development.

Technical requirements

The technical requirements for this chapter are as follows:

- Access to the Microsoft Power Apps platform
- Familiarity with canvas apps, Power Pages, and model-driven apps
- A basic understanding of app development concepts

Understanding when to use a canvas app

In this section, we'll delve into the dynamic world of canvas apps within Microsoft Power Apps. You're about to embark on a journey that will equip you with the knowledge and skills to effectively leverage canvas apps so that you can build powerful and customizable apps tailored to your specific business needs. Through practical examples and hands-on exercises, we'll explore the versatility of canvas apps, learning how to design, build, and deploy apps that empower users and drive business efficiency. Along the way, we'll understand when to use canvas apps as front-facing apps for end users within an organization and for small use cases, in comparison to Power Pages and model-driven apps. This understanding will provide you with a comprehensive perspective on the diverse app development options available within Power Apps.

Creating a canvas app – understanding the why and how

Let's delve into the world of crafting canvas apps, where we'll explore their purpose and mechanics in depth. Reflecting on the data model introduced in the previous chapter, we established several tables, with the Event table as our focal point. Our goal now is to showcase this primary Event table within our canvas app, enabling end users to input data seamlessly.

Canvas apps prove invaluable for engaging end users, particularly within our organization, by facilitating data entry without granting extensive backend access. Beyond mere form exposure, they offer extensive customization options, allowing us to tailor the app's appearance so that it matches our organization's branding effortlessly. Additionally, canvas apps boast seamless integration capabilities with various data sources, unlike model-driven apps or Power Pages apps, which are limited to pulling data directly from Dataverse. We will cover integration with various data sources in *Chapter 5*.

Familiarity with Excel formulas can significantly ease the learning curve for canvas app formulas, something we'll explore further in upcoming chapters. Moreover, the beauty of canvas apps lies in their accessibility; no prior coding knowledge is required for basic apps. With their intuitive drag-and-drop interface, creating a basic functional app tailored to our business needs becomes a straightforward endeavor. However, it is important to set realistic expectations – achieving functionality beyond basic apps can be challenging without at least some coding expertise.

Hands-on exercise – building your first canvas app

In this exercise, we aim to create a simple canvas app for managing event data. The app will allow users to do the following:

- Input new event data

- View and edit existing event data

- Ensure data integration with a sample data source, such as Excel or SharePoint

By the end of this exercise, you will have a functional canvas app that can be used to manage event data within your organization. You'll gain hands-on experience with designing the user interface, adding functionality, and deploying the app so that it can be used by your team.

Step-by-step guide

Here are the steps for building your canvas app:

1. Open the Power Apps maker portal at `https://make.preview.powerapps.com/`.

2. Ensure you are in the **Event Planning Project Development** environment on the top purple ribbon.

3. Click on the **Event Planning Project** solution.

4. Click **New** from the top ribbon, click **App**, then click **Canvas app**, as shown in the following screenshot:

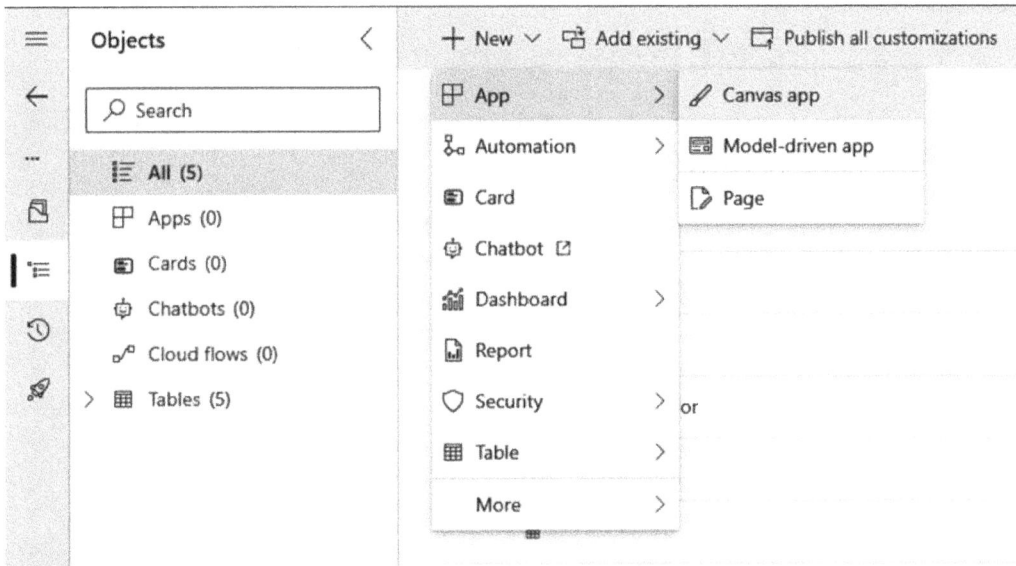

Figure 4.1 – Where to find the Canvas app option

Creating a canvas app

Follow these steps to create the canvas app:

1. Type Event Planning Canvas App in the **App name** field. Then, select the **Tablet** format and click **Create**:

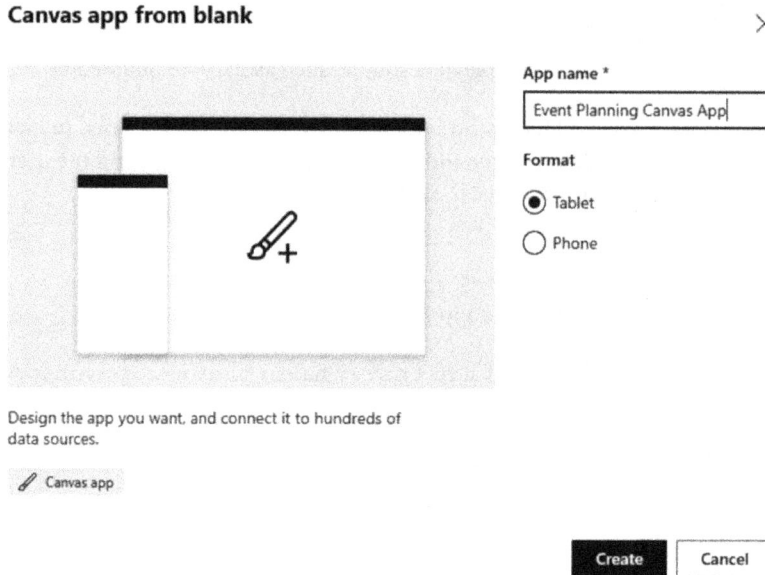

Figure 4.2 – The popup for creating a new canvas app

2. Once the canvas app designer opens, click **Skip** in the **Welcome to Power Apps Studio** popup since we will be creating it together:

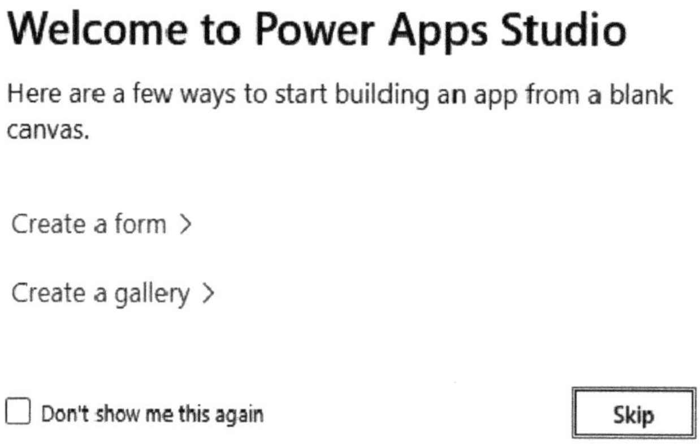

Figure 4.3 – The Welcome to Power Apps Studio popup

> **Note**
>
> For this example, we will create our canvas app using the tables from Dataverse. In *Chapter 5*, we will learn how to integrate other data sources into our canvas app.

Connect to data

Follow these steps to learn how to connect to data:

1. Click on the **connect to data** hyperlink option on the blank screen:

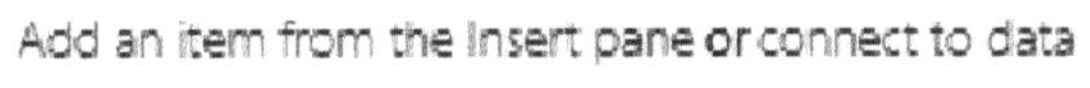

Add an item from the Insert pane or connect to data

Figure 4.4 – The connect to data hyperlink

2. The left panel will look as follows:

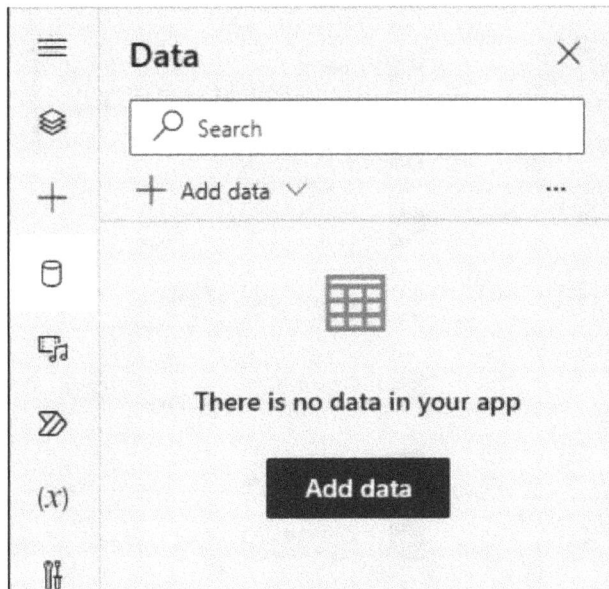

Figure 4.5 – The left panel

3. Click on the **Add data** button and type Events in the popup's search bar:

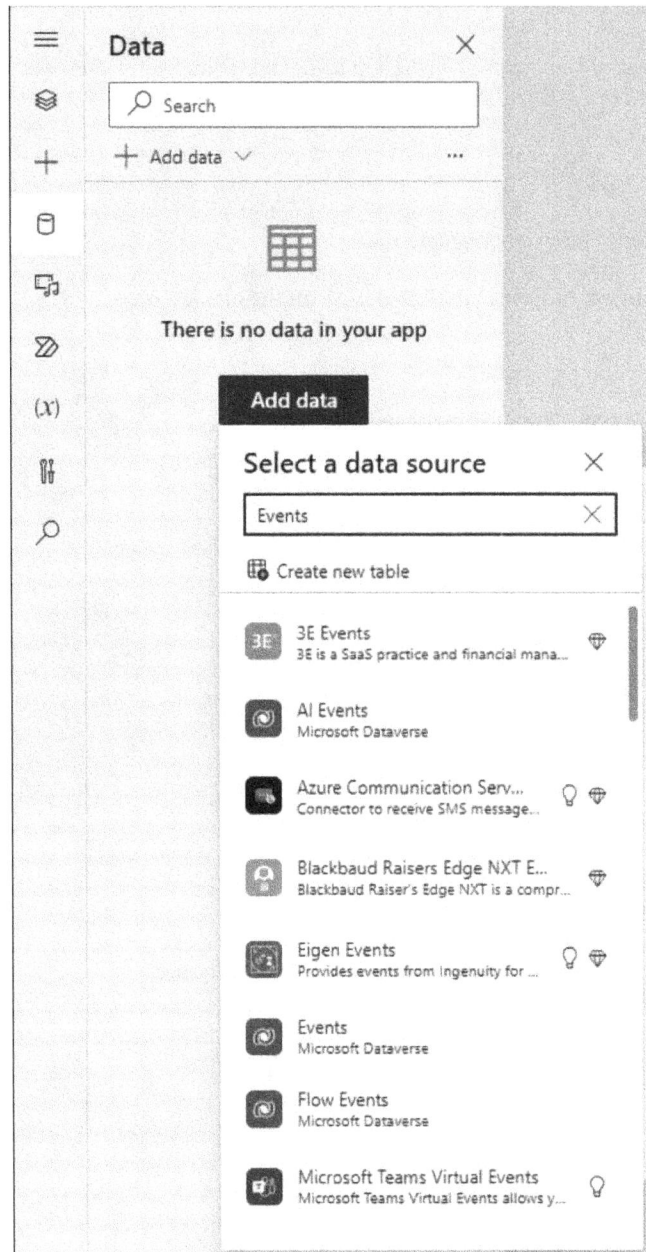

Figure 4.6 – The Add data button in the left panel

Note
You will notice that many different data sources have Events as a keyword.

4. Click on the **Events** option that has **Microsoft Dataverse** as the subtext.

5. The left panel should look as follows and say **Microsoft Dataverse – Current environment**:

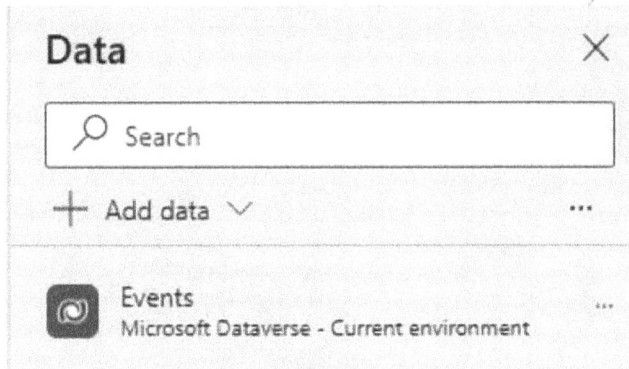

Figure 4.7 – The Microsoft Dataverse – Current environment option

Next, we'll add a form so that the user can input data.

Using forms to input data

In a canvas app, a form serves as the primary means for users to input and edit data. Forms are essential components that facilitate interaction between the app and its data sources. They allow users to enter, modify, and view data in a structured and user-friendly manner. By using forms, you can ensure data consistency, validation, and integration with backend data sources, making the app more reliable and efficient.

Let's explore how to efficiently input data using forms in a canvas app. Follow these steps:

1. From the left panel, click on the plus (+) icon. It should say **Insert** when you hover over it.

2. Click on the **Edit Form** option.

3. From the right panel, select **Data source** and from the popup, select **Events**:

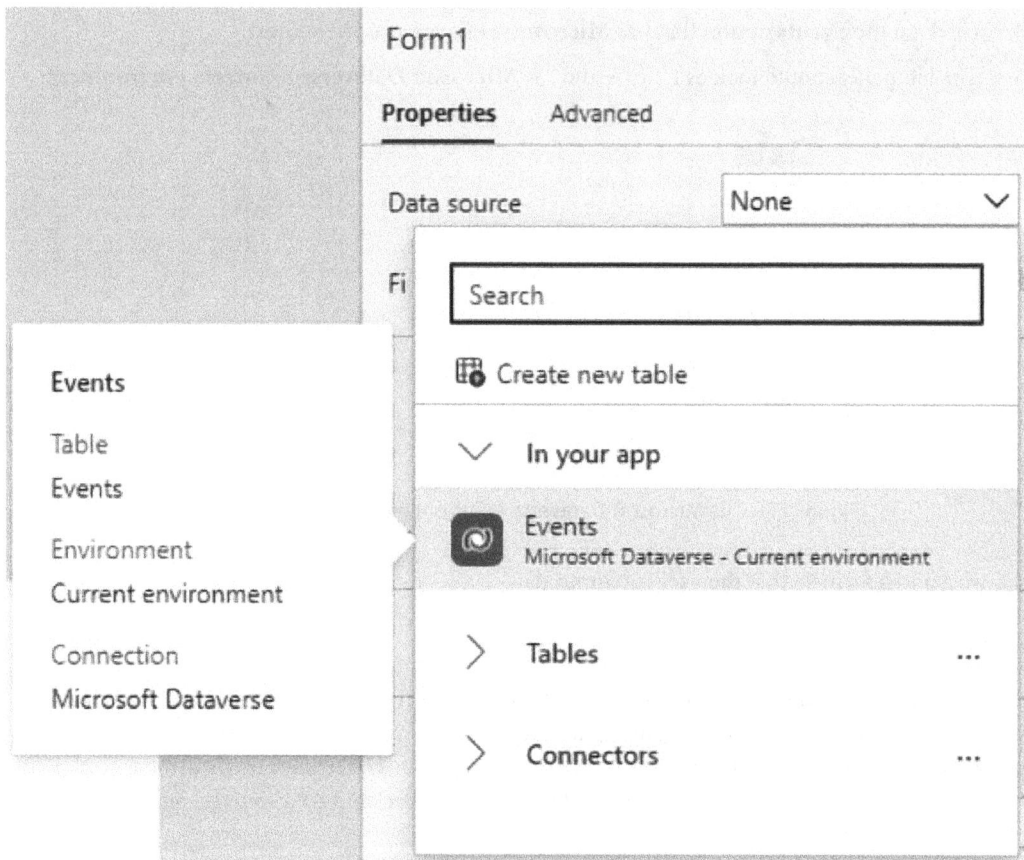

Figure 4.8 – Data source | Events

4.   Under **Data source** on the right panel, click on the **Fields** option to open up the fields.

5.   Click **Add field** and select the following fields:

    A.   **Event Name**

    B.   **Event Description**

    C.   **Event Start Date**

    D.   **Event End Date**

    E.   **Event LocationID**

    F.   **Event Type**

    G.   **Venue**

6. Hover over the **Created On** field, click on the three dots on the right, and click **Remove**:

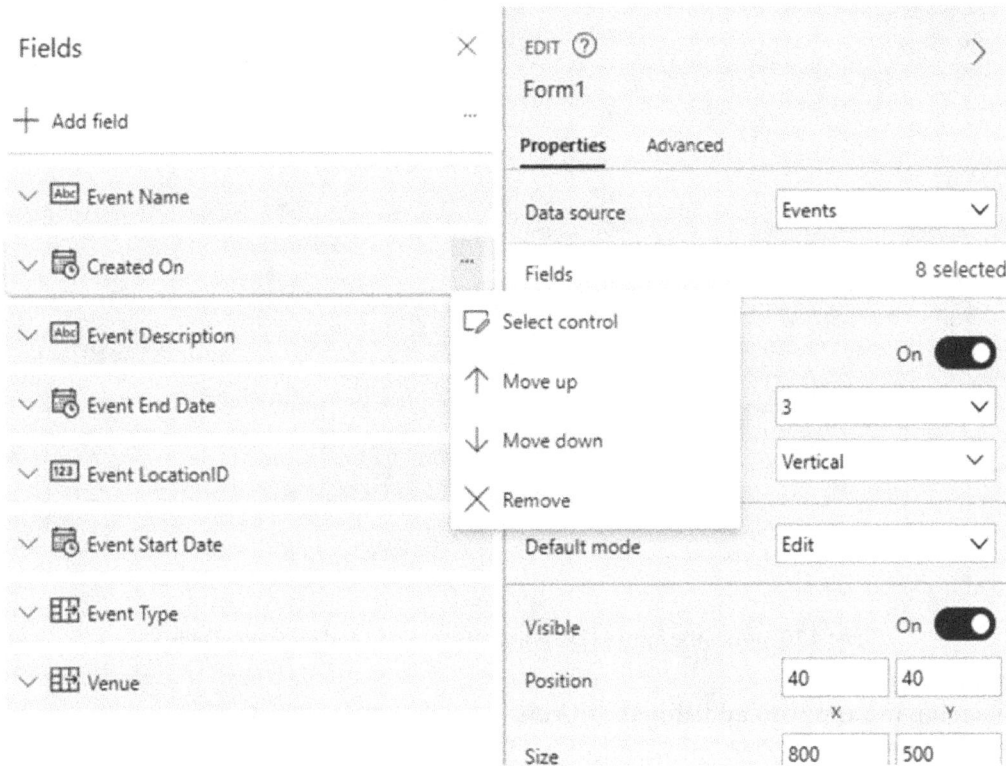

Fields	×	EDIT ⑦	
		Form1	
＋ Add field	...	**Properties** Advanced	
⌄ [Abc] Event Name		Data source	Events ⌄
⌄ [📅] Created On	...	Fields	8 selected
⌄ [Abc] Event Description		☐ Select control	On 🔘
⌄ [📅] Event End Date		↑ Move up	3 ⌄
⌄ [123] Event LocationID		↓ Move down	Vertical ⌄
⌄ [📅] Event Start Date		✕ Remove	
		Default mode	Edit ⌄
⌄ [⊞] Event Type		Visible	On 🔘
⌄ [⊞] Venue		Position	40 40
			X Y
		Size	800 500

Figure 4.9 – Where to find the Remove option under Fields

7. Using the same three dots, move **Event Start Date** above **Event End Date**. Also, move **Event Type** above **Event Description**. We do not necessarily need to move or remove these, but this is a great way to learn how!

8. The designer should now look like this:

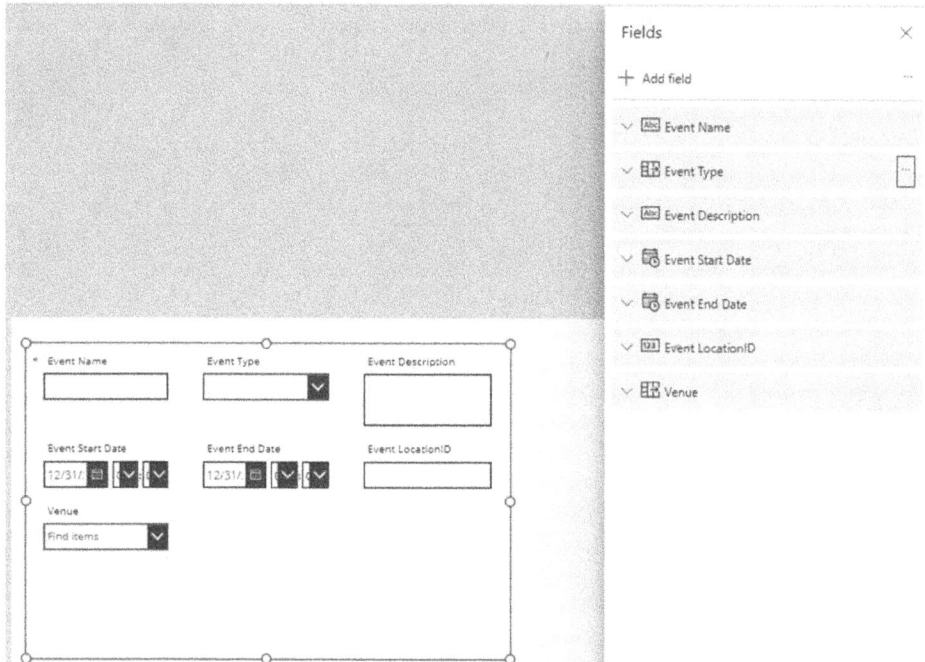

Figure 4.10 – What the form will look like after moving and removing fields

Enhancing the app with additional controls

To make the app more functional, we'll add some additional controls:

1. Click on the plus (+) icon in the left panel and select **Button**.

2. Place the button below the form and set its **Text** property to Submit.

3. Set the **OnSelect** property of the button to SubmitForm(Form1), where Form1 is the name of your form control.

   This will enable users to submit their data entries by clicking the button.

Saving and publishing

Here are the steps to save and publish:

> **Saving**
>
> This action saves your work in progress. It ensures that all the changes you have made to the app are stored securely, allowing you to continue working on it at a later time without losing any data.

1. To save your app, click the **Save** button in the top-right corner of the designer.

> **Publishing**
>
> This action makes the latest version of your app available to end users. Publishing is essential for deploying your app to a wider audience after you have finished making and testing your changes.

2. To publish your app, after saving, click on the **Publish** button located to the right of the **Save** button.

3. Add a description to the publish pop-up window, ideally noting what has been added since the last publish.

4. Finally, click **Publish this version**:

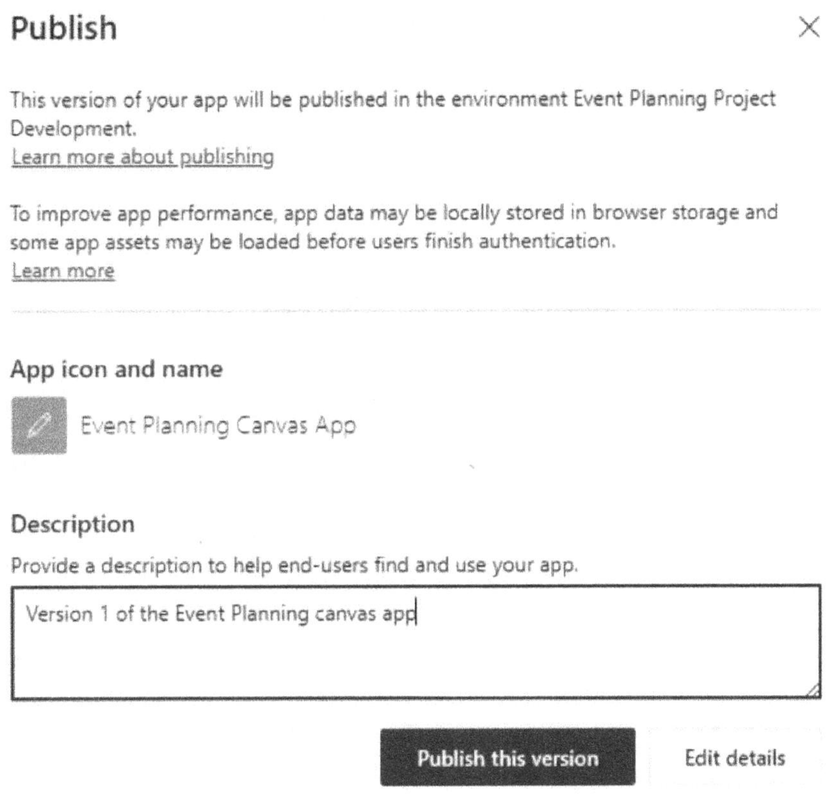

Publish   ✕

This version of your app will be published in the environment Event Planning Project Development.
Learn more about publishing

To improve app performance, app data may be locally stored in browser storage and some app assets may be loaded before users finish authentication.
Learn more

App icon and name

Event Planning Canvas App

Description

Provide a description to help end-users find and use your app.

Version 1 of the Event Planning canvas app

[Publish this version]   Edit details

Figure 4.11 – An example of what to put in the Description box before clicking Publish this version

Feel free to continue to add to this page using the designer. We will continue to build this canvas app in *Chapter 7*.

5. Click on the **Back** button in the top-left corner of the designer.

You'll notice that **Event Planning Canvas app** has been added to the **Event Planning Project** solution.

Commendation and summary

Congratulations on making it this far! You've successfully taken the first steps in building a functional canvas app. By following this hands-on exercise, you've learned how to do the following:

1. **Set up a canvas app environment**: You navigated to the Power Apps maker portal and created a new canvas app tailored to the Event Planning Project.

2. **Connect to data sources**: You connected your app to the Dataverse and integrated the Events table, setting a solid foundation for data management.

3. **Design a user input form**: You added and customized a form, enabling users to input and edit event data. You also learned how to organize form fields for a better user experience.

4. **Add interactive controls**: You enhanced the app by adding a **Submit** button, allowing users to submit their data entries seamlessly.

5. **Save and publish**: You understood the importance of saving your work and publishing your app, ensuring that your changes are preserved and made available to end users.

In our exploration of crafting canvas apps, we've harnessed their versatility and user-friendly features to enhance our data management processes. Leveraging the data model established in the preceding chapter, we've seamlessly integrated our primary Event table into the canvas app, empowering end users to input data effortlessly. The foundation laid by this data model ensures data consistency and reliability throughout the app.

Canvas apps shine as a solution for engaging end users within our organization, offering a streamlined data entry experience without the need for extensive backend access. Beyond their functional utility, canvas apps excel in customization, allowing us to align the app's appearance seamlessly with our organization's branding. Moreover, their ability to integrate with diverse data sources sets them apart from model-driven or Power Pages apps, which are constrained to Dataverse. As we delve deeper into canvas app development, we'll explore the initial ease of using canvas app formulas, especially for those familiar with Excel functions. While getting started and creating basic functionalities can be straightforward, it's important to note that as you progress, you'll encounter nuances and intricacies that require careful consideration and deeper understanding. Formulas in canvas apps can empower you to build sophisticated solutions, but mastering their full potential may present challenges that demand attention to detail and practice.

Keep up the great work, and get ready to explore more exciting features and capabilities in the upcoming chapters!

When to use Power Pages

Entering the realm of Power Pages, formerly known as Power Apps portals and Dynamics 365 portals, opens a world of possibilities for creating dynamic front-facing apps tailored to the needs of end users, both within and outside the organization. Unlike canvas apps, which are often used for smaller-scale deployments, Power Pages shine in scenarios demanding extensive customization and a responsive web interface. While Power Pages are particularly well-suited for complex use cases that require robust scalability and advanced customization, it's important to note that canvas apps and model-driven apps also offer robust and scalable solutions, depending on the specific requirements and scope of the app.

As we explore the intricacies of Power Pages, we'll explore their effectiveness in engaging end users across various scenarios, both internal and external. Whether you're delivering a seamless user experience within the organization or reaching out to external stakeholders, Power Pages provides a versatile platform for building impactful apps that meet diverse user needs.

Let's delve into the practical steps of leveraging Power Pages to create event apps. Follow these steps:

1. In the **Event Planning Project** solution, click **New**. Next, click **App**, and then **Page**.

2. Once you're in the **Pages** designer, click **With data**:

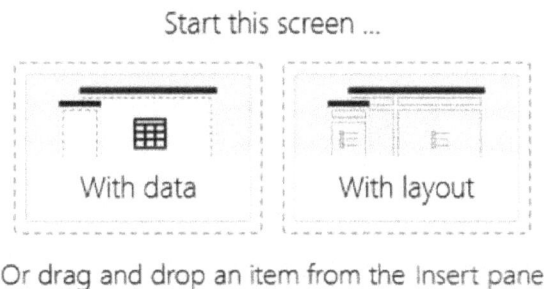

Figure 4.12 – The With data button

3. Search for the **Events** table, similar to what you did with your canvas app.

4. Select the **Events** option that includes **Microsoft Dataverse** as the subtext.

    Similar to canvas apps, the **Pages** designer allows you to customize using drag and drop.

5. Click the form within the design window and ensure **EditForm1** is selected in the left panel:

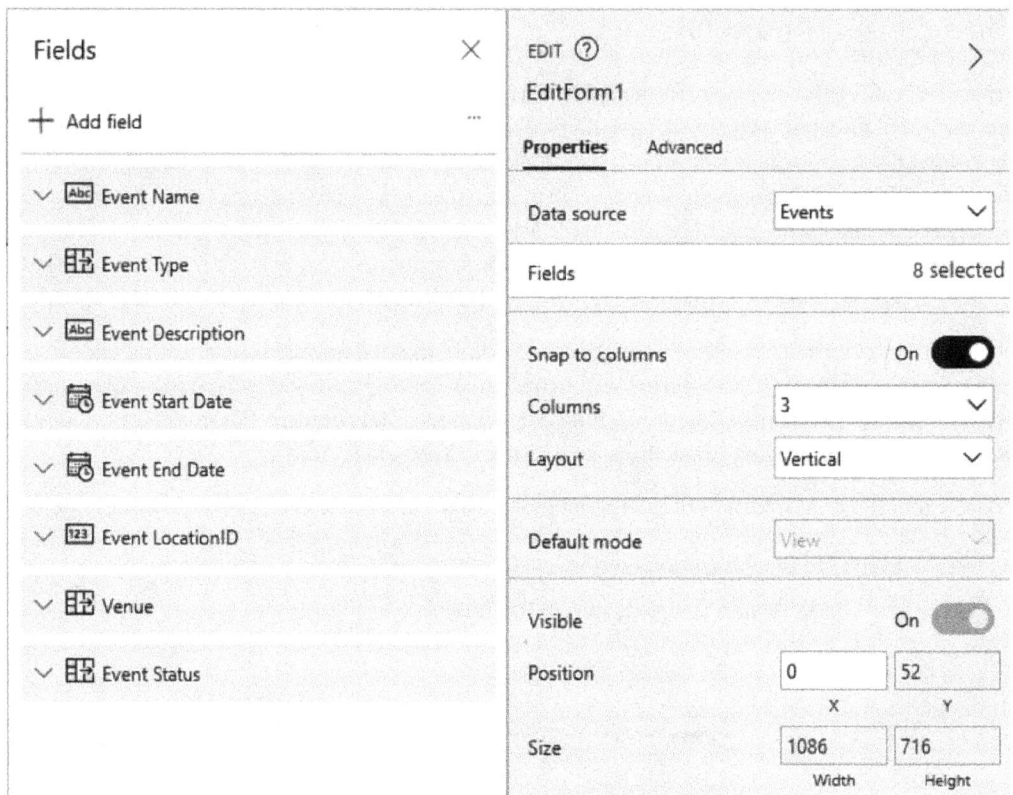

Figure 4.13 – The EditForm1 popup

6.   Under **Properties** in the right panel, ensure **Events** is selected for **Data source**.

7.   Add the following fields by using the **Fields** field in the right panel:

   A.   **Event Description**

   B.   **Event End Date**

   C.   **Event LocationID**

   D.   **Event Start Date**

   E.   **Event Status**

   F.   **Event Type**

   G.   **Venue**

Make the necessary adjustments to ensure your screen looks as follows. These field adjustments are similar to the ones we made for the canvas app:

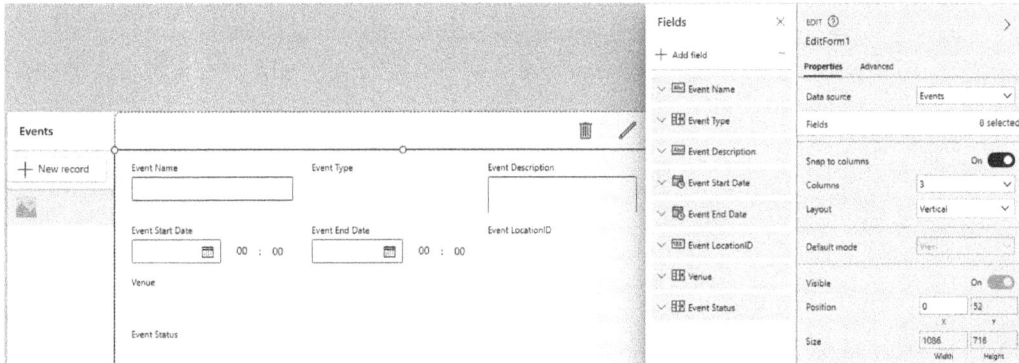

Figure 4.14 – Moving and removing fields in Power Pages

8. Click the **Save** button in the top-right corner of the designer.

9. Set **Name** to `Event Planning Pages App` in the **Save as** pop-up window:

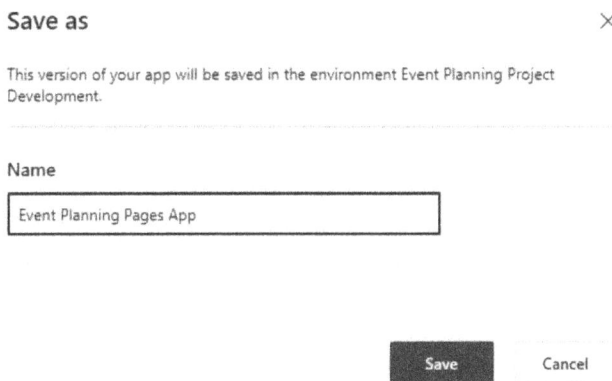

Figure 4.15 – The Save as popup

10. Once you've saved it, click on the **Publish** button located to the right of the **Save** button:

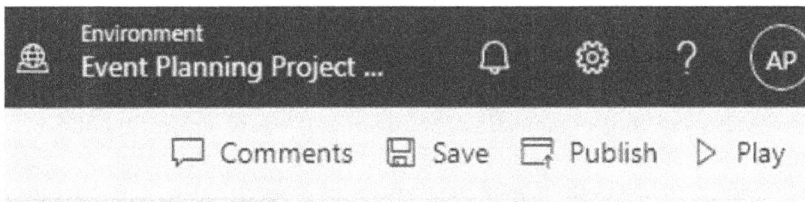

Figure 4.16 – The Publish button to the right of the Save button

11. Add a description in the **Publish** pop-up window and click **Publish this version**.

12. Click the **Play** button in the top-right corner to see it come to life!

13. Click on **New record** from the left panel. You should see the following form:

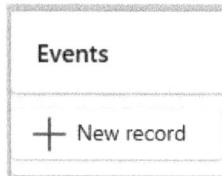

Figure 4.17 – The New record button

14. Add data and then click on the checkmark at the top right of the form:

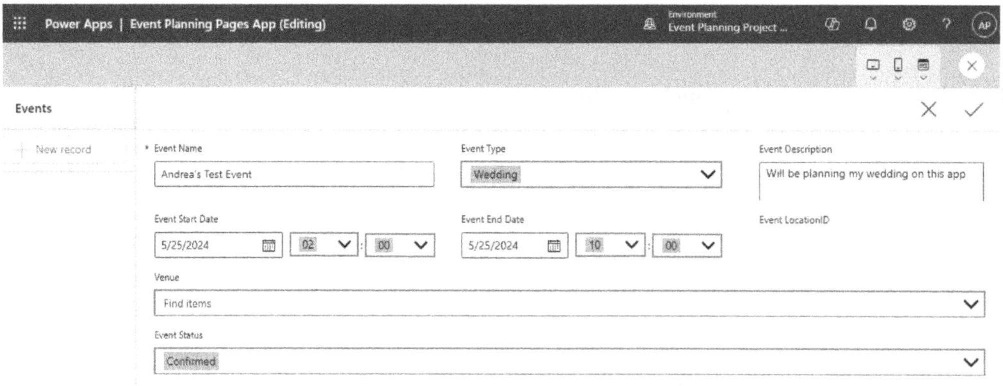

Figure 4.18 – An example of where to find the checkmark on the right-hand side of the screen

15. Click on the **X** button at the top right of the designer. Feel free to continue to add to this page using the designer. We will continue to build on this in *Chapter 7*.

16. Click on the **Back** button at the top left of the designer.

17. Notice that the **Event Planning Pages** app has been added to the **Event Planning Project** solution!

When considering whether to use Power Pages apps or canvas apps for your app development needs, several factors come into play. Power Pages apps and canvas apps are both part of the Microsoft Power Platform suite, but they serve different purposes and offer distinct advantages. Power Pages, formerly known as Power Apps portals and Dynamics 365 portals, offers a compelling solution for larger-scale projects within and beyond your organization's boundaries. If you're aiming to build an extensive app or one intended for external users, Power Pages provides the ideal platform. With Power Pages, you can seamlessly integrate forms from your Dataverse, expose specific data, and share the app with individuals outside your organization – an option not available with canvas apps. Furthermore, Power Pages offers a full-width view and a website-like appearance, enhancing user experience and engagement.

While canvas apps are suitable for specific use cases within the organization, Power Pages excels in handling more complex projects requiring additional logic and multiple pages. Although canvas apps can mimic the functionality of a website and accommodate various data exposures, it's prudent to reserve them for smaller-scale initiatives to avoid potential limitations.

In the next chapter, we'll delve into the rationale behind choosing a model-driven app over canvas apps or Power Pages apps, underscoring the importance of understanding the distinctions between these solutions to make informed development decisions. By understanding the distinctive strengths and capabilities of each platform, you can enhance your app development process and steer clear of potential challenges, thus achieving optimal efficiency and effectiveness.

When to use a model-driven app

In this section, we'll step into the world of model-driven apps, which are designed to streamline data management for your internal business teams, admins, IT professionals, and developers. Unlike canvas apps, where designers have free rein over layout and design, model-driven apps provide a more structured interface determined by the components you add. This structured approach offers robust capabilities in managing data at scale and facilitating complex business processes.

Both canvas apps and model-driven apps have distinct strengths that cater to different app development needs within the Microsoft Power Platform suite. While canvas apps emphasize highly customizable app design, model-driven apps prioritize quick data access and decision-making within a standardized user interface across different devices. This ensures a consistent experience for all users and facilitates efficient data management within a structured framework.

In contrast to Power Pages apps, which are geared toward creating modern, external-facing websites, model-driven apps are specifically tailored for internal business operations. While Power Pages apps enable the creation of external-facing websites using low-code tools, model-driven apps focus on structured data management and process automation to enhance operational efficiency within your organization.

Essentially, while canvas apps focus on intricate designs and Power Pages apps cater to external-facing websites, model-driven apps serve as the optimal solution for structured data management and process automation within your organization.

Summary and comparison

Consider the following comparison to guide your decision-making process when choosing between canvas apps, Power Pages apps, and model-driven apps within Microsoft Power Platform. Understanding these distinctions will help you make informed decisions when you're choosing the right solution for your specific app development needs.

Now, let's put this knowledge into practice by creating a model-driven app tailored for the **Event Planning Project** solution:

1. In the **Event Planning Project** solution, click **New** | **App** | **Model-Driven app** and name the app `Event Planning Model-Driven App`.

2. Next, give it a short description and click **Create**:

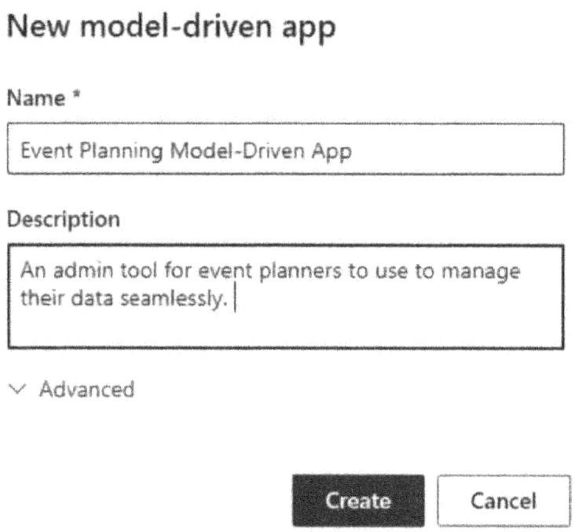

Figure 4.19 – An example of what to add to the Description box before clicking the Create button

3. Click on **Data** in the left panel, search for `Events`, click **Event**, and click the three dots. Finally, click **Add to app**:

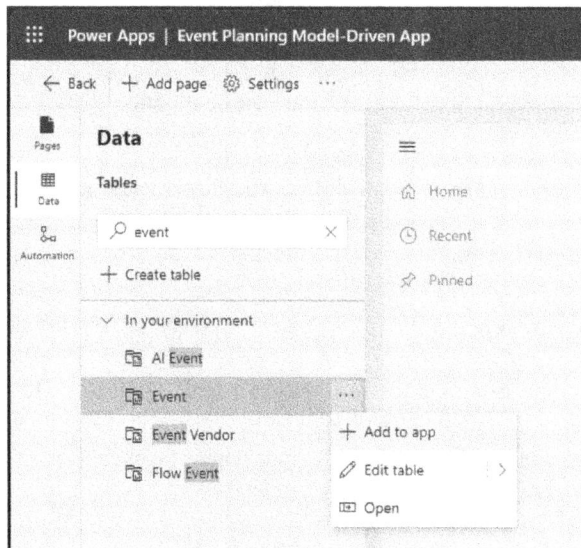

Figure 4.20 – An example showing which data option to choose

4. Notice the **Active Events** view:

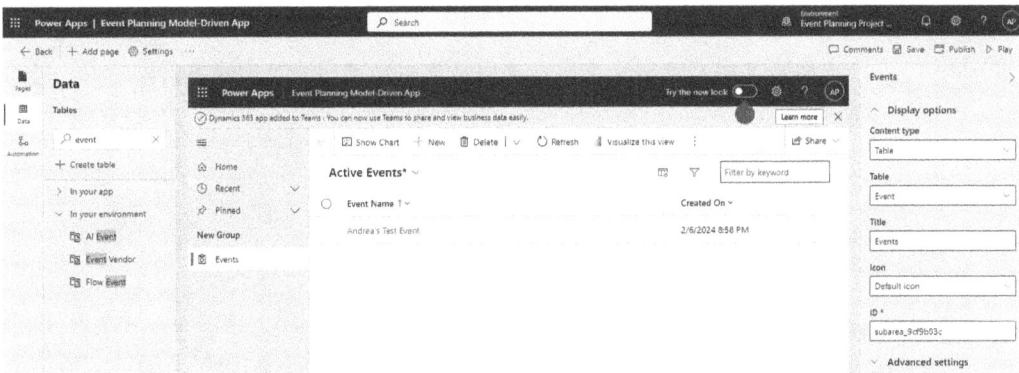

Figure 4.21 – The Active Events view

5. Click on the **Andrea's Test Event** record that we created in the Power Pages app.

6. This will open the main form that we edited in the **Solutions** chapter:

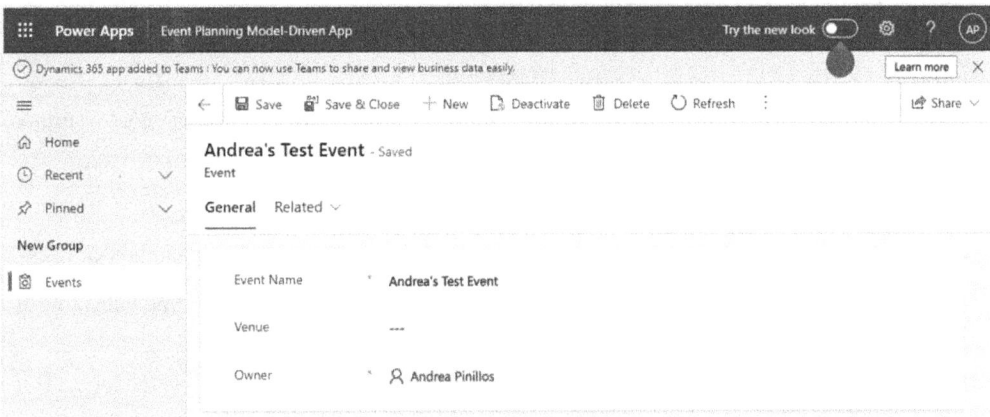

Figure 4.22 – The main form created in the Solutions chapter

7. Click **Save**, then **Publish**, and finally click **Play**.

Welcome to the world of model-driven apps! (They are my personal favorite Power App.)

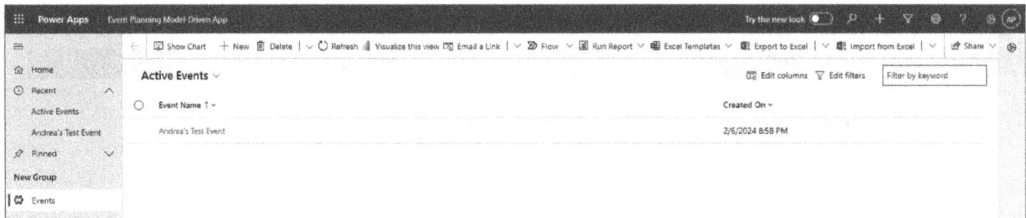

Figure 4.23 – A blank model-driven app

Exploring the model-driven app

Now that you can create a model-driven app, all of your business needs can be met. We'll explore the capabilities of the app in this subsection!

Starting from the left-hand side, we have the following:

1. First, we have the **Recent** tab, which allows us to view tables that we have recently clicked on. This makes it easy to find things when we start to have many different events that we're working on. Notice that **Andrea's Test Event** is under there. Once we start adding various events, it will be easier to find things that we recently worked on.

2. We can also pin something from our **Recent** tab if we're continuously working on a few things and want to ensure it doesn't get lost in the **Recent** tab.

Now, let's consider the middle section's top ribbon:

1. The **Show Chart** button allows us to create a chart to display the data being presented for that table. In this instance, it would be the Event table. This provides a nice visual that will allow us to visualize the data versus just seeing the data in a list.

2. The **New** button allows us to create a new record, similar to what we saw in Power Pages. This will open up the **Events** form so that we can create a new record.

3. The **Delete** button will permanently delete the record(s) we select from the list.

4. The **Refresh** button will do what its name suggests.

5. This Visualize this **View** button will open Power BI and give you a more dynamic view of our data, similar to **Show Chart**:

    I.   Once we add more data, it'll make things more interesting in this view:

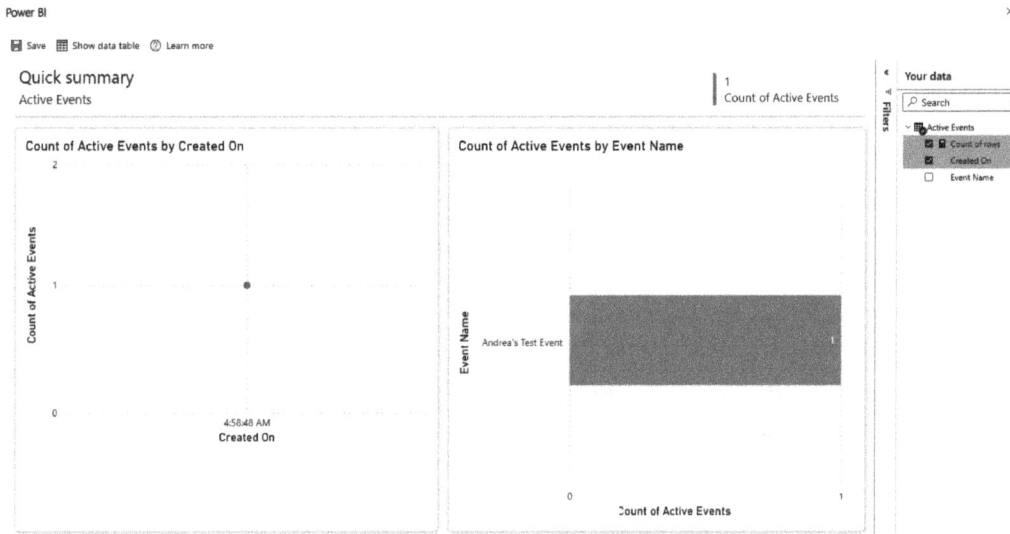

Figure 4.24 – Power BI dashboard

6. Select the **Andrea's Test Event** record from the left select button and click **Email a Link**. This will open Outlook and paste the direct link to the record:

I. Similarly, we can click on the **Andrea's Test Event** record and copy the URL:

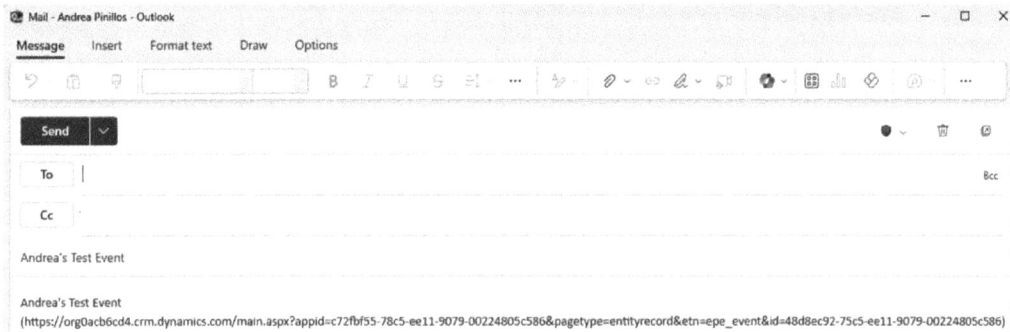

Figure 4.25 – Outlook with a link in the body of the email

7. Click on the dropdown next to **Email a Link** and select **Of Current View**:

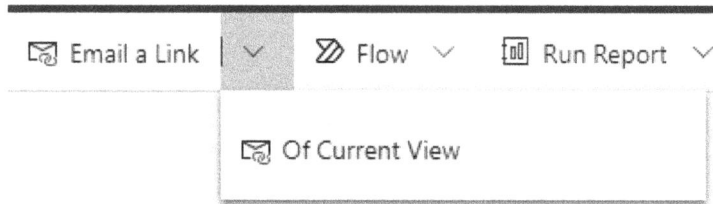

Figure 4.26 – Email a link | Of Current View

8. This will open Outlook and paste the direct link to the **Active Events** view. The **Flow** button will display any flows associated with the **Events** view. This can be created using Power Automate.

9. Utilize Excel templates in model-driven apps to easily extract, format, and share data, streamlining the import process and facilitating efficient analysis within your organization:

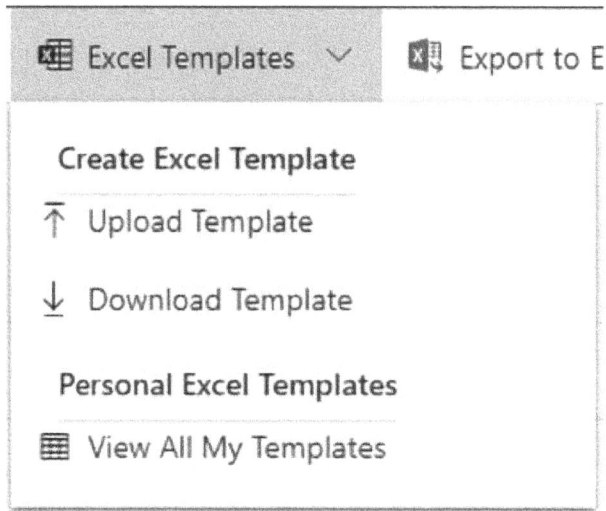

Figure 4.27 – Excel Templates options

10. Export the entire list of active events to Excel:

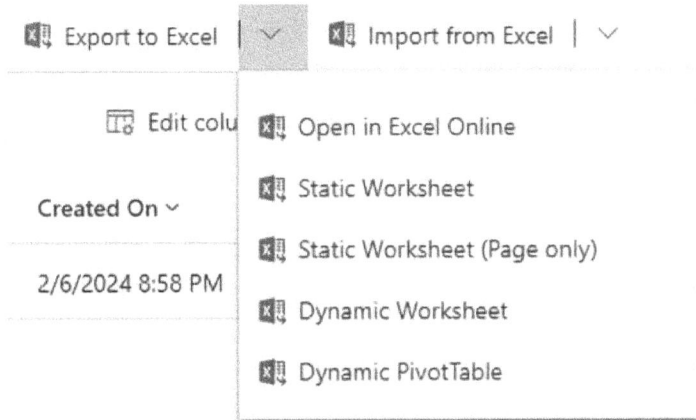

Figure 4.28 – The Import from Excel options

11. If you want to filter some of the data that will be exported, simply click on the column name and click **Filter By**:

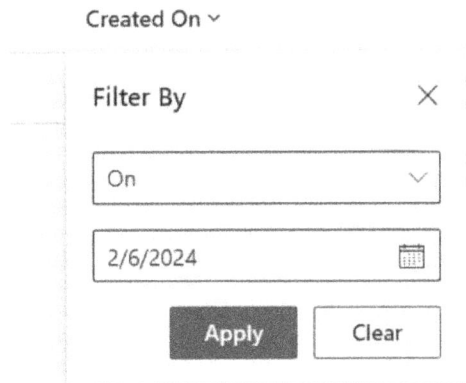

Figure 4.29 – Filter By

12. Then, click **Apply** and click **Export to Excel**.

13. Similarly, you can import data into your model-driven app in bulk by clicking the **Import from Excel** button. Ensure the import columns match what is in the view in the model-driven app.

Personalized settings in the Power Apps ribbon

Follow these steps to personalize settings in the Power Apps ribbon:

1. Click on the gear icon in the top ribbon.

2. Click **Personalization Settings**.

3. Ensure your time zone is correct so that you see the correct date and times in your model-driven app.

4. The **Default Pane** and **Default** tabs will come in handy once we have more tables exposed in our model-driven app. Click **OK**:

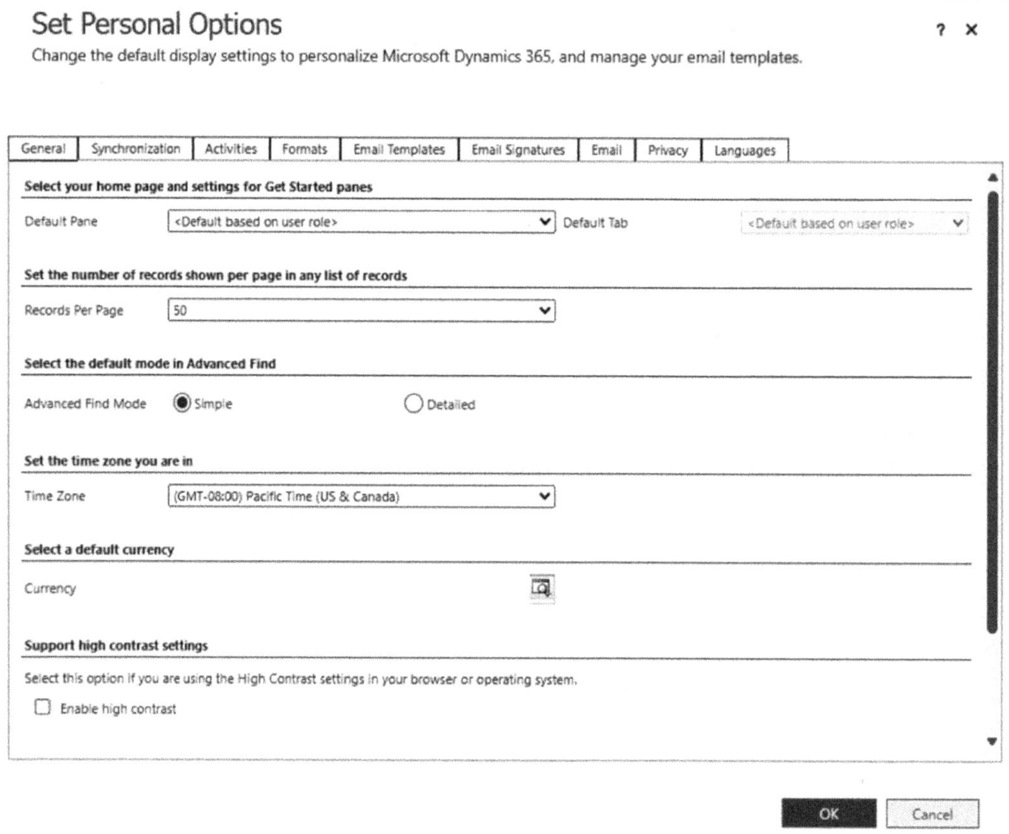

Set Personal Options

? ✕

Change the default display settings to personalize Microsoft Dynamics 365, and manage your email templates.

General	Synchronization	Activities	Formats	Email Templates	Email Signatures	Email	Privacy	Languages

Select your home page and settings for Get Started panes

Default Pane   `<Default based on user role>`   Default Tab   `<Default based on user role>`

Set the number of records shown per page in any list of records

Records Per Page   `50`

Select the default mode in Advanced Find

Advanced Find Mode   ◉ Simple   ◯ Detailed

Set the time zone you are in

Time Zone   (GMT-08:00) Pacific Time (US & Canada)

Select a default currency

Currency

Support high contrast settings

Select this option if you are using the High Contrast settings in your browser or operating system.

☐ Enable high contrast

OK   Cancel

Figure 4.30 – Set Personal Options

5. Click on the link for **Andrea's Test Event**. This will take you to the main form.

6. Click on the empty field for **Venue**.

7. Click **New Venue**:

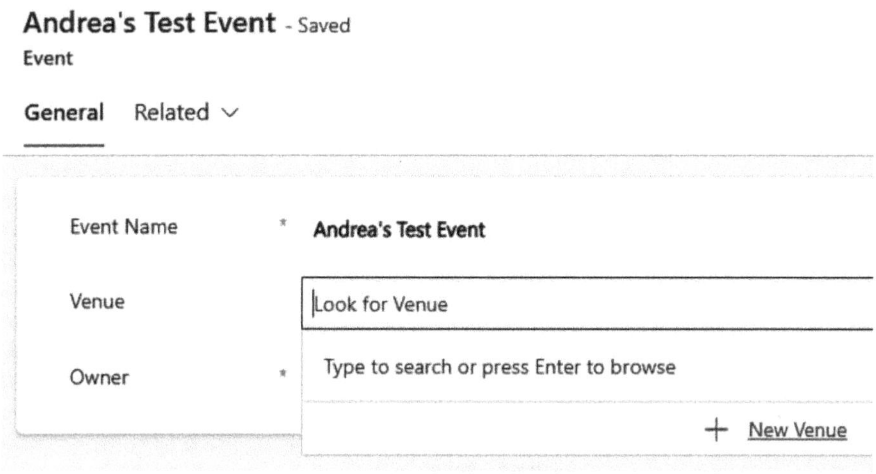

Andrea's Test Event - Saved
Event

General   Related ⌄

Event Name	*	**Andrea's Test Event**
Venue		Look for Venue
Owner	*	Type to search or press Enter to browse
		＋ New Venue

Figure 4.31 – The New Venue form

This is the lookup field we added for the Venue table. This means we can add data to the Venue table without having to expose the Venue table on the left navigation as we did for Events.

8. Let's add a venue called Garden Center:

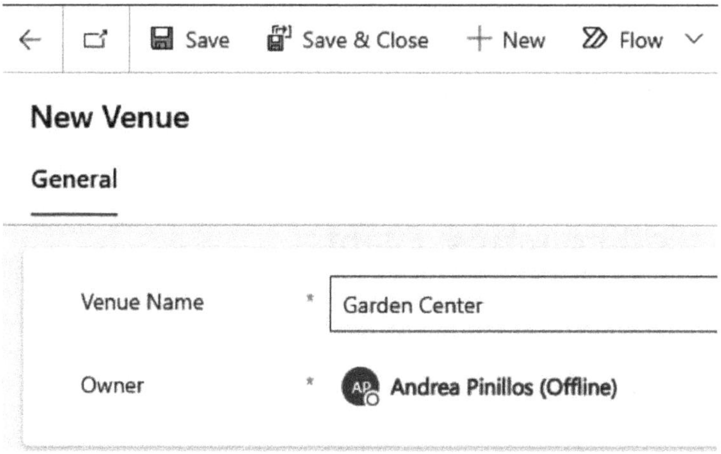

← ⬀   💾 Save   Save & Close   ＋ New   ⟫ Flow ⌄

New Venue

General

Venue Name	*	Garden Center
Owner	*	AP Andrea Pinillos (Offline)

Figure 4.32 – Naming the venue Garden Center

9. Click **Save** and close (**X**) at the top. This will take you back to **Andrea's Test Event**.
10. Hover over the blank field for **Venue** and click on the magnifying glass.
11. Select **Garden Center** from the dropdown:

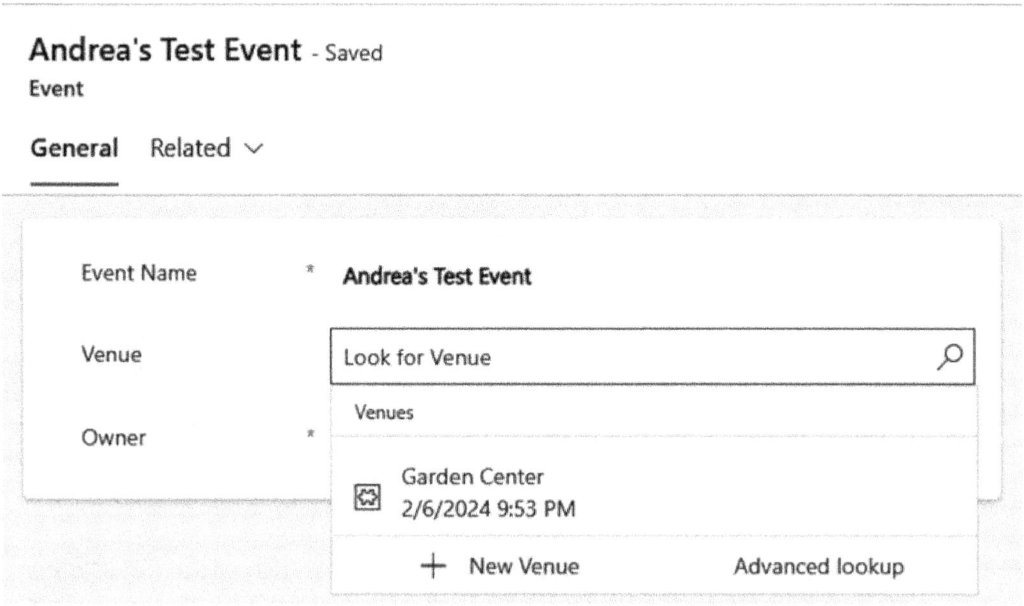

Andrea's Test Event - Saved
Event

General   Related ∨

Event Name   *   **Andrea's Test Event**

Venue   Look for Venue   🔍

                    Venues

Owner   *   Garden Center
                    2/6/2024 9:53 PM

                +   New Venue   Advanced lookup

Figure 4.33 – Choosing Garden Center from the Venue lookup

12. Click **Save & Close**.

13. Click on the **Edit columns** button next to the **Edit filters** button. Then, click **Add columns** from the right slide-out pane. Click **Venue** and then **Close**. Next, click and drag **Venue** above **Created On**. Finally, click **Apply**:

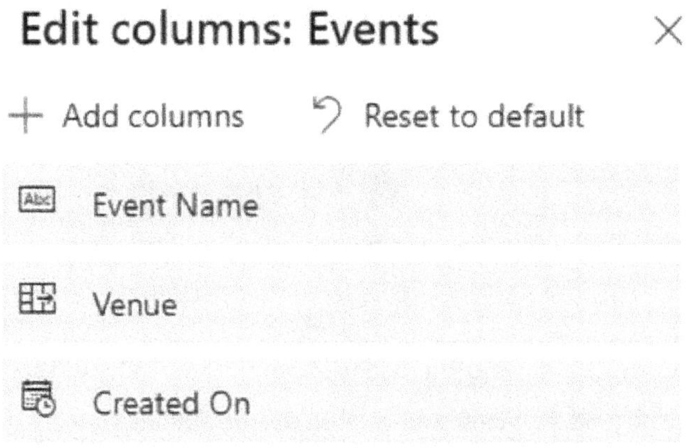

Edit columns: Events   ✕

+ Add columns   ↺ Reset to default

[Abc]   Event Name

[⊞]   Venue

[⊡]   Created On

Figure 4.34 – Edit columns: Events | Venue

14. Now, we can see **Venue** in our **Active Events** view:

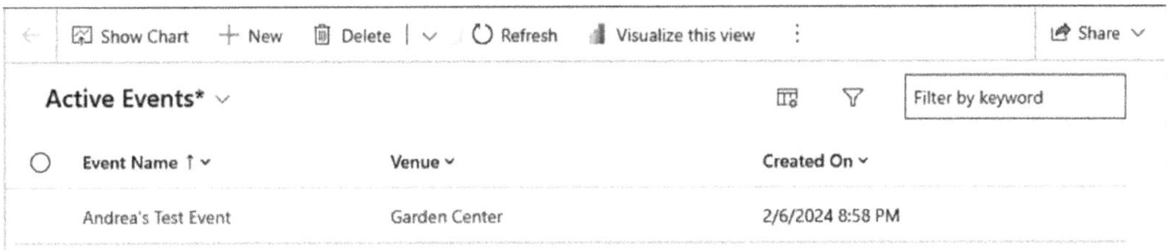

Figure 4.35 – Andrea's Test Event

Notice the asterisk next to the **Active Events** label. This means that this view hasn't been saved and will go back to the default **Active Events** view after the browser refreshes or we move away from the screen.

15. Click the drop-down button next to **Active Events**.

16. Click **Save as new view**:

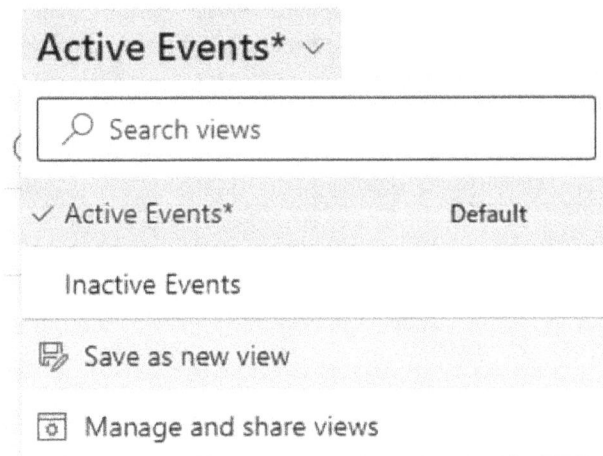

Figure 4.36 – Save as new view

17. Name this new view Events with Venue and give it a description.

18. Click **Save**:

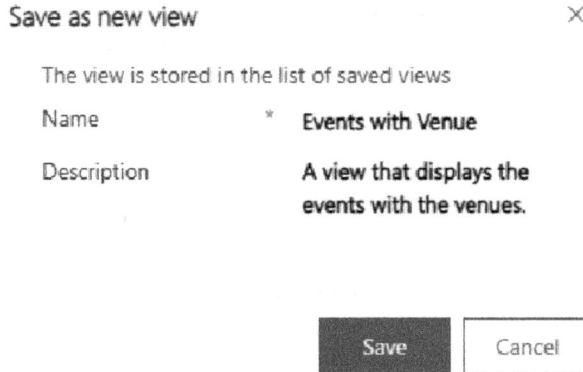

Figure 4.37 – An example of what should be added to the Save as new view popup

Notice that your new view has replaced the **Active Events** view. This view you just created is only visible to you unless you share it with others.

19. Next, click on the dropdown next to **Events with Venue**. Then, click on **Manage and share views**. Finally, click on the three dots next to **Events with Venue** and click **Share**:

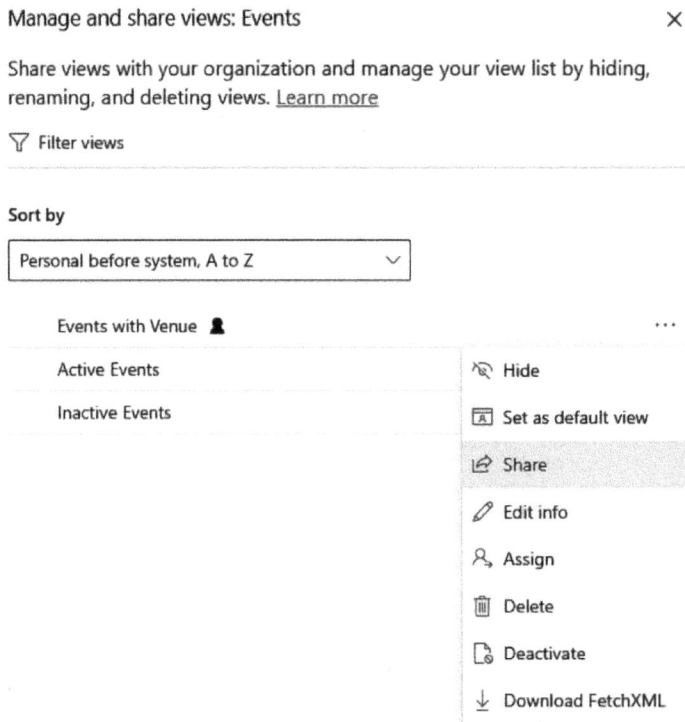

Figure 4.38 – The Share button

20. Start typing the name of someone you know who is in your organization and has access to the model-driven app:

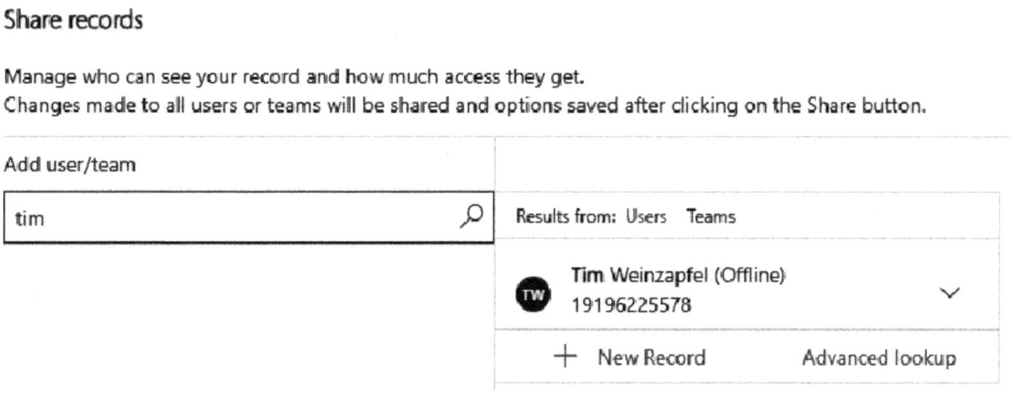

Share records

Manage who can see your record and how much access they get.
Changes made to all users or teams will be shared and options saved after clicking on the Share button.

Add user/team	
tim 🔍	Results from: Users Teams
	Tim Weinzapfel (Offline) 🔽
	TW 19196225578
	＋ New Record Advanced lookup

Figure 4.39 – Adding a name to share the view

21. Feel free to assign that person specific permissions. Click **Share**:

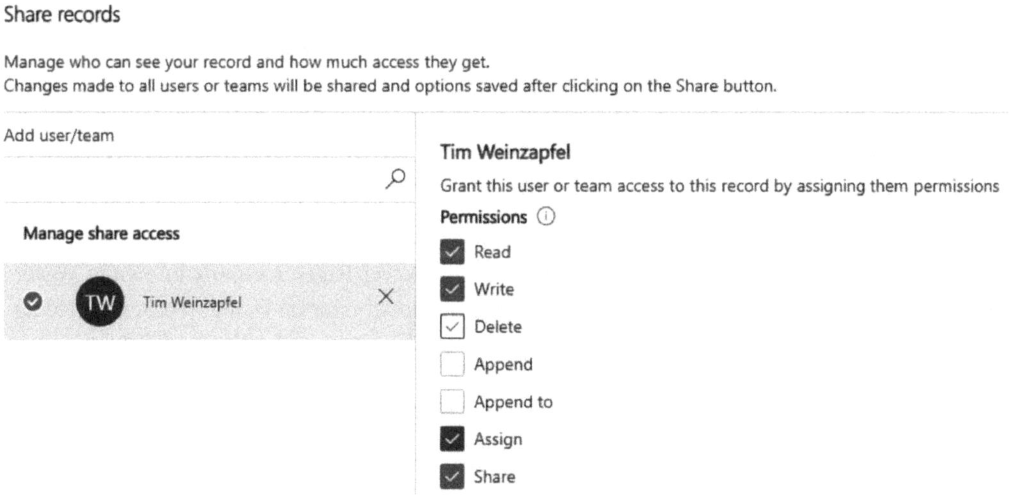

Share records

Manage who can see your record and how much access they get.
Changes made to all users or teams will be shared and options saved after clicking on the Share button.

Add user/team	**Tim Weinzapfel**
🔍	Grant this user or team access to this record by assigning them permissions
Manage share access	Permissions ⓘ
	✅ Read
✅ TW Tim Weinzapfel ✕	✅ Write
	☑ Delete
	☐ Append
	☐ Append to
	✅ Assign
	✅ Share

Figure 4.40 – Applying permissions

In conclusion, model-driven apps serve as the backbone for internal business operations, offering structured data management and process automation for teams, admins, IT professionals, and developers. Unlike canvas apps, where designers have full control over layout and design, model-driven apps provide a standardized interface across devices, ensuring consistency and ease of use.

While Power Pages apps are ideal for creating external-facing websites, model-driven apps excel in managing data at scale within an organization. With their rapid build process, consistent user interface, and seamless migration between environments, model-driven apps streamline app development and enhance operational efficiency.

Choosing the right tool

Consider the following tips when you need to select the right tool for the job:

- Use **canvas apps** if you need a quick, low-code solution for internal apps with a focus on mobile and tablet accessibility

- Choose **model-driven apps** if your app is data-centric, requiring robust data handling and integration with Dataverse or Dynamics 365

- Opt for **Power Pages apps** if you need to create a public-facing or external user-accessible web app with extensive customization capabilities

Summary

In this chapter, we explored the fundamental differences between canvas apps, Power Pages, and model-driven apps within the Microsoft Power Apps ecosystem. We began by understanding when to use each type of app and considered a comprehensive set of factors, including primary use case, development style (low code versus full web development), target audience, integration capabilities, customization options, user experience, accessibility, data management capabilities, security features, ease of use, deployment speed, scalability, cost implications, learning curve, and potential disadvantages.

Canvas apps emerged as the go-to choice for creating front-facing apps for end users within an organization, offering flexibility and ease of development. Power Pages, formerly known as Power Apps portals or Dynamics 365 portals, proved valuable for building external-facing websites, providing a secure, low-code platform for rapid website creation and deployment. Lastly, model-driven apps emerged as the preferred solution for internal business teams, admins, and IT professionals, offering structured data management and process automation at scale.

By discerning the unique strengths and capabilities of each platform, you have gained valuable insights into optimizing your app development process and avoiding potential pitfalls, ensuring efficient and effective outcomes.

The next chapter will build upon this foundation by guiding you through the process of connecting data from various sources to your Power Apps, including SharePoint integration, data mapping using dataflows, and leveraging Dataverse. This practical demonstration will provide you with invaluable skills so that you can enhance the functionality and usability of your Power Apps through seamless data integration.

5

Data Connections

In this chapter, we are going to cover different data connections that you can use when connecting to your Power App. Power Apps allows you to create a seamless connection to a wide array of data sources. Whether it's a simple task-tracking app leveraging data from Excel stored in OneDrive, a customer engagement app drawing on Dynamics 365, or a complex project management tool integrated with SQL Server, data connections are the lifelines that infuse apps with power and purpose.

This chapter will explore three of the more popular data sources that Power Apps can connect to – Excel, SharePoint, and Dataverse. We'll cover the details of each connection type, offering insights into their unique characteristics, advantages, and considerations. We will also touch on other data connections.

Specifically, we are going to cover the following topics:

- Using Excel as a data source
- Using SharePoint Lists as a data source
- Using Dataverse as a data source
- Comparing Excel, SharePoint Lists, and Dataverse
- Other data connections

Technical requirements

To follow the subsequent chapters in this book, you will need to have access to Microsoft Power Apps. If you currently do not have access to Power Apps, such as through your employer, you can still sign up for a Power Apps Developer plan. You can find the files used in this chapter on GitHub: `https://github.com/PacktPublishing/Power-Apps-Tips-Tricks-and-Best-Practices/tree/main/Chapter05`

Using Excel as a data source

Excel can be a popular choice as a data source for your app due to its widespread use and familiarity. When creating a new app, creating one based on an existing Excel file is easy and there are various ways to go about this. The first and simplest approach is to create a new app by uploading an Excel file. This approach allows you to create an app with minimal to no coding or design as Power Apps does most of the work for you. Another approach is to start with a blank canvas app and then connect to Excel. We'll cover both approaches in this section.

Creating a new Power App by uploading an Excel file

If you have an existing Excel file, Power Apps makes it very easy to quickly create a Power App with minimal coding. In this example, we will use an Excel file entitled `Events Planning Data - Chapter 5.xlsx`. This file is available on this book's GitHub repository if you would like to download and follow along. A subset of the data is shown here:

	A	B	C
1	Title	Vendor Type	Description
2	Bob's Catering	Catering and Food Services	Offers overall food and catering services
3	Rocking DJ	Music and Entertainment	Offers DJ and music
4	All is Safe Security	Security	Security Services
5	Green Thumb Florist	Floral and Decor	A full service floral business
6	ABC Beer and Liquors	Catering and Food Services	They provide full bartender including beer, wine, and spirits
7	Louie's Limousine Services	Transportation	Offers up limousine and other transportation services
8	Flashy Lighting Technologies	Music and Entertainment	Provides lighting and other visual effects.

Figure 5.1 – Sample Excel file data

To create the app, simply follow these steps:

1. Start at the Power Apps home page and select **Start with data**, as shown in *Figure 5.2*:

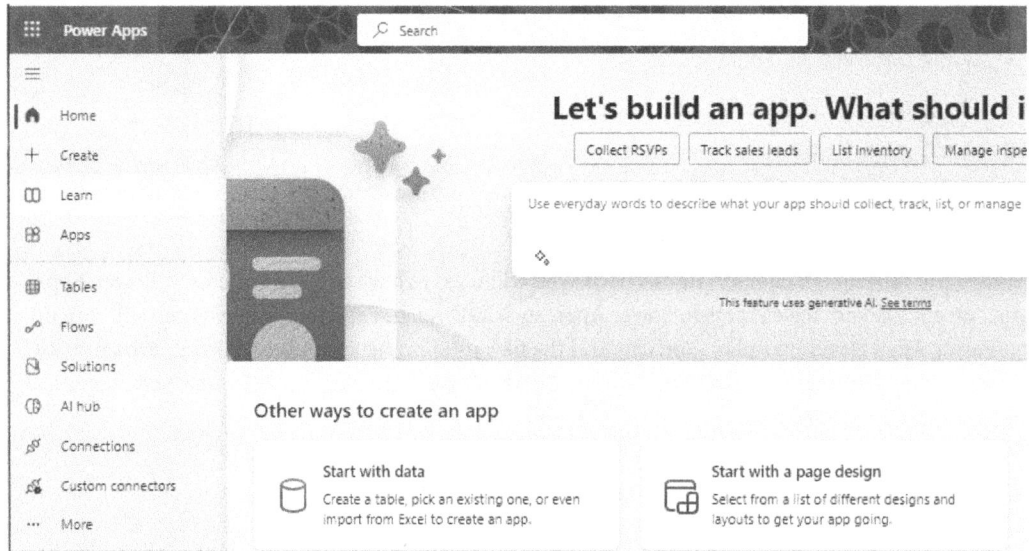

Figure 5.2 – Creating an app with data

2. Then, select the **Upload an Excel file** option, as shown in *Figure 5.3*:

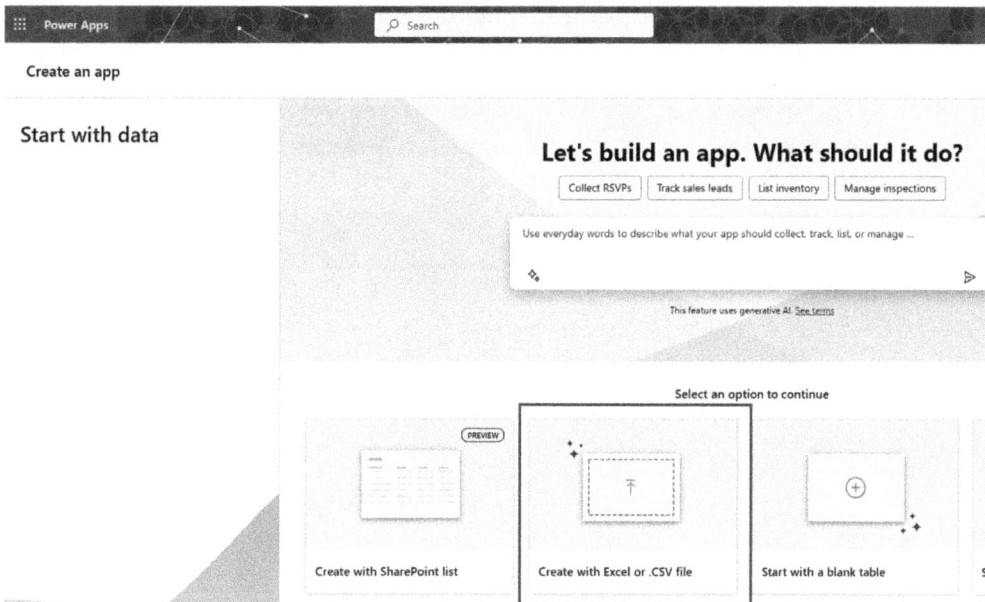

Figure 5.3 – Uploading an Excel file

3. After selecting the applicable Excel file, Power Apps will provide a sample preview of the data. Then, click on the **Create app** button at the bottom right of the screen, as shown in *Figure 5.4*:

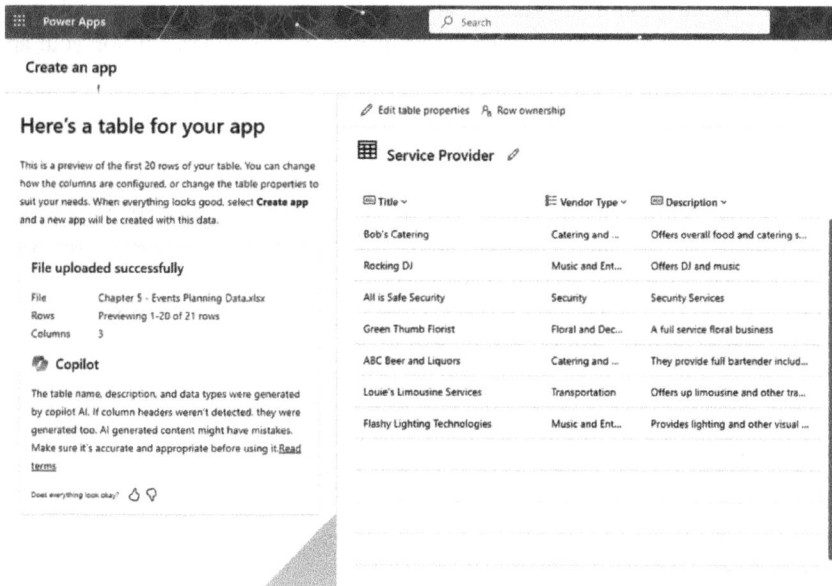

Figure 5.4 – Sample data preview (note the option to use the first row as column headers)

4. That's it! Power Apps will automatically create the app without any involvement, coding, or design work necessary. This app will be fully functional and complete and can add new data, edit existing data, or even delete data. *Figure 5.5* provides an example of the fully completed app:

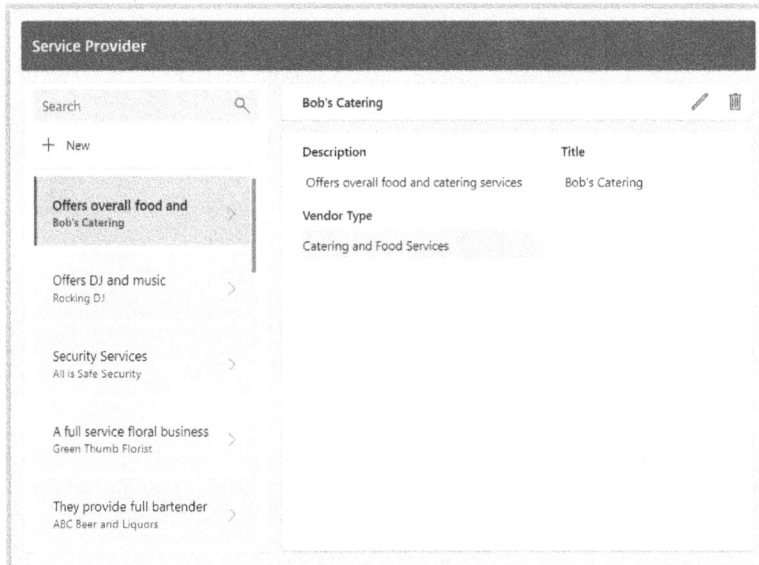

Figure 5.5 – Completed Power App

This approach allows you to create a fully functional app within minutes without any coding or design work required. However, you may want to start with a blank canvas app and develop your app from scratch. We will cover that in the next section.

Creating a blank canvas app with Excel

In the previous example, we showed how you can quickly create an app by uploading an Excel file. However, you can also create a blank canvas app and connect it to Excel. *Chapter 4* covered how to create a blank canvas app. However, instead of using and connecting to the Dataverse that was used in the example in *Chapter 4*, we will select Excel as our data. However, there are several connection approaches that we have available, as shown in *Figure 5.6*:

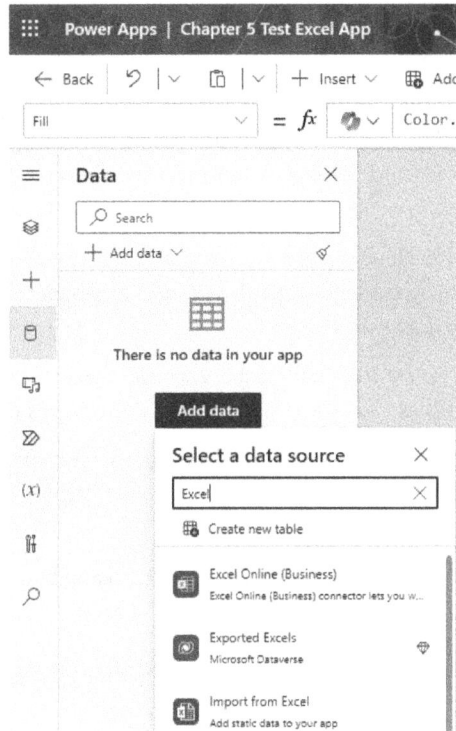

Figure 5.6 – Excel connection options

Let's take a closer look at these approaches:

- **Excel Online (Business)**: This option allows you to connect to Excel files in document libraries supported by Microsoft Graphs. This includes OneDrive for Business, SharePoint Sites, and Office 365 Groups. This option not only allows you to import the data but also to write data back to your Excel file. Please be aware that your Excel data must be formatted as a table.

- **Import from Excel**: This option imports the data as static data. There is no writing data back to the original Excel file. This option does not require the data to be formatted as a table.

Note

There is a third option for Excel called **Exported Excel**. This data source involves connecting to data exported from Microsoft Dataverse to Excel. We will only focus on the first two here.

Using Excel as a data source for Power Apps comes with its own set of advantages and disadvantages. It's a popular choice for many users due to Excel's widespread use and familiarity, but there are considerations to keep in mind, depending on the complexity and scale of the app. We will have a detailed look at the pros and cons next.

Advantages of Excel

There are several advantages to using Excel as a data source. These include the following:

- **Ease of use and accessibility**: Excel is a widely used tool with a familiar interface for many users. This makes it easy to set up and manage data for Power Apps without requiring specialized database knowledge.

- **Rapid prototyping**: For developers and business analysts, Excel allows you to quickly test and prototype Power Apps solutions. Changes can be made swiftly in the Excel file, and the effects can be seen almost immediately in the app.

- **Flexibility**: Excel's formula and calculation capabilities make it a versatile tool for data manipulation and analysis. Such capabilities can be leveraged within Power Apps for dynamic data handling.

- **Integration with Microsoft 365**: Being part of the Microsoft ecosystem, Excel files stored in OneDrive or SharePoint can be easily connected to Power Apps, enabling seamless data flow and updates.

- **Cost-effective**: For small businesses or individual users, using Excel as a data source can be a cost-effective solution since it avoids the need for additional database software or infrastructure.

Now that we've covered the advantages, let's cover some of the disadvantages of using Excel.

Disadvantages of Excel

The following are some key disadvantages you should consider when using Excel as a data source:

- **Scalability issues**: Excel is not designed to handle large datasets or serve as a database for high-concurrency apps. Performance can degrade significantly as the amount of data grows, affecting app responsiveness.

- **Data integrity**: Unlike databases that offer transactional integrity and robust data validation mechanisms, Excel has limited capabilities in this area. This can lead to data inconsistencies and errors, especially with multiple users editing the file simultaneously.

- **Security concerns**: Managing access and securing data in Excel can be challenging, particularly for sensitive information. Excel lacks the sophisticated security features of database management systems.

- **Complexity in collaboration**: While Excel files can be shared and edited by multiple users (especially when hosted on OneDrive or SharePoint), managing versions and ensuring data consistency can become complex in collaborative scenarios.

- **Limited relational data support**: Excel is not inherently designed to handle relational data efficiently. Designing complex data models that require relationships between tables is more cumbersome and less intuitive than in dedicated database systems.

- **Dependency on Microsoft 365 environment**: For Excel files to be used effectively as data sources in Power Apps, they typically need to be stored in OneDrive or SharePoint. This creates a dependency on the Microsoft 365 environment, which might not always be desirable or feasible.

In summary, while Excel can be a convenient and accessible option for small-scale or prototype Power Apps projects, its limitations in terms of scalability, data integrity, and security make it less suitable for more complex, high-volume, or sensitive apps. It's essential to evaluate the specific needs and constraints of your project before deciding on Excel as your data source. Let's explore another more popular option: **SharePoint Lists**.

Using SharePoint Lists as a data source

SharePoint Lists are a very popular data source for Power Apps and for good reason. SharePoint offers not only robust document management but SharePoint Lists also connect seamlessly with Power Apps.

Let's dig into the specific steps around creating a Power App from SharePoint.

Creating a Power App from SharePoint

There are several ways to create a Power App from a SharePoint list. The first – and easiest – way is to create an app directly from your SharePoint list. While in your SharePoint list, you have the option to integrate with Power Apps. This can be done as follows:

1. View your list directly in SharePoint.
2. Select **Integrate** above the list. This will open an additional menu.
3. Select **Power Apps**, then **Create an app**, as shown in *Figure 5.7*:

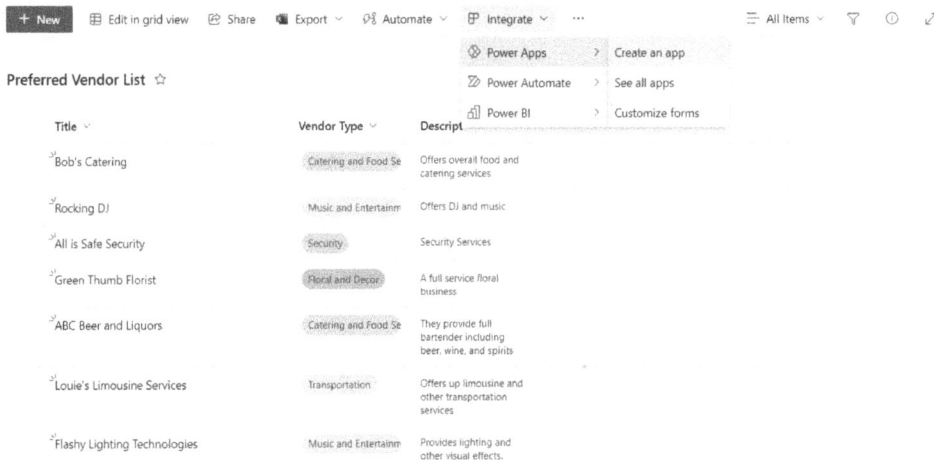

Figure 5.7 – Creating a Power App from a SharePoint list

4. Enter a name for your app. In our example, we'll use `Preferred Vendor List App`:

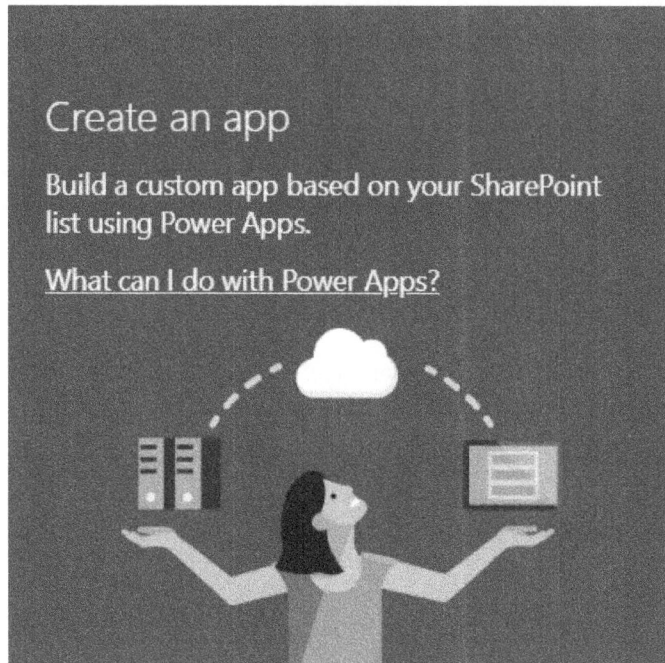

Figure 5.8 – Naming the app

5. Power Apps will automatically create a three-screen app that will consist of a main gallery page, a page to view a specific record, and an edit screen to add a new item or edit an existing item. This app is now connected to your SharePoint list, is fully functional, and is ready to be used. No coding is required:

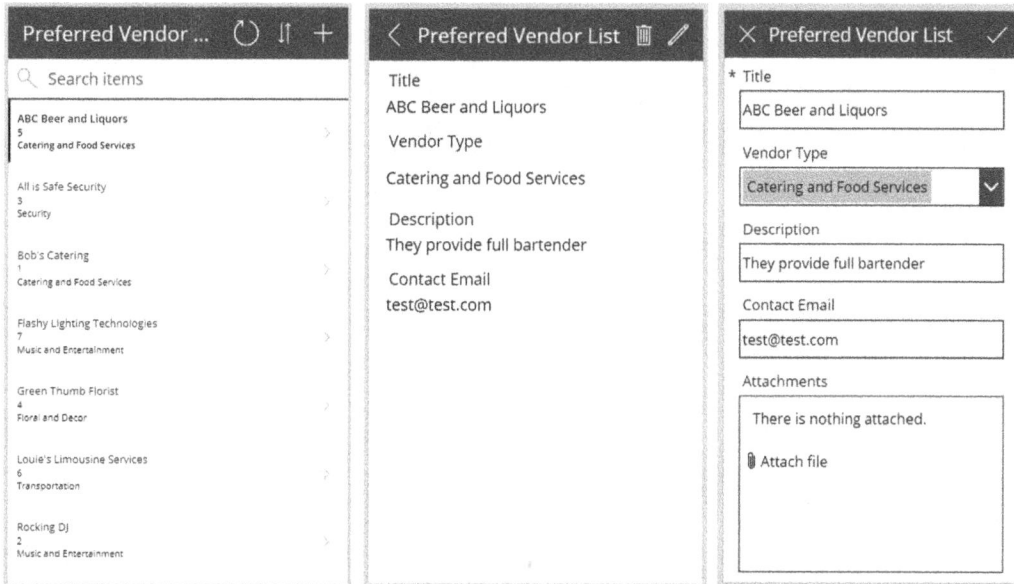

Figure 5.9 – View of the three screens created in the app

> **Note**
>
> When creating an app directly from a SharePoint list, the layout will always be in a mobile phone layout and always consist of the same three pages. If you prefer to use a tablet or custom-size app, then this approach should not be used. Furthermore, the app will be saved to the default environment. Therefore, the app will need to be moved if a different environment and/or solution is used.

Creating a new Power App from a SharePoint list

We can automatically create an app from a SharePoint list by following a process similar to what was described in the *Creating a blank canvas app with Excel* section. Let's walk through this process:

1. Go to the Power Apps home page and select **Start with data**. Then, select **Select external data**, as shown in *Figure 5.10*:

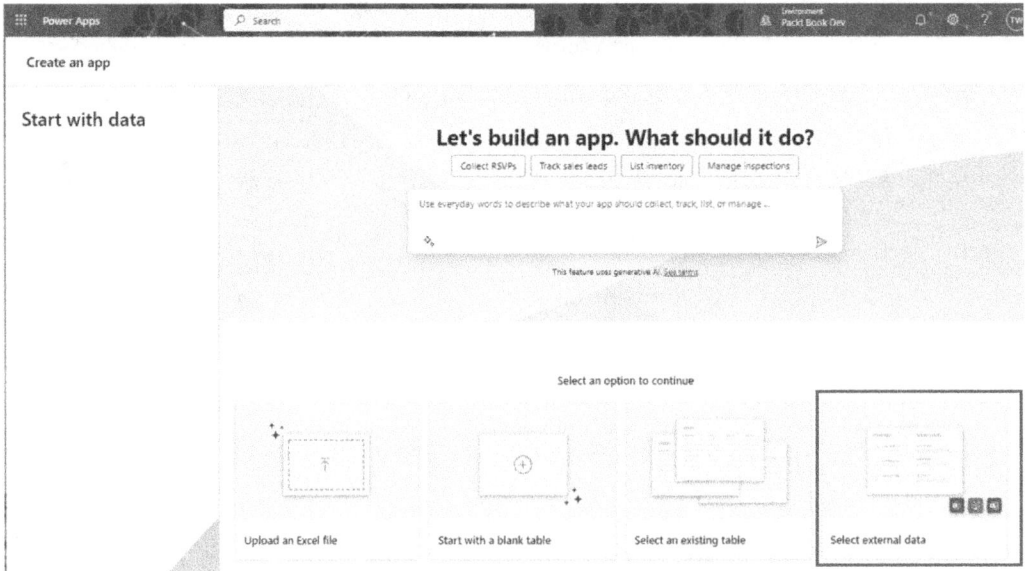

Figure 5.10 – Select external data

2. Select the SharePoint location and list from the sections shown in *Figure 5.11*:

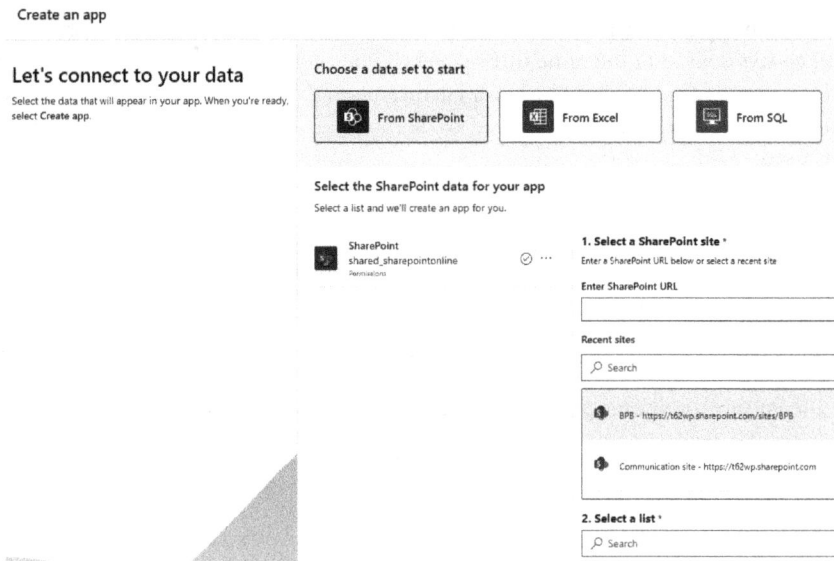

Figure 5.11 – Select a SharePoint site

3. This will create an app, similar to what happened in the *Creating a blank canvas app with Excel* section.

Similar to Excel, using SharePoint Lists also has its advantages and disadvantages. Let's take a look.

Advantages of SharePoint Lists

The following are some of the advantages of using a SharePoint list as your Power App data source:

- **Version control and data integrity**: SharePoint Lists offer better version control over data versus Excel. With Excel, while users can connect to the spreadsheet directly, it is also possible that copies are downloaded onto their drives, with data being added. This results in data integrity issues as data is no longer in one central location. Furthermore, in SharePoint Lists, it is much easier to add data validation rules so that specific columns contain the correct type of data, such as numerical information, only the available choices, and so on.

- **Customization and rich data types**: SharePoint Lists supports a wide range of data types and custom columns, including choice, lookup, and person or group types, allowing for detailed and structured data collection that can be directly used in Power Apps. Even rich text format can be used in SharePoint Lists.

- **Security and permission management**: SharePoint provides granular control over permissions, enabling secure data access and management at both the list and item levels. This ensures that sensitive information can be protected and only made accessible to authorized users.

- **Scalability**: SharePoint Lists can handle large amounts of data and support up to 30 million items per list, making it suitable for a wide range of apps, from small departmental tools to enterprise-level solutions.

Disadvantages of SharePoint Lists

The following are some areas to consider when using a SharePoint list as your data source:

- **Complexity for large datasets**: While SharePoint Lists can store large volumes of data, performance can become an issue with very large datasets, especially when complex queries or multiple views are involved.

- **User experience and customization limits**: Although SharePoint and Power Apps offer a range of customization options, there may be limitations in achieving the exact look, feel, and functionality compared to custom-developed apps.

- **Learning curve**: Properly utilizing SharePoint Lists as a data source in Power Apps requires familiarity with both platforms. Users new to SharePoint or Power Apps may face a learning curve to effectively use these tools together. However, note that SharePoint's lists are very easy to implement and maintain.

- **Dependency on SharePoint infrastructure**: Relying on SharePoint Lists as a data source means that your Power Apps are tied to the SharePoint infrastructure, which might not always align with an organization's IT strategy or external data storage preferences.

- **Data structure constraints**: SharePoint's lists are essentially flat, with limited support for complex relational data models. Designing apps that require intricate relationships between data entities can be challenging.

In summary, SharePoint Lists offers a robust and integrated solution for storing and managing data within the Microsoft ecosystem, providing significant advantages for collaboration, data integrity, and security. However, considerations regarding performance, customization, and complexity must be addressed to ensure that the use of SharePoint Lists as a data source aligns with the specific needs and capabilities of your Power Apps project.

Now that we've covered both Excel and SharePoint, we will consider using Dataverse as a data source.

Using Dataverse as a data source

Using Microsoft Dataverse as a data source for Power Apps provides a robust and secure platform for building and deploying professional-grade apps. Dataverse, formerly known as the **Common Data Service** (**CDS**), is an integral part of Microsoft Power Platform. It offers a cloud-based storage and data management solution that is not only powerful but also highly flexible, enabling organizations to store and manage data used by business apps in a secure environment.

In this section, we'll explore using Dataverse as a data source for Power Apps. We'll start by examining its key features and benefits, followed by important considerations for its use. Finally, we'll provide a step-by-step guide on how to add data via Dataverse, ensuring a smooth integration process for our Power Apps solutions.

Overview of Dataverse

Microsoft Dataverse is designed to facilitate the development of apps by providing a unified and standardized data schema, thus enabling data to be shared and used across apps seamlessly. It supports a wide range of standard and custom data types, relationships, and business logic, making it an excellent choice for complex business apps that require integration with various Microsoft services and external data sources.

Key features and benefits

Dataverse offers a plethora of key features and benefits that make it a powerful data source for Power Apps. Some of these include its rich data model, seamless integration with the Microsoft ecosystem, global accessibility, robust security measures, support for logic and validation, and its suitability for both no-code and low-code development approaches. Let's explore each of these aspects to understand how they contribute to creating robust and versatile Power Apps solutions:

- **Rich data model**: Dataverse provides a comprehensive data model that supports complex entities, relationships, and business rules, enabling developers to model real-world data and processes effectively.

- **Integration with the Microsoft ecosystem**: Dataverse is deeply integrated with other Microsoft services, including Fabric, Power Apps, Power Automate, Power BI, and Dynamics 365. This integration provides a seamless experience for users and developers, allowing for the easy creation of end-to-end business solutions.

- **Global accessibility and security**: Hosted in Microsoft's cloud infrastructure, Dataverse benefits from global availability, built-in data security, and various compliance features that adhere to Microsoft's strict security standards.

- **Logic and validation**: With support for server-side logic and validation rules, Dataverse ensures data integrity and consistency across all apps that access the data, minimizing the potential for errors.

- **No-code/low-code development**: Dataverse is designed to be accessible to both professional developers and business analysts, supporting a no-code/low-code approach to app development. This democratizes the app development process, enabling users without deep technical skills to build and customize apps.

Considerations for using Dataverse

When considering the use of Dataverse as a data source for your Power Apps, there are several important factors to keep in mind. These include the learning curve associated with understanding Dataverse's data model and capabilities, considerations regarding cost and licensing, strategies for optimizing performance, and the potential complexities of customization. We'll explore each of these considerations in detail to ensure you can implement Dataverse in your Power Apps solutions successfully:

- **Learning curve**: While Dataverse is designed to be user-friendly, there is still a learning curve associated with understanding its data model, capabilities, and best practices for design and development.

- **Cost and licensing**: Access to Dataverse and its features is subject to Microsoft's licensing terms, which can vary based on the organization's needs and the scale of the deployment. It's important to consider the cost implications when planning to use Dataverse as a data source.

- **Performance optimization**: As with any database platform, designing efficient data models and queries is crucial to ensuring good performance. Overly complex data models or inefficient queries can impact app responsiveness.

- **Customization and complexity**: While Dataverse's flexibility and customization options are significant advantages, they can also lead to complexity, especially in large-scale implementations. Proper planning and governance are essential to maintain manageability.

In summary, Microsoft Dataverse offers a powerful and secure platform for data management, making it an excellent choice for Power Apps development. Its integration with the broader Microsoft ecosystem, combined with its extensive data modeling capabilities, provides a solid foundation for building enterprise-grade apps. However, organizations should carefully consider the platform's learning curve, cost, and governance requirements to ensure successful implementation.

A step-by-step guide on how to add data via Dataverse

As we continue to explore the practical aspects of utilizing Dataverse as a data source, we'll walk through a detailed step-by-step guide on how to add data directly within the Dataverse environment. From navigating to the appropriate tables within the **Event Planning Project** solution to inputting and saving new data entries, each action will be outlined to ensure a smooth and seamless process. By following these instructions meticulously, you'll gain firsthand experience in leveraging Dataverse for efficient data management and integration within your Power Apps solutions:

1. Start by navigating to the **Event Planning Project** solution and opening the **Vendor** table.

2. Under **Vendor columns and data**, click **Edit**.

3. Click on the +**14 more** dropdown on the far right:

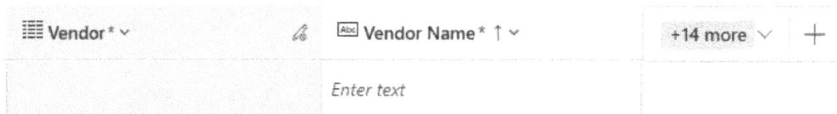

Figure 5.12 – The +14 more dropdown

4. Click on the checkboxes next to **Created On** and **Created By (Delegate)** to remove those from our grid, then click **Save**:

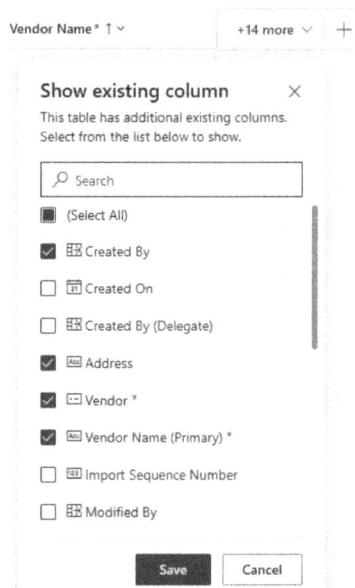

Figure 5.13 – Selecting existing columns

5. Double-click the blank box under **Vendor Name** and type `Restaurant`:

    A. Notice that the Vendor GUID auto-fills after you click out of the first box.

    B. Notice that **Created By** auto-fills as well.

6. Double-click the blank box under **Address** and type `123 Main St.`

7. Your grid will look as follows. Feel free to add more vendors to this list:

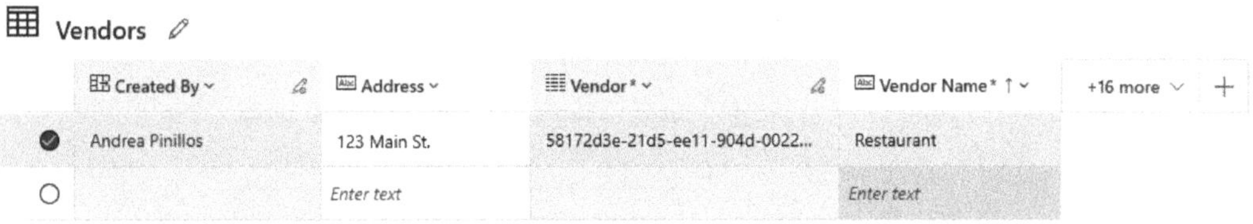

⊞ Vendors 🖉

⊞ Created By ˅	🔏	🗛 Address ˅	⊟ Vendor * ˅	🔏	🗛 Vendor Name * ↑ ˅	+16 more ˅	+
✅ Andrea Pinillos		123 Main St.	58172d3e-21d5-ee11-904d-0022...		Restaurant		
○		*Enter text*			*Enter text*		

Figure 5.14 – Vendor table data

8. Click the **Back** button and click on **Apps**, then click **Event Planning Model-Driven App**. We're doing this to validate that the data is populating in both the backend and the frontend.

9. Click **Play**.

10. Click **New in the Event table**, which can be found in the top ribbon of the model-driven app, and name the event `Data Event`.

11. Click on the magnifying glass in the **Venue** field and click **Garden Center**.

12. Click on the **Related** tab and click **Event Vendors**.

13. Click **New Event Vendor**.

14. Next to the **Event Vendor Name** field, type **Data Vendor**. This is the field in the form where you will input data.

15. Next to **Vendor**, hover over the empty field and click the magnifying glass to show all options.

16. You should see **Restaurant** as an option; click it:

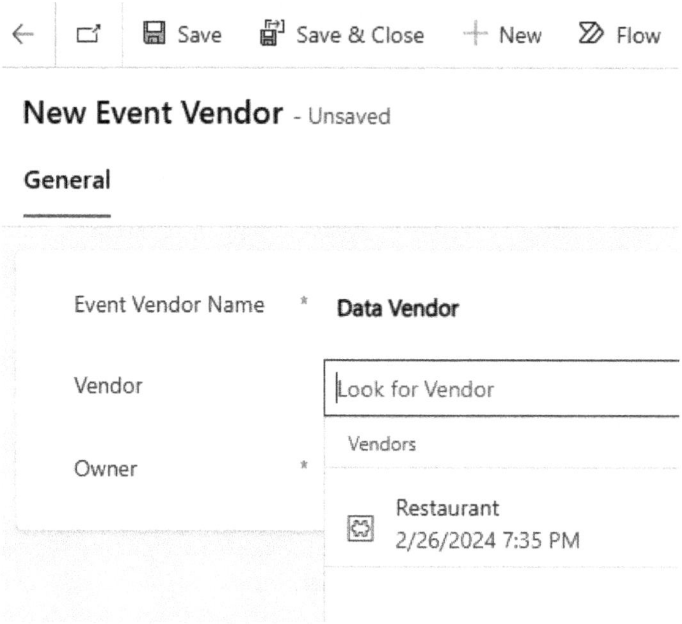

Figure 5.15 – The Event Vendor Name field in the model-driven app form

17. Click **Save**.

18. Then, click **Restaurant** to see the **Vendor** form with the data we added in Dataverse:

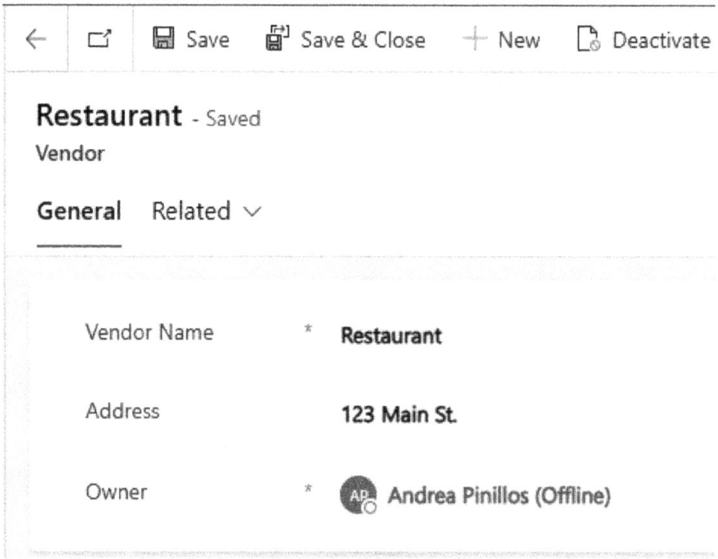

Figure 5.16 – The Vendor form within the model-driven app

This section provided a comprehensive step-by-step process for adding data to Dataverse and accessing it within a model-driven app. From configuring dataflows and importing data to troubleshooting potential issues, each stage has been covered meticulously. Additionally, we explored how to navigate the model-driven app interface to locate and interact with the imported data seamlessly. By following these instructions, you now possess the knowledge and skills necessary to effectively manage data within the Dataverse ecosystem, empowering you to harness its full potential for enhanced data integration and analysis.

Now that we've covered using Excel, SharePoint, and Dataverse as potential data sources for your Power App, let's provide a comparison.

Comparing Excel, SharePoint Lists, and Dataverse

So far, we've covered three popular data sources for Power Apps: Excel, SharePoint Lists, and Dataverse. In this section, we'll provide a summary of each, including their benefits and any considerations. We'll consider the following areas:

- Ease of use
- Cost
- Scalability
- Data integrity and security
- Collaboration and integration
- Complexity and learning curve
- Customization and flexibility

The following table provides this comparison:

Area	Excel	SharePoint Lists	Dataverse
Ease of use	High ease of use; familiar interface and rapid prototyping capabilities	Moderate; requires some familiarity with SharePoint but generally user-friendly	Moderate to high; requires an understanding of the data model but accessible through a no-code/low-code approach
Cost	Cost-effective; included in most Microsoft 365 plans, with no additional software needed	Cost-effective; included in Microsoft 365 subscriptions	Higher cost; requires additional licensing, which can vary based on usage and features

Area	Excel	SharePoint Lists	Dataverse
Scalability	Limited scalability and performance issues with large datasets	Moderate scalability; supports up to 30 million items but may have performance issues with complex queries	High scalability; designed for large and complex datasets with efficient data handling
Data integrity and security	Limited; lacks robust data validation and security features	Good; offers version control, data validation, and granular permissions	Excellent; strong data integrity and built-in security and compliance features
Collaboration and integration	Moderate; easy to share but can have issues with data integrity and version control	Good; supports collaboration with version control and integration within Microsoft 365	Excellent; seamless integration with the Microsoft ecosystem, including Power Apps, Power Automate, and Power BI
Complexity and learning curve	Low complexity; minimal learning curve, especially for existing Excel users	Moderate; requires learning SharePoint basics but this is manageable for most users	Higher; involves learning Dataverse's data model and capabilities but is designed to be approachable
Customization and flexibility	High flexibility with formulas and calculations, but limited in terms of relational data support	Moderate; supports rich data types and custom columns but is limited in complex relational models	High; supports complex data models, relationships, and business rules, allowing for significant customization

Table 5.1 – A comparison of Excel, SharePoint Lists, and Dataverse

Other data connections

So far, we have covered three popular data source connections. However, there are significantly more data sources that can be used within your app. This includes more native options, from SQL Server and Azure SQL Database to third-party connectors. Providing an in-depth walkthrough on these connectors is beyond the scope of this book. In addition, many connectors require additional licenses for access.

The important point to remember is that Power Apps can connect to a significant number of data sources, depending on what apps are used within your organization (as well as any licensing requirements). Furthermore, as you learn how to connect to the more common data sources to build out your app, it may build a foundation for working with other apps.

Using dataflows as a data source

Using a dataflow as a data source in Power Apps opens up advanced scenarios for data preparation, transformation, and integration, allowing developers and business users to create more sophisticated and data-centric apps. Dataflows are a feature within Microsoft's Power Platform that enables users to ingest, transform, cleanse, and enrich data from a wide variety of sources, making it readily available for apps, automated workflows, and analytics.

In this section, we will explore dataflows in Power Apps, focusing on their key features and benefits, considerations for their use, and a step-by-step guide on utilizing dataflows as data sources. These subsections will provide you with comprehensive insights and practical guidance for leveraging dataflows effectively in your Power Apps development journey.

Overview of dataflows in Power Apps

Dataflows are designed to bring the power of data warehousing and **extract, transform, and load** (ETL) processes directly into the hands of app developers and business analysts without them requiring deep technical expertise in data engineering. By using a familiar Power Query-based interface, users can define and execute data preparation steps that consolidate data from diverse sources into a common format, store it in the CDS (now part of Microsoft Dataverse), and make it accessible to Power Apps and other components of Power Platform.

Key features and benefits

In this section, we'll explore the key features and benefits of using dataflows in Power Apps, highlighting their pivotal role in data integration, the Power Query experience they offer, their automated refresh capabilities, and their contribution to centralized data management:

- **Data integration**: Dataflows can connect to a broad range of data sources, including cloud-based services, databases, Excel files, and more, enabling users to bring data together from across their digital landscape.

- **Power Query experience**: Leveraging the Power Query tool, dataflows offer a rich set of capabilities for data transformation and cleaning, such as filtering, aggregation, and merging of data, all through a user-friendly interface.

- **Automated refresh**: Dataflows can be configured to refresh data at scheduled intervals, ensuring that Power Apps always has access to the latest information without manual intervention.

- **Centralized data management**: By storing transformed data in Microsoft Dataverse, dataflows facilitate a centralized approach to data management, improving data consistency and security across apps.

Considerations for using dataflows

As we explore using dataflows, it's important to consider various factors. These include complexity, performance, licensing, and data governance, each influencing the effectiveness of dataflows in Power Apps development:

- **Complexity and learning curve**: While dataflows abstract away much of the complexity associated with traditional ETL processes, there is still a learning curve to use them effectively, especially for users unfamiliar with data transformation concepts.

- **Performance and scalability**: The performance of Power Apps relying on dataflows can be influenced by the size of the datasets, the complexity of the transformations, and the frequency of data refreshes. It's important to plan and optimize dataflows to meet the app's performance requirements.

- **Licensing and cost**: Access to dataflows and the volume of data storage and processing may be subject to licensing restrictions and additional costs, depending on the organization's subscription to Power Platform and Azure services.

- **Data governance**: As with any data integration and transformation tool, proper data governance practices are crucial to ensure data quality, security, and compliance with regulatory requirements.

In summary, dataflows offer a powerful way to enhance Power Apps with sophisticated data processing capabilities, making them an excellent choice for scenarios requiring complex data integration and transformation. However, leveraging dataflows effectively requires planning carefully and considering factors such as complexity, performance, and cost.

A step-by-step guide – utilizing dataflows as data sources

In this section, we'll learn how to integrate a dataflow as a vital component within the Power Apps ecosystem. It's important to note that Power Apps cannot directly connect to a dataflow. Instead, dataflows transform and load data into Dataverse or other target data sources, which Power Apps can then connect to. Follow along with our comprehensive step-by-step guide to understand how to use dataflows to seamlessly incorporate external data sources into your Dataverse, and subsequently integrate this data into your Power Apps environment.

Part 1 – setting up an Excel sheet

Before we dive into creating dataflows in Power Apps, we'll start by setting up our data source. Open an Excel sheet and define the necessary column headers while leveraging the provided sample data for guidance:

1. Open an Excel sheet on your local machine and add two column headers:

    C. Column header 1: **Attendee Name**

    D. Column header 2: **Attendee Email**

2. Use the following Excel sheet as a reference and use the data, if needed:

Attendee Name	Attendee Email
John Smith	john.smith@example.com
Emily Johnson	emily.johnson@example.com
Michael Williams	michael.williams@example.com
Sarah Brown	sarah.brown@example.com
David Jones	david.jones@example.com
Jessica Davis	jessica.davis@example.com
Christopher Miller	christopher.miller@example.com
Amanda Wilson	amanda.wilson@example.com
Matthew Taylor	matthew.taylor@example.com
Jennifer Martinez	jennifer.martinez@example.com
Andrew Anderson	andrew.anderson@example.com
Elizabeth Garcia	elizabeth.garcia@example.com
Ryan Lopez	ryan.lopez@example.com
Megan Lee	megan.lee@example.com
Joshua Hernandez	joshua.hernandez@example.com
Samantha Young	samantha.young@example.com
Daniel Scott	daniel.scott@example.com
Ashley King	ashley.king@example.com
Kevin Green	kevin.green@example.com
Nicole Adams	nicole.adams@example.com

Table 5.2 – The sample data for the Event Attendee Excel sheet

> **Note**
> In typical organizations, we usually work with large datasets and manipulate the data from different tables.

Part 2 – creating a dataflow in Power Apps

Now that our Excel sheet is ready, let's move on to Power Apps so that we can create a dataflow. We'll navigate the interface to initiate the process of importing our Excel data into Power Apps:

1. Ensure you are in the **Event Planning Project Development** environment.

2.   Click the **More** button in the left panel navigation.

3.   Click **Dataflows**:

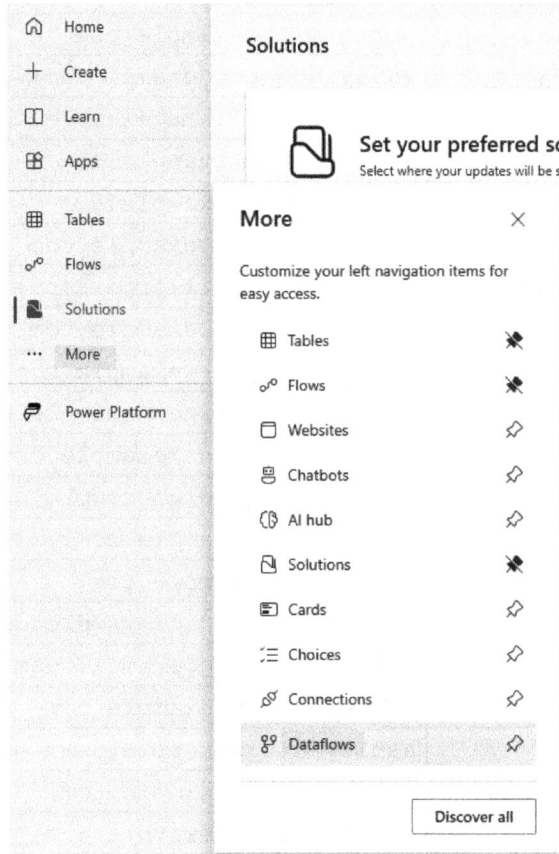

Figure 5.17 – The Dataflows option

4.   Click the **Create a dataflow** button located in the middle of the screen or from the top gray ribbon:

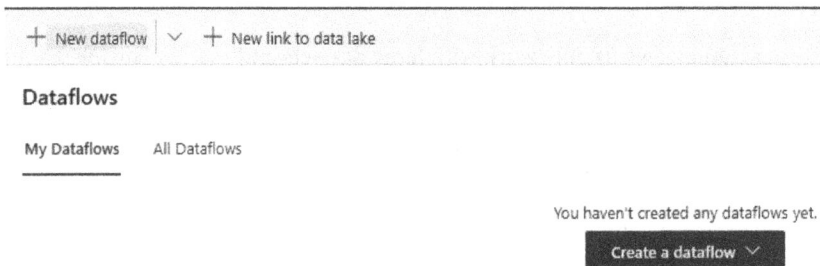

Figure 5.18 – The Create a dataflow button

5. Click on the **Start from blank** option.

6. In the **New dataflow** pop-up window, name the flow Event Attendees:

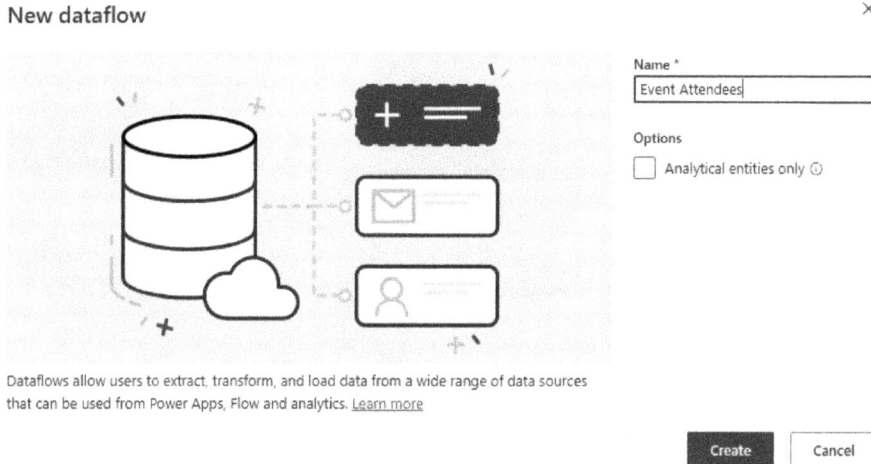

Figure 5.19 – The New dataflow popup

7. Click **Create**.

Note

This Power Query window works similarly to a Power BI query. Here, you will be able to choose from a plethora of data sources, including (but not limited to) a SQL Server database, SharePoint, an Azure SQL database, an Excel workbook, and more!

---

8. Click **Excel workbook**.

9. Click **Upload file (Preview)**.

10. Click the **Browse...** button:

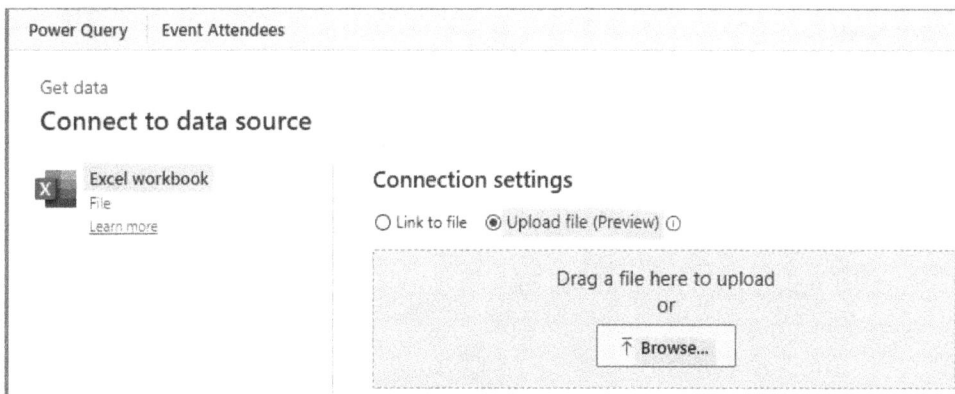

Figure 5.20 – The Power Query window after an Excel workbook has been selected

11. Select the Event Attendees file.

12. Click **Sign in** to create the connection:

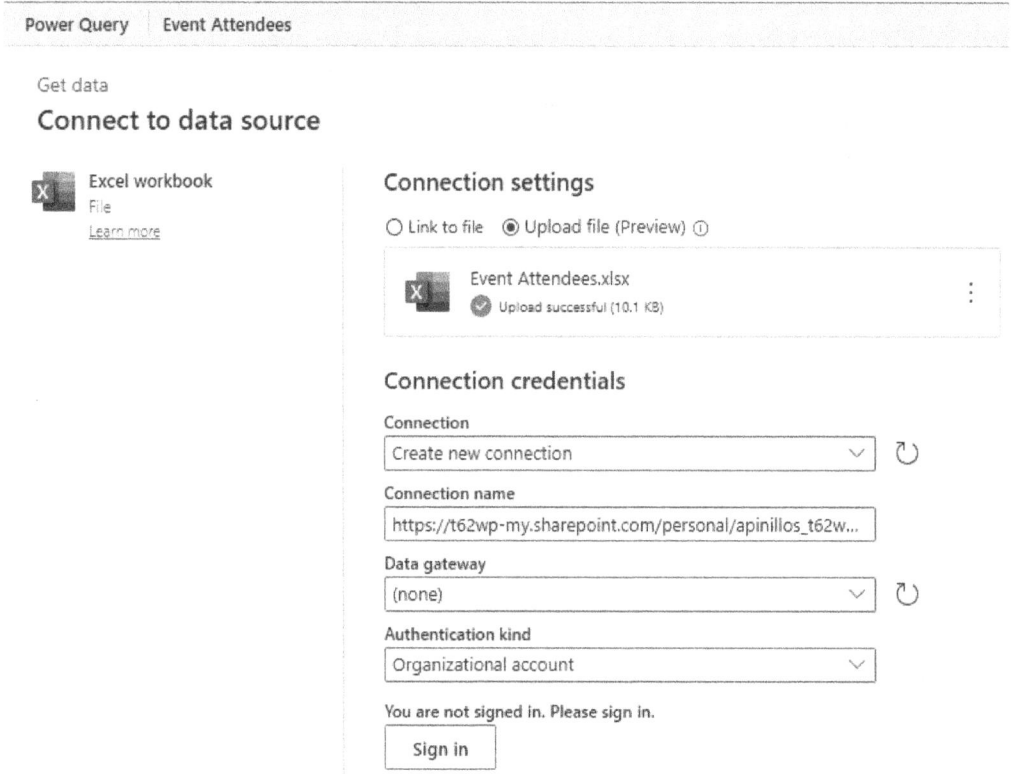

Power Query	Event Attendees

Get data
Connect to data source

Excel workbook
File
Learn more

Connection settings

○ Link to file   ⦿ Upload file (Preview) ⓘ

Event Attendees.xlsx
✓ Upload successful (10.1 KB)   ⋮

Connection credentials

Connection
Create new connection   ∨   ↻

Connection name
https://t62wp-my.sharepoint.com/personal/apinillos_t62w...

Data gateway
(none)   ∨   ↻

Authentication kind
Organizational account   ∨

You are not signed in. Please sign in.
Sign in

Figure 5.21 – Successfully uploading the Event Attendees Excel file

13. Once you've signed in, the **Next** button will no longer be disabled at the bottom right of your screen.

14. Click **Next**:

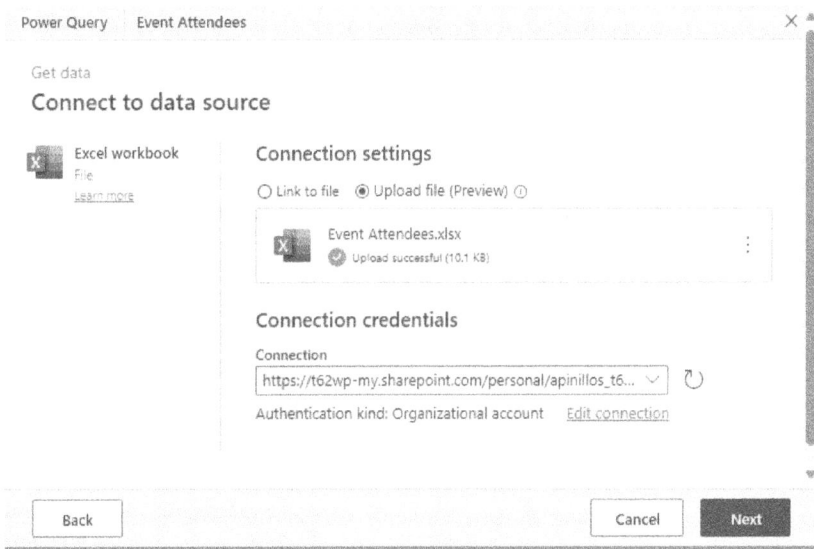

Figure 5.22 – Signing up successfully and the Next button

15. On the **Choose data** screen, click **Sheet1** from the left panel to display the data from the Excel sheet.

16. Once you see your data, click **Transform data**:

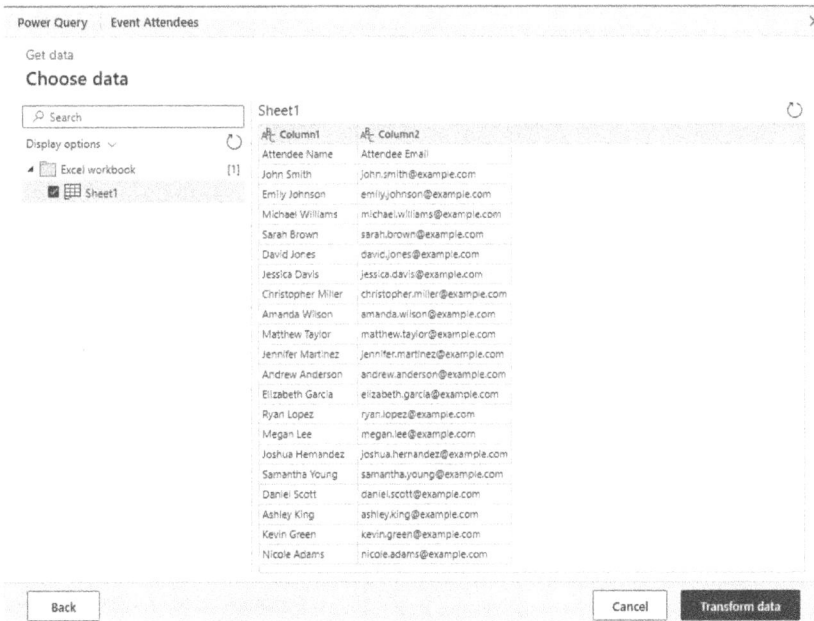

Figure 5.23 – Sheet1 selected, the data, and the Transform data button

17. Both of our columns will remain as Text fields. You can click on the icon next to **Column2** to see the other options.

Part 3 – data transformation

With our dataflow framework in place, it's time to transform our raw data into a structured format suitable for our app. We'll leverage Power Query tools to refine and clean our dataset:

1. Under the **Home** tab, above the **Transform** section, click on **Use first row as headers** to use the first row as a header instead of using the default Column1, Column2, and Column3 options:

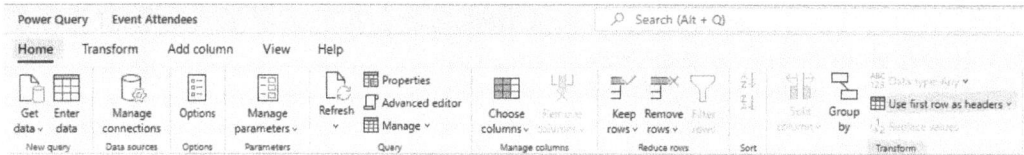

Figure 5.24 – Where to find the Use first row as headers button in the ribbon

2. The column headers should now appear as Full Name, Attendee Email, and Attendee ID.

3. The **Applied steps** area under **Query settings** in the right panel should have an extra step called **Promoted headers**:

> **Tip**
> You can click on the **X** icon next to **Promoted headers** if you need to remove the **Promoted headers** action at any time.

All runs Failed runs				
Start time	**Duration**	**Type**	**Status**	**Actions**
2/20/2024, 9:49:10 PM	00:00:48	On demand	Failed	↓

Figure 5.25 – What the column headers should look like and where to find the new step in Applied steps

4. Click **Next** at the bottom right of the screen.

Part 4 – mapping and publishing

Once our data has been transformed, we can map it to the corresponding fields in our Power Apps environment and publish the dataflow. This step ensures seamless integration and data consistency across the platform:

1. On the next screen, click **Load to existing table** under **Load settings**.
2. In the **Destination table** dropdown, scroll down to epe_Attendee.

It will be easy to find since we gave our solution a unique publisher.

1. Under the **Column mapping** section, select epe_AttendeeId under the **Select key (optional)** dropdown.

> **Tip**
> This will be useful if we want to merge queries or add more data later and want to ensure we don't create duplicate data from a previous upload.

2. Map the following source columns to their respective destination columns:

    E. Source column: Attendee Email; destination column: epe_AttendeeId
    F. Source column: Attendee Email; destination column: epe_AttendeeEmail
    G. Source column: Attendee Name; destination column: epe_AttendeeName

3. Click **Next**:

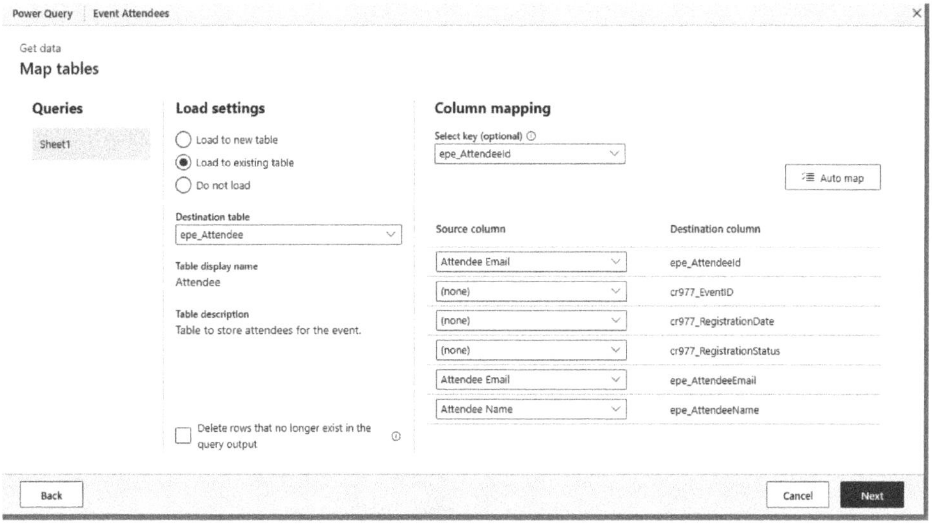

Figure 5.26 – The Map tables window before clicking Next

4. In the **Refresh settings** window, click **Refresh manually**.

5. Select **Send refresh failure notification to the dataflow owner**.

> **Tip**
>
> Clicking **Refresh** automatically will allow us to set an interval to pull the data and update our database. This is great if we need to constantly pull for a database that is constantly being updated and we need to keep our data updated from that data source.

6. Click **Publish**.

7. Wait until the **Next refresh** column is no longer In Progress:

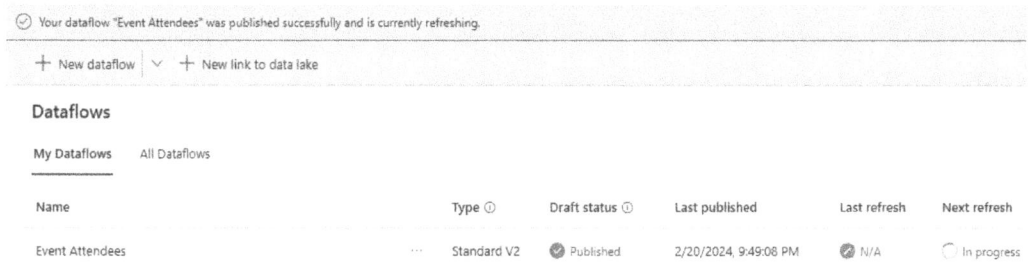

✓ Your dataflow "Event Attendees" was published successfully and is currently refreshing.

╋ New dataflow ∨   ╋ New link to data lake

Dataflows

My Dataflows   All Dataflows

Name		Type ⓘ	Draft status ⓘ	Last published	Last refresh	Next refresh
Event Attendees	⋯	Standard V2	✓ Published	2/20/2024, 9:49:08 PM	⊘ N/A	◯ In progress

Figure 5.27 – A successful dataflow will display a green ribbon and a Draft status of Published

8. If your dataflow fails, it will have a red **X** icon under the **Last refresh** column:

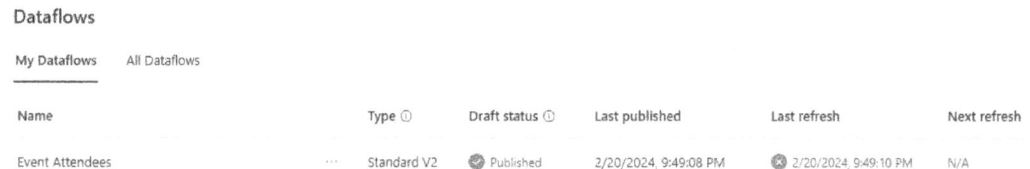

Dataflows

My Dataflows   All Dataflows

Name		Type ⓘ	Draft status ⓘ	Last published	Last refresh	Next refresh
Event Attendees	⋯	Standard V2	✓ Published	2/20/2024, 9:49:08 PM	✗ 2/20/2024, 9:49:10 PM	N/A

Figure 5.28 – Example of the red X under the Last refresh column

9. Click on the three dots next to the name of the dataflow and click **Show refresh history** to see details about the error:

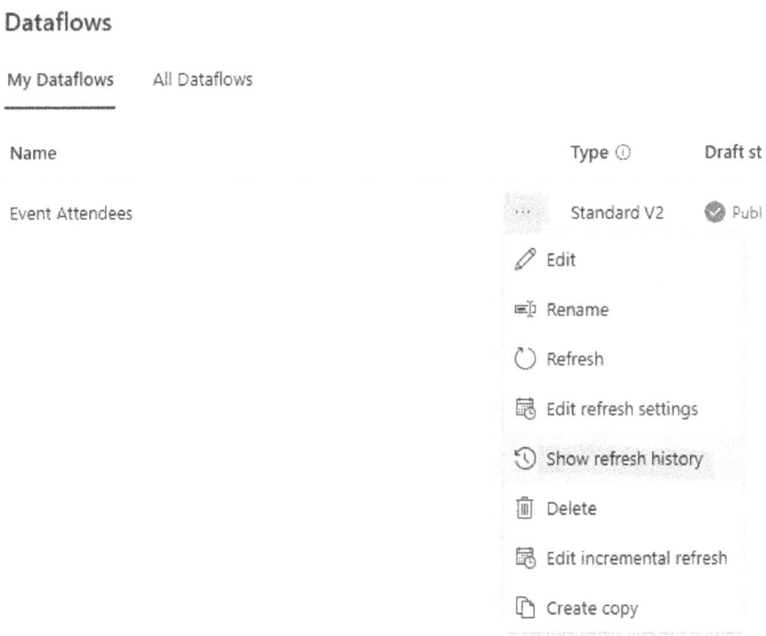

Dataflows

My Dataflows   All Dataflows

Name	Type ⓘ	Draft st
Event Attendees	··· Standard V2	✅ Publ

✏️ Edit

⌨️ Rename

🔄 Refresh

🗂️ Edit refresh settings

🕑 Show refresh history

🗑️ Delete

🗂️ Edit incremental refresh

📋 Create copy

Figure 5.29 – The Show refresh history button

10. Click on the **Download** button under the **Actions** column name:

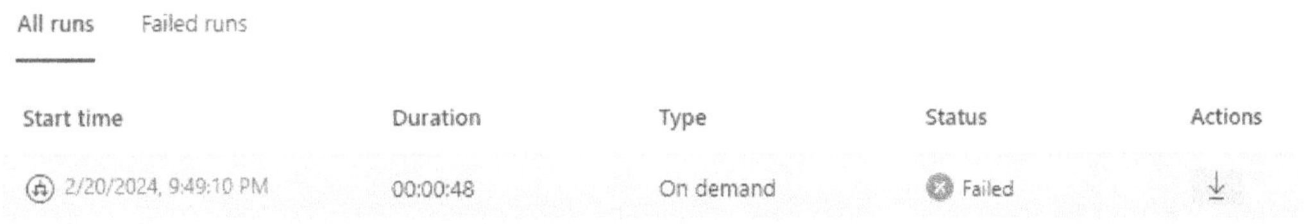

All runs   Failed runs

Start time	Duration	Type	Status	Actions
🔄 2/20/2024, 9:49:10 PM	00:00:48	On demand	❌ Failed	↓

Figure 5.30 – The Download button

11. In this instance, the dataflow did not accept **Attendee Email** as the key and gave the following error message for all 20 rows. This error message states that the `epe_attendeeid` column is not being accepted as a key field, so we need to create a key field for this instead of trying to use the attendee email:

```
{"message":"Syntax error at position 45 in '(epe_
attendeeid=michael.williams@example.com)'.","cdsErrorCode":"0x80
060888","cdsRequestId":"02e46aec-a4f1-40b2-a374-76e8cb953b97","c
dsRequestUtcTime":"2/21/2024 5:49:45 AM"}
```

12. Let's go to the **Attendee** table and add a whole number column for **Attendee ID**.

13. Click **Close** in the **Refresh History – Event Attendees** window (the screen you are currently on).

14. Click **Solutions**, then **Event Planning Project**, and finally **Attendee**.

15. Click **New**, then **Column**. Add the following details:

    H.   **Display name**: `Attendee Key`

    I.   **Data type**: **Whole number**

    J.   **Format**: **None**

    K.   **Schema name**: `AttendeeKey`

16. Click **Save**:

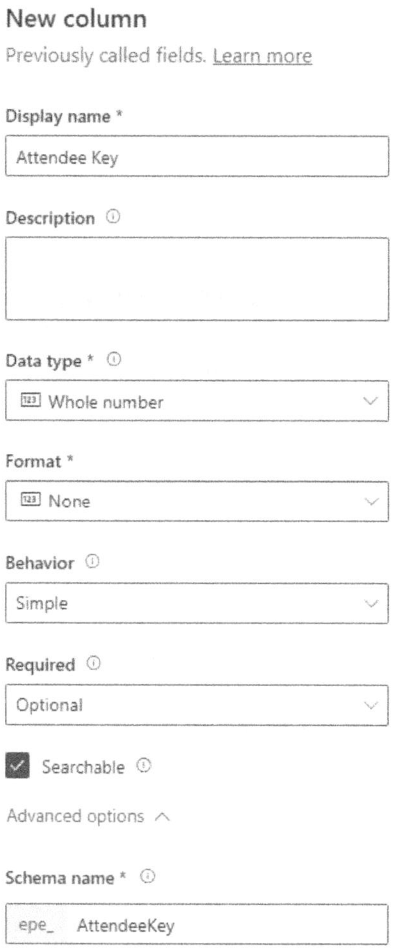

New column

Previously called fields. Learn more

Display name *

Attendee Key

Description ⓘ

Data type * ⓘ

🔢 Whole number ⌄

Format *

🔢 None ⌄

Behavior ⓘ

Simple ⌄

Required ⓘ

Optional ⌄

☑ Searchable ⓘ

Advanced options ⌃

Schema name * ⓘ

epe_   AttendeeKey

Figure 5.31 – What the New column page for Attendee Key should look like

17. Click on **Key** under the **Schema** section of the `Attendee` table:

   L. **Display name**: `Attendee Key ID`

   M. **Name**: `AttendeeKeyID`

   N. **Columns: Attendee Key**

18. Click **Save**:

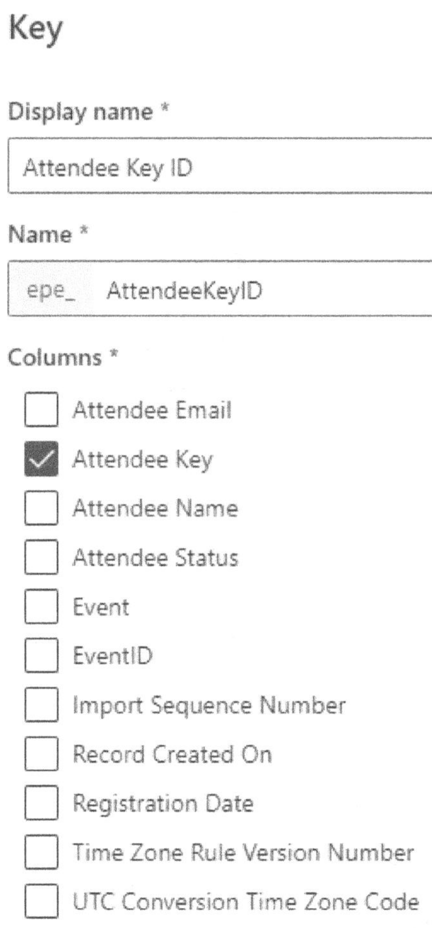

Key   ✕

Display name *

Attendee Key ID

Name *

epe_   AttendeeKeyID

Columns *

☐ Attendee Email

☑ Attendee Key

☐ Attendee Name

☐ Attendee Status

☐ Event

☐ EventID

☐ Import Sequence Number

☐ Record Created On

☐ Registration Date

☐ Time Zone Rule Version Number

☐ UTC Conversion Time Zone Code

Figure 5.32 – The completed Key form

19. Now, we will need to update our Excel file with a new column for **Attendee ID** and add numbers 1-20 for all attendees.

20. Add a new column called `Attendee ID` in the **Event Attendees** Excel sheet.

21. Add numbers 1-20, as shown in *Figure 5.28*:

    O. **Column header 1**: `Full Name`

    P. **Column header 2**: `Email`

    Q. **Column header 3**: `Attendee ID`:

Full Name	Email	Attendee ID
John Smith	john.smith@example.com	1
Emily Johnson	emily.johnson@example.com	2
Michael Williams	michael.williams@example.com	3
Sarah Brown	sarah.brown@example.com	4
David Jones	david.jones@example.com	5
Jessica Davis	jessica.davis@example.com	6
Christopher Miller	christopher.miller@example.com	7
Amanda Wilson	amanda.wilson@example.com	8
Matthew Taylor	matthew.taylor@example.com	9
Jennifer Martinez	jennifer.martinez@example.com	10
Andrew Anderson	andrew.anderson@example.com	11
Elizabeth Garcia	elizabeth.garcia@example.com	12
Ryan Lopez	ryan.lopez@example.com	13
Megan Lee	megan.lee@example.com	14
Joshua Hernandez	joshua.hernandez@example.com	15
Samantha Young	samantha.young@example.com	16
Daniel Scott	daniel.scott@example.com	17
Ashley King	ashley.king@example.com	18
Kevin Green	kevin.green@example.com	19
Nicole Adams	nicole.adams@example.com	20

Figure 5.33 – Attendee ID column added

22. Go back to Power Apps and click **More** via the left panel, then **Dataflows**.

23. Since we added a new field, let's create a new Dataflow so that the **Key** option will populate.

24. Click **New dataflow** and name it `Event Attendees`.

25. Click **Excel workbook**, click on **Upload file (Preview)**, click **Browse**, select **Event Attendees** from your documents, and click **Next**. This is important so that you can set up the dataflow so that it passes data from our Excel sheet to Dataverse.

26. Click **Sheet 1** on the left, then **Transform data**.

27. Click **Use first row as headers**, then **Next**.

28. Click **Load to existing table** and select the **epe_Attendees** table.

29. Map the following source columns to their respective destination columns:

    R. Under the **Column mapping** section, select `Attendee Key ID (epe_AttendeeKey)` under the **Select key (optional)** dropdown.

    S. Source column: `Email`; destination column: `epe_AttendeeEmail`

    T. Source column: `Attendee ID`; destination column: `epe_AttendeeKey`

    U. Source column: `Full Name`; destination column: `epe_AttendeeName`

    V. Source column: `(none)`; destination column: `epe_AttendeeStatus`

    W. Source column: `(none)`; destination column: `epe_EventID`

    X. Source column: `(none)`; destination column: `epe_RegistrationDate`

30. Click **Next**, then **Publish**. This will initiate the data transfer from the Excel sheet to Dataverse.

31. Wait for **Last Refresh** to finish and be set to `In Progress`.

Part 5 – error handling and resolution

Sometimes, things don't go as planned. In this section, we'll learn how to troubleshoot errors that may arise during the dataflow process and take corrective actions to ensure successful data transfer:

1. You should see a green check mark, indicating that your dataflow succeeded and uploaded the data:

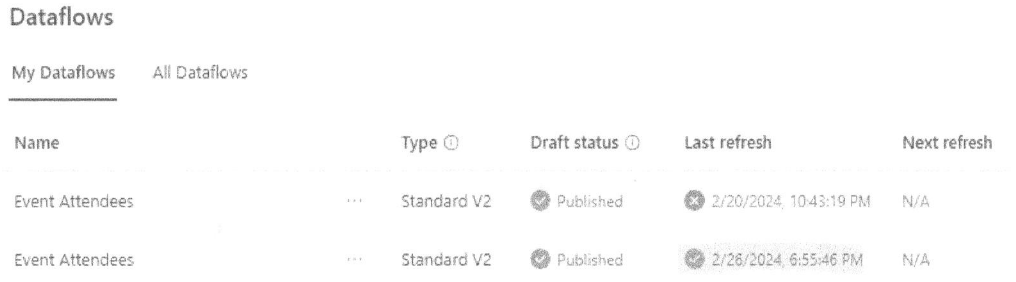

Dataflows

My Dataflows   All Dataflows

Name		Type ⓘ	Draft status ⓘ	Last refresh	Next refresh
Event Attendees	···	Standard V2	✅ Published	❌ 2/20/2024, 10:43:19 PM	N/A
Event Attendees	···	Standard V2	✅ Published	✅ 2/26/2024, 6:55:46 PM	N/A

Figure 5.34 – Data was uploaded successfully

2. Go to the **Event Planning Project** solution to check your data.

Part 6 – verification

Finally, we'll verify the data upload by checking the **Attendee** table in our Power Apps solution. This step ensures that our data has been imported accurately and is ready for use in our apps:

1. Click on the **Attendee** table.
2. Under **Attendee columns and data**, you should see the first 10 rows of your attendee data:

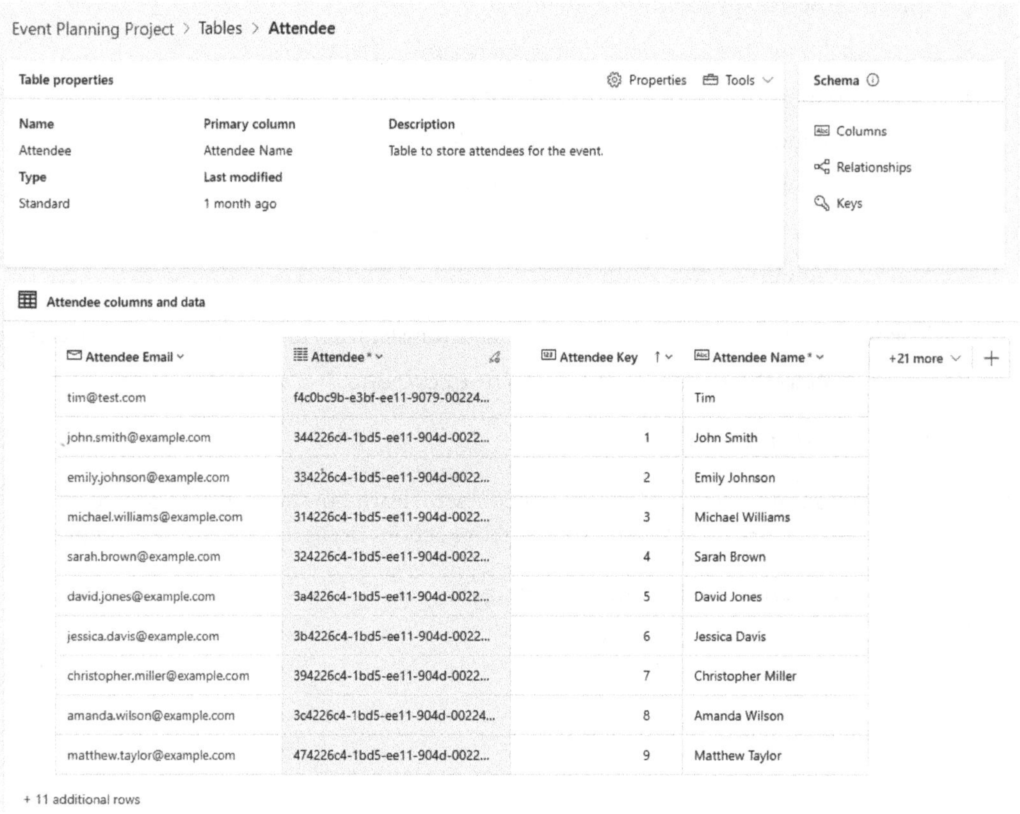

Event Planning Project > Tables > **Attendee**

Table properties				Properties	Tools ∨	Schema ⓘ

Name	Primary column	Description				Columns
Attendee	Attendee Name	Table to store attendees for the event.				Relationships
Type	Last modified					Keys
Standard	1 month ago					

Attendee columns and data

Attendee Email ∨	Attendee * ∨	Attendee Key ↑ ∨	Attendee Name * ∨	+21 more ∨ +
tim@test.com	f4c0bc9b-e3bf-ee11-9079-00224...		Tim	
john.smith@example.com	344226c4-1bd5-ee11-904d-0022...	1	John Smith	
emily.johnson@example.com	334226c4-1bd5-ee11-904d-0022...	2	Emily Johnson	
michael.williams@example.com	314226c4-1bd5-ee11-904d-0022...	3	Michael Williams	
sarah.brown@example.com	324226c4-1bd5-ee11-904d-0022...	4	Sarah Brown	
david.jones@example.com	3a4226c4-1bd5-ee11-904d-0022...	5	David Jones	
jessica.davis@example.com	3b4226c4-1bd5-ee11-904d-0022...	6	Jessica Davis	
christopher.miller@example.com	394226c4-1bd5-ee11-904d-0022...	7	Christopher Miller	
amanda.wilson@example.com	3c4226c4-1bd5-ee11-904d-00224...	8	Amanda Wilson	
matthew.taylor@example.com	474226c4-1bd5-ee11-904d-0022...	9	Matthew Taylor	

+ 11 additional rows

Figure 5.35 – Uploaded attendee data in Dataverse

This section explored the process of adding data to Dataverse using Excel, supplemented by troubleshooting examples to identify and resolve common issues. By delving into the nuances of problem-solving within the Dataverse environment, we've equipped you with valuable skills to navigate challenges effectively. Through practical examples and expert guidance, you now have the tools to address any potential issues that may arise during the data integration process, ensuring smooth and efficient operations within Microsoft Dataverse.

Summary

In this chapter, we covered different sources that can be used as data for your Power Apps. First, we covered how Excel can be used as a data source and provided different ways to connect. Then, we showed a similar approach using SharePoint Lists. In both situations, we highlighted the advantages and disadvantages of each. After, we explored using Dataverse. This provided you with a broad overview of the various connectors that are available when you're creating your app. Lastly, we introduced the concept of using dataflows.

This chapter completes *Part 1*, which is designed to cover some important overall aspects of your project app. This includes discussing the importance of using solutions, how to effectively use environments, the difference between a canvas app and a model-driven app, and various data connections. In *Part 2*, we will dive deeper into more advanced Power Apps techniques.

Part 2:
Advanced Power App
Techniques

In this part, you'll get a comprehensive overview of three key areas essential for enhancing your Power Apps development skills - variables, collections, and data filtering. You'll start by exploring variables, collections, and data filtering, which are necessary for managing and temporarily storing data, enabling the creation of more complex apps. You'll also learn how to apply data filtering techniques to refine and display data more effectively. Next, you'll dive into canvas app formulas, where you'll discover important functions for form submission, app navigation, and connecting forms to Dataverse tables—critical for building interactive and data-driven apps. Finally, you'll explore conditional formatting and URL deep linking, two capabilities that significantly enhance user interaction improving overall user experience and interface navigation. Together, these topics provide the tools and knowledge to create more dynamic and user-friendly Power Apps.

This part has the following chapters:

- *Chapter 6, Variables, Collections, and Data Filtering*
- *Chapter 7, Canvas App Formulas*
- *Chapter 8, Conditional Formatting and URL Deep Linking*

6

Variables, Collections, and Data Filtering

When building out your Power App, efficiency and functionality are very important. In this chapter, we will focus on core components that elevate Power Apps beyond mere app creation. This includes covering variables, collections, and data filtering. Understanding and mastering these elements are crucial for crafting interactive, responsive, and data-driven applications that stand out.

Variables in Power Apps are fundamental. They store data temporarily and can be thought of as containers that hold values that your app can manipulate. Whether you're managing user inputs, controlling app flow, or just temporarily storing data for calculations, variables play an important role. In addition, they can make development significantly more efficient.

Collections take data management a step further. These are more complex structures capable of storing tables of data within your app. Collections are powerful for handling multiple pieces of data simultaneously, allowing for sophisticated data manipulation, storage, and retrieval operations. This is important for apps that require working with large datasets or complex data structures.

Lastly, **data filtering** is the process of refining data, an essential practice for managing large datasets. By implementing filtering, you can design apps that present users with the most relevant information, improving user experience by making navigation and data interaction intuitive and efficient.

In this chapter, we're going to cover the following main topics:

- Variables and variable scope
- Collections
- Filtering data

Technical requirements

In order to follow all the subsequent chapters in this book, you will need to have access to Microsoft Power Apps. If you currently do not have access to Power Apps, such as through your employer, you can still sign up for a Power Apps Developer plan.

Variables and variable scope

The use of variables is core to any programming or application development. Overall, variables are used to store and maintain data dynamically. In Power Apps, variables are indispensable tools that allow you to hold data temporarily for various purposes, such as manipulating values, controlling the flow of operations, or storing information to be used across the app. Their use is important for making apps interactive and responsive, adapting to user inputs, and ensuring a seamless user experience. Furthermore, using variables can make your overall development easier and more efficient.

Importance of variables

Using variables provides several advantages:

- **Data management**: Variables offer a flexible way to store and manipulate data within your app. Whether you're capturing user inputs, calculating values, or storing results of operations, variables make these tasks manageable and straightforward.

- **Control flow**: By utilizing variables, developers can direct the flow of an application, making decisions based on stored values. This capability is vital for creating dynamic and logical app behaviors that respond to user interactions or external data changes.

- **State management**: Variables help to maintain the state of an application. For instance, you can use variables to keep track of whether a user is logged in, what step they are on in a multi-step process, or to hold temporary settings that can be toggled by the user.

- **Enhanced user experience**: With variables, you can personalize the app experience for users by displaying content or options based on their preferences, previous selections, or inputs. This level of personalization makes the app more engaging and user-friendly.

- **Efficiency and performance**: Variables can improve the efficiency and performance of your app by minimizing the need for repetitive data retrieval operations. By storing data in variables, you can reduce the number of calls to data sources, which can be especially beneficial in scenarios with limited bandwidth or expensive data operations.

Types of variables

Power Apps supports several types of variables, each suited for different scenarios:

- **Global variables**: These variables are accessible throughout the entire app, from any screen or function. They are ideal for storing information that needs to be accessed or modified from multiple places within the app.

- **Context variables**: Context variables are specific to a screen and can only be accessed or modified within that screen. They are perfect for managing the state or data of individual screens without affecting the rest of the app.

- **With() function and named formulas**: The `With()` function allows you to create named values that are essentially variables that can be used within the function. Named Formulas are very similar to global variables because they allow you to define reusable formulas that are accessible across the app.

- **Collections**: Although not variables in the traditional sense, collections are important to mention in the context of Power Apps variables. Collections can store tables of data and are powerful tools for managing complex datasets, enabling operations such as sorting, filtering, and aggregating data across multiple records.

In the following sections, we'll dive deeper into each of these categories, exploring their specific uses, syntax, and practical examples to illustrate how they can be effectively utilized in Power Apps development. This exploration will empower you to harness the full potential of variables in your applications, enabling you to create more dynamic, efficient, and user-centric apps.

Global variables

Global variables in Power Apps are pivotal for storing data that needs to be accessible across the entire application, regardless of the screen the user is currently interacting with. They provide a powerful means to share and manipulate data throughout your app, ensuring that key pieces of information are consistently available wherever needed.

Creating global variables

To create a global variable, you use the Set (Variable Name, Value) function. There are two examples in the following figure. In the first one, a variable named gblAppName, is set with the text Event Planning App. The second variable, gblDate, is set with today's date.

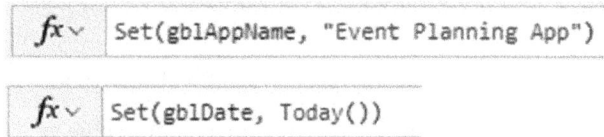

fx ⌄   Set(gblAppName, "Event Planning App")

fx ⌄   Set(gblDate, Today())

Figure 6.1 – Creating a global variable

TIP

When creating variables, there are two recommended practices to follow. First, use prefixes, such as gbl for global variables and loc for context variables. This helps differentiate between them and helps to prevent you from using the same variable name twice. In addition, use a name that indicates the intent or purpose of the variable.

Variables can store a variety of data types, including text, number, Boolean (true/false), or even complex data types such as records and tables.

Accessing global variables

To access the variable, you simply reference the name of the variable in any formula or property within your app. For example, to display the variable that tracks the current date, the variable name is shown in the formula bar. *Figure 6.2* provides an example of this:

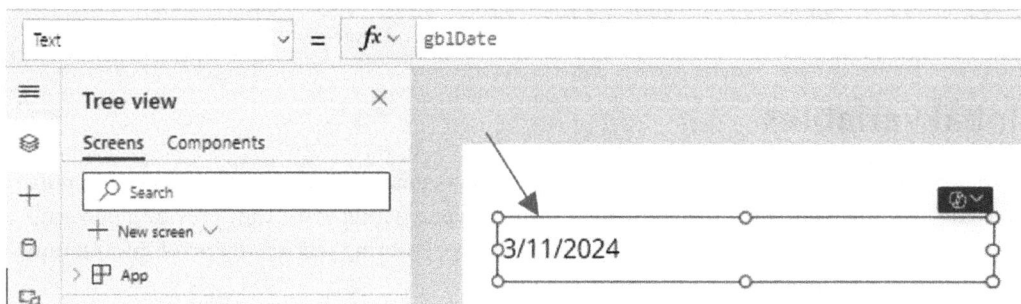

Figure 6.2 – Displaying a global variable

Examples of global variables

Global variables are particularly useful for maintaining information such as the following:

- **User preferences**: Store and retrieve user settings or preferences across the app, allowing a personalized experience on every screen

- **Session data**: Keep track of session-specific data, such as login status, access tokens, or the user's current role, that might determine what content is accessible to them

- **App state**: Maintain the state of the app, such as which tab or section is currently active, enabling consistent navigation and user interface states across different screens

- **Dynamic content**: Dynamically update content based on user interactions or data changes by referencing global variables in formulas that control visibility, text, or other properties of screen components

Let's provide a practical example of creating global variables. In this scenario, we want to capture the user of the app, including their name and their email address. Storing this information in global variables will allow us to use this information in other areas, simplifying our code. We will define these in the app's OnStart property, as shown in *Figure 6.3*:

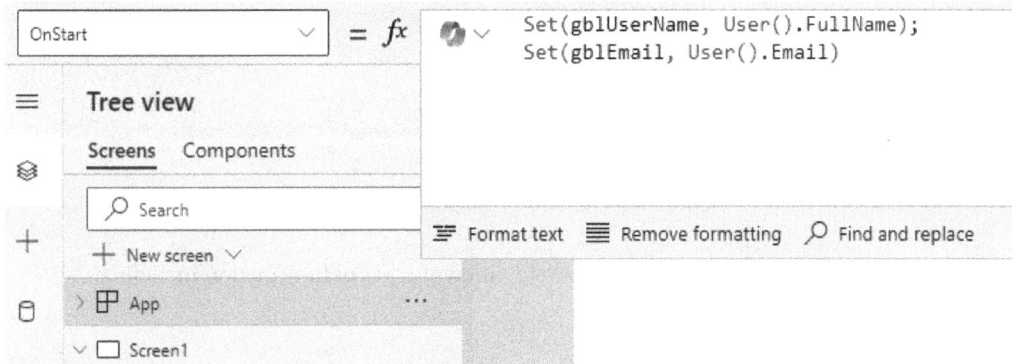

Figure 6.3 – Creating two global variables

> **Tip**
>
> It is common to create global variables in the app's OnStart property to ensure all variables are defined when the app is loaded. This is shown in *Figure 6.3*. While this is a viable option, be aware that all formulas contained in this property will execute before the application starts. This can result in the application taking longer to open. An alternative to defining variables at startup is using *Named Formulas*, which we will cover later in this chapter.

By understanding and implementing global variables effectively, you can significantly enhance the data handling capabilities of your Power Apps applications, making them more dynamic, efficient, and user-friendly. In the next section, we will cover context variables.

Context variables

Context variables provide a way to store and use data within a specific screen in Power Apps. Unlike global variables, which are accessible throughout the entire app, context variables are local to the screen where they are created. This scoping makes context variables particularly useful for managing data, states, and behaviors that are relevant only within a single-screen context.

Creating context variables

To create a context variable, you use the `UpdateContext({Variable name: Value})` function. Note that this is slightly different than creating a global variable in that the variable name and associated value are contained within `{}`. This allows you to create multiple context variables with one function. Coming up are two examples. In the first one, a variable named `locTitle` is set with the text `Home Screen`. In the second example, two variables are set including both `locTitle` as well as `locHeader`. See *Figure 6.4* for two examples of creating local variables:

```
UpdateContext({locTitle: "Home Screen"})
```

```
UpdateContext({locTitle: "Second Screen", locHeader: "Screen Header"})
```

Figure 6.4 – Creating a context variable

Like global variables, context variables can also hold different types of information including:

- A single value
- A table
- A record
- An object reference
- Any result from a formula

Accessing context variables

Context variables are accessed the same way as global variables and can be done by simply referencing the variable name. *Figure 6.5* provides an example of using a local variable as the title of the screen.

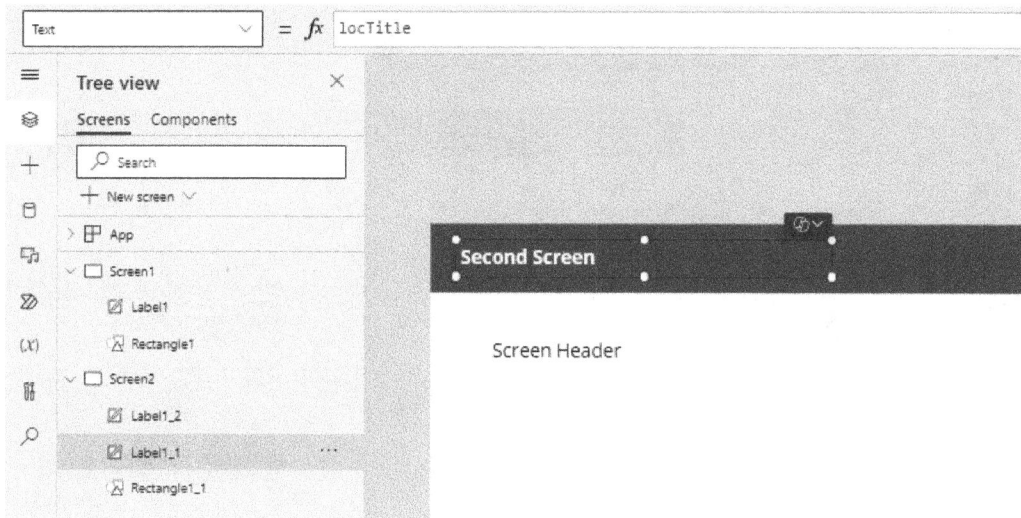

Figure 6.5 – Displaying a context variable

Examples of context variables

Context variables are particularly useful for maintaining information or situations such as the following:

- **Screen-specific data**: Storing information that is only relevant to the current screen, such as the title of the screen, as shown in *Figure 6.5*.

- **Performance optimization**: Reducing app complexity and potentially enhancing performance by limiting the scope of variables to where they are needed.

- **Local state management**: Managing the state of user interface components or data within the screen that would not be relevant to other screens. Examples include buttons, toggles, and input fields.

Let's provide a practical example of creating a context variable. Using the example provided in *Figure 6.5*, we'll create two context variables to display the title of the screen being viewed and a label on the screen.

To do this, we'll use the **OnVisible** property of the screen to create the two context variables. Then, two labels are added to the screen with their **Text** properties referencing the respective variable. This is shown in *Figure 6.6*.

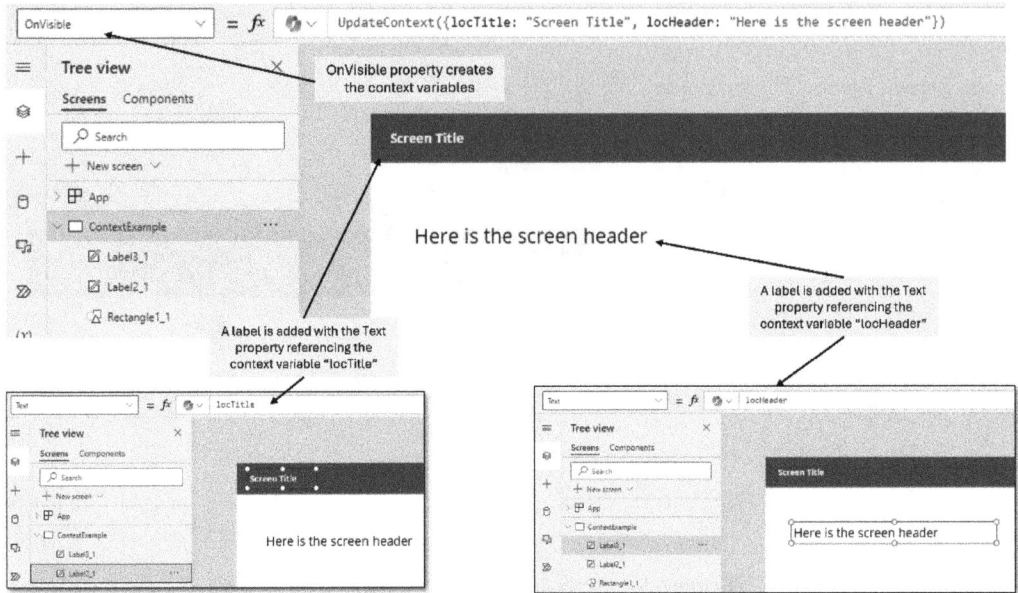

Figure 6.6 – Creating and using context variables

Like global variables, by using context variables effectively, you can continue to enhance the data handling capabilities of your Power Apps applications, making them more dynamic, efficient, and user-friendly.

> **TIP**
>
> It is recommended to use a context variable over a global variable when the information is only needed for a specific screen. Since context variables are screen-specific they can help optimize performance by only being used when necessary.

In the next section, we will delve into two areas that are very similar to variables and this includes using the `With()` function as well as Named Formulas.

With() function and Named Formulas

When covering variables, it is important to cover two additional areas that are related to variables. These include using the `With()` function and Named Formulas. We will cover each of these briefly.

Using the With() function

The `With()` function allows you to create temporary named values that can be used within the scope of specific code. Unlike context variables where the variable information is maintained for the specific screen, named values defined within the `With()` function are only available within that specific function. The context of the `With()` function is similar to the `UpdateContext()` with the addition of a formula to be evaluated against.

The `With()` function uses the following format:

```
With( {Record1, Record2,…}, Formula )
```

The parameters of each function are as follows:

- `Record1…`: This is where each named values are defined using the syntax `{name1: value1, name2: value2, …}`.
- *Condition*: The formula that is used to evaluate. All records meeting the condition are removed.

An example of using the function is as follows:

```
With(
    {Value1: 10, Value2: 20},
    Value1 * Value2
)
```

The output of this function would be 200 (10 * 20). *Figure 6.7* provides an example of this:

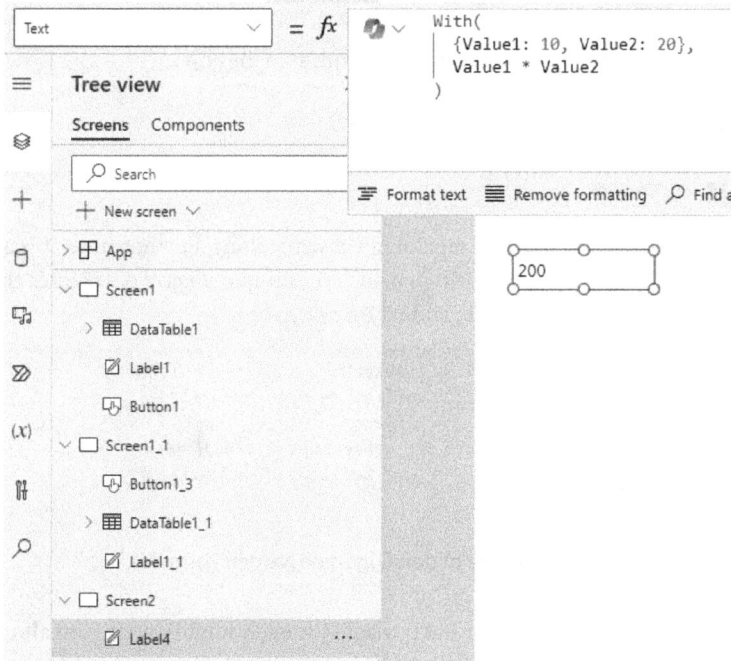

Figure 6.7 – Using the With() function in a Text label

As shown in this example, we have defined two named values, `Value1` and `Value2`, and then used that information within the scope of the function. We will now cover Named Formulas.

Using Named Formulas

Named Formulas are very similar to global variables in that they are available across the entire app. However, they are only created in the **Formulas** property for the overall app. See *Figure 6.8* to find the location of the **Formulas** property:

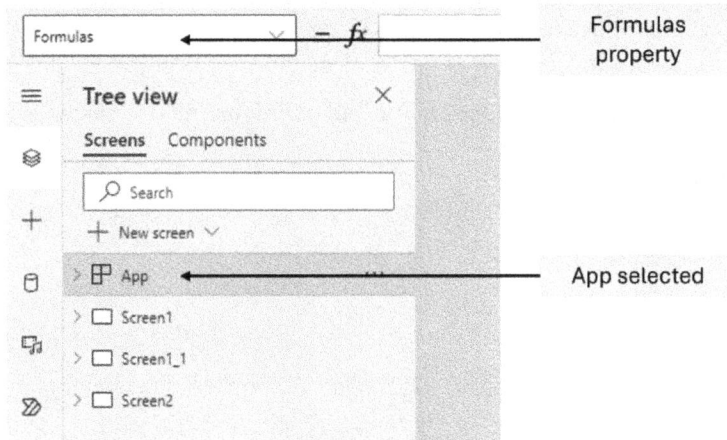

Figure 6.8 – Location of Formulas in the app

Creating Named Formulas is done by using the format:

```
Named formula = value
```

Multiple items can be created and must be separated via semicolons. Furthermore, it is also possible to store records within each named value. To provide an example, *Figure 6.9* creates three Named Formulas called `nfUserName`, `nfEmail`, and `nfPerson`:

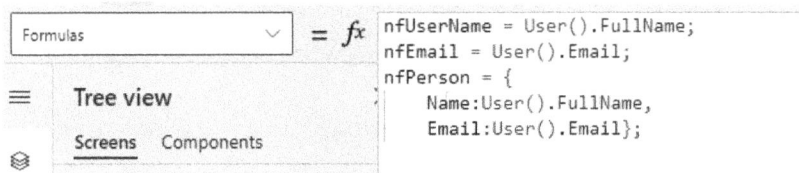

Figure 6.9 - Example of defining three Named Formulas

In this example, the first two are Named Formulas with each having a single value. The third, `nfPerson`, contains a record with two items. Calling `nfUserName` will provide the same result as calling `nfPerson.Name`.

As you can see, Named Formulas are very similar to global variables. This results in a logical question – is it better to use one than the other? Let's compare them in the next section.

Comparing global variables and Named Formulas

Global variables and Named Formulas provide similar functionality in that they allow you to store values that are accessible across the app. In *Figure 6.10*, you can see that the output of the global variable, gblUserName, is the same as the Named Formula, nfUserName.

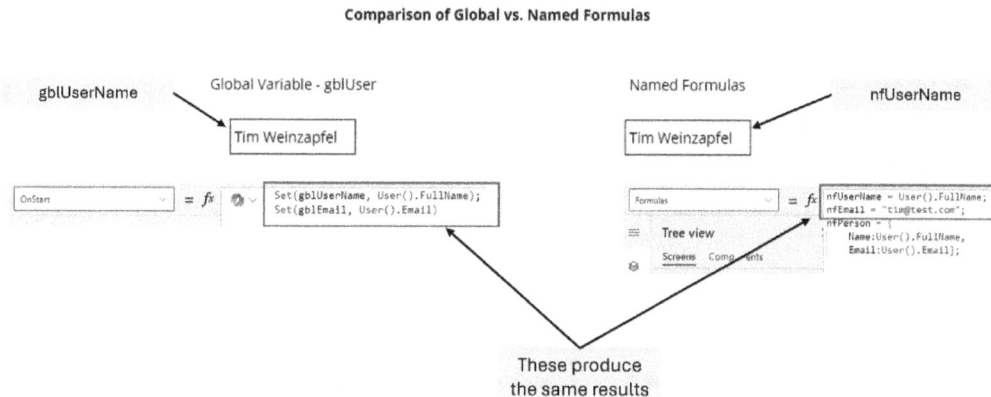

Figure 6.10 – Comparing a global variable to a same-named formula

This raises two questions. Are they the same, and is one better than the other?

Let's address the first question: are global variables and Named Formulas the same? The simple answer is no. While they have similar functionality, there are differences. Notably, global variables can be updated easily and anywhere in the app. For example, a global variable can be defined in the App. OnStart property, as shown in *Figure 6.10*, but can be easily defined or updated by other areas of the app, such as clicking on a button or changing screens. Named Formulas, on the other hand, can only be created in the App.Formulas property and cannot be changed in other areas.

Is one better than the other? The best answer is – it depends. Global variables are better when you want to be able to change the value, whereas Named Formulas are better for more static data or calculations. However, global variables can consume more memory use, especially in large apps, thus potentially impacting performance. Named Formulas are only calculated when needed and thus are better from an app performance perspective.

The important thing is understanding the functionality of them both and then determining which approach may be best for your given situation. We will give a summary of these different approaches shortly. However, we want to move to one more important area of storing data, and that is collections.

Collections

Collections in Power Apps are powerful data structures that allow you to store and manipulate sets of data as tables. Unlike variables that hold single values, collections can hold multiple records, each containing one or more fields. This capability makes collections incredibly useful for handling complex data operations, such as sorting, filtering, aggregating, or displaying data in galleries or tables.

Importance of using collections

Collections are particularly beneficial in scenarios where you need to work with multiple pieces of data collectively:

- **Temporary data storage**: Collections can temporarily hold data fetched from a data source, allowing for offline use or manipulation within the app

- **Bulk operations**: Collections allow you to perform operations on multiple records at once, such as bulk updates, deletions, or complex calculations

- **Dynamic UI elements**: Power dynamic UI components, such as galleries, tables, and dropdowns, where multiple records need to be displayed or selected by the user

- **Data manipulation**: Easily sort, filter, and aggregate data within a collection to display it according to specific criteria or user inputs

Creating collections

Collections can be created by using one of two functions: `Collect()` and `ClearCollect()`. Three other useful functions are `Clear()`, `Remove()`, and `RemoveIf()`. Let's dig into each of these functions.

Using the Collect() Function

`Collect()` adds records to an existing collection or creates a new collection if one does not already exist. However, this function does not clear out any existing data. It uses the following format:

```
Collect(Collection Name, Value1, Value 2, … )
```

An example of this is as follows:

```
Collect(Vendors, "Acme Beverage Company", "XYZ Catering Services")
```

This function creates a collection called `Vendors` and will either create a new one if it doesn't already exist or add those records to the existing list.

When adding records to a collection, they can be added the following:

- A single value
- A record
- A table

Let's look at an example of each of these. The format for adding a *single value* is as follows:

```
Collect(Vendors, "Acme Beverage Company")
```

This function adds a single value to the collection, `Vendors`:

Value

Acme Beverage Company

Figure 6.11 – Data in a collection

Figure 6.12 provides an example of adding the `Collect()` function to the `OnSelect` property of a button. Each time the button is pushed, the `Vendors` collection will add a new record of Acme Beverage Company. Next to the button is a Gallery that uses the `Vendors` collection in the `Items` property.

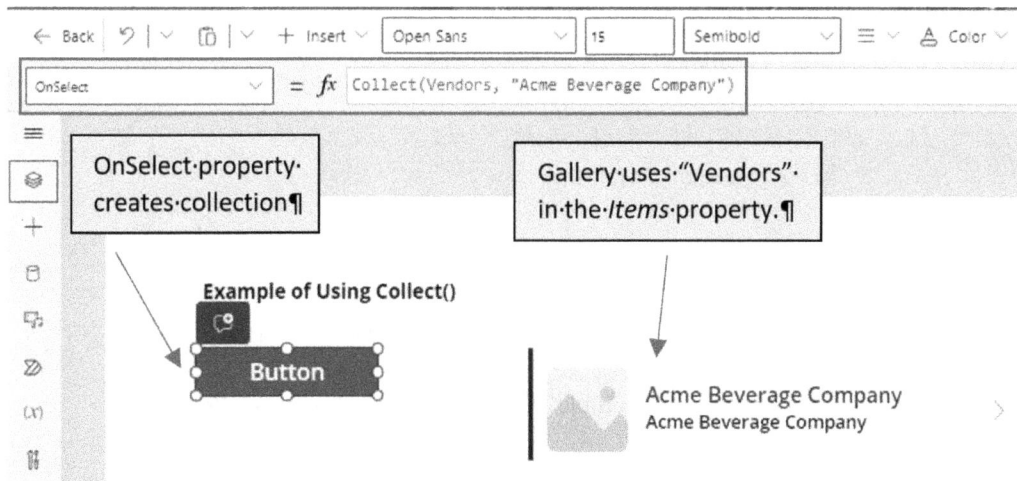

Figure 6.12 – Use of Collect() in a button

Calling the `Collect()` function additional times and referring to new values will add them to the existing table.

In addition to adding a single value, `Collect()` also allows you to add **records** with their corresponding column. Be aware that each record must have the applicable column names and corresponding values contained with brackets, { }, and each record is separated by a comma. As an example, the following function creates two records, each with a `Vendor` and a `VendorID`:

```
Collect(Vendors, {Vendor: "Acme Beverage Company", VendorID: 100},
{Vendor: "XYZ Catering Services", VendorID: 200} )
```

Vendor	VendorID
Acme Beverage Company	100
XYZ Catering Services	200

Figure 6.13 – Calling the Collect() function to add records

A third approach is to add these items as a single table by wrapping the records around a `Table()` function as follows:

```
Collect(Vendors, Table({Vendor: "Acme Beverage Company", VendorID:
100}, {Vendor: "XYZ Catering Services", VendorID: 200} ))
```

The output of this will be similar to the previous output:

Vendor	VendorID
Acme Beverage Company	100
XYZ Catering Services	200

Figure 6.14 – Calling the Collect() function to add records from a table

However, it is then also possible to add multiple tables into individual cells, as follows:

```
Collect(Vendors,
{PrimaryVendors: Table({Vendor: "Acme Beverage Company", VendorID:
100}, {Vendor: "XYZ Catering Services", VendorID: 200} ) },
{AltVendors: Table({Vendor: "123 Company", VendorID: 300}, {Vendor:
"No good ", VendorID: 400} ) }
)
```

AltVendors   PrimaryVendors

                                          [Table]

[Table]

Figure 6.15 – Calling the Collect() function to add records as table values

Thus, each table is stored in an individual cell and the values are not extracted. While there may be specific uses for this approach, this is beyond the scope of this book. However, these examples have been provided to demonstrate that there are different ways to store data within collections.

Using the ClearCollect() function

ClearCollect() is very similar to Collect() but will clear out any existing data and create a new collection. It uses a similar format:

ClearCollect(Collection Name, Value1, Value 2, …)

An example of this is as follows:

```
ClearCollect(Vendors, "Acme Beverage Company", "XYZ Catering
Services")
```

This function creates a collection called Vendors but will delete any records that were previously created. In addition, all of the methods described in the Collect() section also apply to the ClearCollect() function. Thus, the primary difference between Collect() and ClearCollect() is whether existing records remain or are deleted.

However, there are also situations where records need to be removed from a collection. This can be done using the Clear(), Remove(), and RemoveIf() functions. Each of these will be discussed next.

Removing records with Clear(), Remove(), and RemoveIf()

There are several functions available to delete records depending on your situation. They are Clear(), Remove(), and RemoveIf(). Let's start with Clear(). This function simply requires the collection name and will clear it out completely. You might quickly recognize that if you were to slice the ClearCollect() function down the middle, you would have Collect(), which creates a collection, and Clear(), which clears a collection. The format for Clear() is as follows:

```
Clear( Collection Name )
```

However, you may certainly run into situations where only specific records need to be deleted from the collection. In this situation, the `Remove()` or `RemoveIf()` functions are more appropriate. Let's start with `Remove()`. It uses the following format:

```
Remove(Collection Name, Record1, Record2, …, RemoveFlags.All)
```

The parameters of each function are as follows:

- *Collection Name*: This is the name of the collection.

- *Record1, Record2, …*: This is the record or records to remove.

- *RemoveFlags.All*: This is an optional parameter. If included and if the collection has duplicates of the same record, all will be removed.

The important issue is specifying which record(s) to remove. There are different ways to approach this depending on the structure of your collection. However, two common ways are to use the `Lookup()` or the `Filter()` function. Both functions are explained in the next section. However, as an advanced look, here are two examples of using each one. First, let's look at an example collection:

Vendor	VendorID
Acme Beverage Company	100
XYZ Catering Services	200

Figure 6.16 – A collection before records are removed

Next, we'll remove `Acme Beverage Company` by applying the following function to a button. In this example, we are using the `LookUp()` function to specify a record based on where `VendorID` equals `100`:

```
Remove(Vendors, LookUp(Vendors, VendorID = 100)
```

The collection now only has `XYZ Catering Services`, as shown in *Figure 6.9*.

Vendor	VendorID
XYZ Catering Services	200

Figure 6.17 – A collection after the Remove() function is applied

As an alternative, we could also use the following approach with the `Filter()` function:

```
Remove(Vendors, Filter(Vendors, VendorID = 100) )
```

It is important to note that different approaches may result in a different number of records being removed. For example, in using the LookUp() function, only one record where the VendorID is equal to 100 will be removed. However, in the second example, all records with a VendorID equal to 100 would be removed. We will cover both Lookup() and Filter() in more detail in the next section. Furthermore, while these are two examples of specifying which records to remove, there are other approaches to use, depending on your situation. However, let's move on to RemoveIf().

The RemoveIf() function uses the following format:

```
RemoveIf( Collection Name, Condition)
```

The parameters of each function are as follows:

- *Collection Name*: This is the name of the collection.

- *Condition*: The formula that is used to evaluate. All records meeting the condition are removed.

Use the same collection as before, the following example removes all records where the VendorID (which is a numeric field) is greater than or equal to 100. This would remove both Acme Beverage Company as well as XYZ Catering Services as each VendorID meets this criterion:

```
RemoveIf(Vendors, VendorID >= 100 )
```

Advantages of using collections

Collections in Power Apps offer several advantages, making them a useful feature for app developers. Here's a summary of the key benefits:

- **Local caching of data**: Collections allow you to store and manipulate data within the user's session locally. This means the data is readily accessible without the need for constant server calls, which can significantly improve the performance and responsiveness of your app, especially in scenarios with slow network conditions.

- **Offline capabilities**: Linked to local caching, collections can be used to create offline capabilities within your app. You can design your app to load data into a collection when online and then work with this data even when the app goes offline, syncing back when the connection is re-established.

- **Data manipulation**: Collections provide a flexible way to manipulate data. You can easily add, remove, and update items within a collection. This is particularly useful for complex data operations that might be cumbersome to perform directly with data sources.

- **Ease of use with visual elements**: Collections can be easily integrated with Power Apps visual elements such as galleries, forms, and lists. This makes it straightforward to display, sort, filter, and interact with the data stored in collections, enhancing the user interface and experience.

- **Complex data structures**: Collections allow you to work with complex data structures, including nested collections. This can be particularly useful when dealing with data that has multiple levels of detail or hierarchical relationships, enabling more sophisticated data models and applications.

- **Temporary data storage**: Collections can act as temporary data storage, which is useful for scenarios where you need to gather and work with data across multiple screens before committing it to a database or another permanent data store. This can simplify data handling and reduce the complexity of app development.

- **Batch processing**: With collections, you can perform batch processing operations on data, such as bulk updates or analyses. This can be more efficient than processing each item individually, especially for large datasets.

Incorporating collections into your Power Apps development strategy can significantly enhance the efficiency, performance, and user experience of your apps. They provide a versatile set of tools for handling data locally, offering benefits across various aspects of app development, from performance optimization to complex data manipulation.

Delegation in collections

Delegation in Power Apps is a crucial concept, especially when working with data sources and operations such as collections. It pertains to the way Power Apps handles data operations and whether these operations are processed locally on the user's device or remotely on the data source itself. Understanding delegation is important because it directly impacts the app's performance and its ability to handle large datasets effectively.

What is delegation?

Delegation is the process by which Power Apps offloads the data processing tasks to the data source rather than performing these tasks locally within the app. When an operation is delegated, it means the data source (such as SQL Server, SharePoint, or Dataverse) executes the query or data operation, returning only the relevant result set to the app. This approach is scalable and efficient, allowing apps to work with very large datasets without downloading all the data to the client device at once.

Determining which operations are delegable will depend on the data source and the operation itself. For example, when working with SharePoint lists, the `Filter()` function is delegable, whereas the `Search()` function is not. Providing a complete list of delegable and non-delegable operations is beyond the scope of this book. However, Microsoft provides a complete list of operations depending on the applicable data source. This can be found at `https://learn.microsoft.com/en-us/power-apps/maker/canvas-apps/delegation-overview#delegable-data-sources`.

Non-delegable operations and collections

While delegation optimizes performance and scalability, not all data operations or functions in Power Apps are delegable. When an operation is non-delegable, Power Apps must load the data into the app's memory to process it locally, which is less efficient and limits the amount of data you can work with. By default, there is a limit of 500 records that will be received. This limit can be increased to 2,000. This limitation is where collections come into play as a workaround in certain scenarios.

Collections are inherently local to the app. When you use a collection, you're working with data that has been loaded into the app's context. Thus, using collections has some advantages:

- **Local processing**: All operations on collections are performed locally, which bypasses the delegation system. This is advantageous for complex data manipulations that are not supported as delegable operations on your data source.

- **Workaround for delegation limits**: By loading data into a collection (within the app's delegation limits), you can perform non-delegable operations on this subset of data. This approach is particularly useful for operations such as complex filters, sorts, or transformations that aren't delegable to your backend data source.

- **Performance considerations**: While collections can sidestep delegation limits, they also inherit the limitations of local processing—primarily, the amount of data you can efficiently work with is constrained by the client device's resources. It's essential to balance the use of collections with the potential impact on app performance.

There are some best practices when working with collections. First, use delegable queries where possible. Delegable operations for initial data retrieval and filtering are preferred because this ensures scalability and efficiency.

In addition, when using collections it is best to consider optimizing the data loaded into the app. Specifically, limit the fields and records to only what is necessary for the user or the app's functionality.

Understanding and effectively managing delegation and the use of collections is key to building high-performing, scalable Power Apps. It requires a balance between using the power of the data source to handle data operations and the flexibility of local data manipulation through collections.

Variable comparisons and inventory

We have covered the different ways to create and use variables across your apps. Let's provide an overall comparison between them. In addition, we'll end this section by showing an easy way to get an inventory of variables within your app.

Summary of descriptions and best uses

Table 6.1 provides a summary of each item covered and the best use cases.

Feature	Description	Scope	Best Uses
Global Variables	Store values that need to be accessed from anywhere within the app.	App-wide	When you need to share data or state between different screens in the app. In addition, when you want the ability to change the value within the app.
Context Variables	Store temporary values that are specific to a single screen in the app.	Screen	When you need to store temporary data that is only relevant to the current screen.
With() Function	Used to calculate values and reduce the number of variables needed by providing temporary values within its scope.	Within function scope	When you need to simplify formulas by reducing the need for multiple variables.
Named Formulas	Define reusable calculations or expressions that can be used throughout the app without being recalculated each time they are used.	App-wide	Complex calculations or expressions that need to be reused across the app to improve performance and maintainability. In addition, you do not need to change the value (or expression) within the app.
Collections	Used to store a table of data that can be accessed globally throughout the app. Can hold multiple records and columns.	App-wide	When you need to create multiple records or a dataset and share it across the app.

Table 6.1 – Summary of each category and best use cases

Now that we have summarized each type of item, let's move to managing all the variables that you have created.

Variable Inventory

At this point, you may have created many variables within your app. The next question is – how do you know what variables have been created? Fortunately, Power Apps provides an easy way to view an inventory of all your variables, including collections and Named Formulas.

On the left navigation bar in Power Apps is an icon, designated as an **(x)** that will display all variables used within the app. *Figure 6.18* shows this:

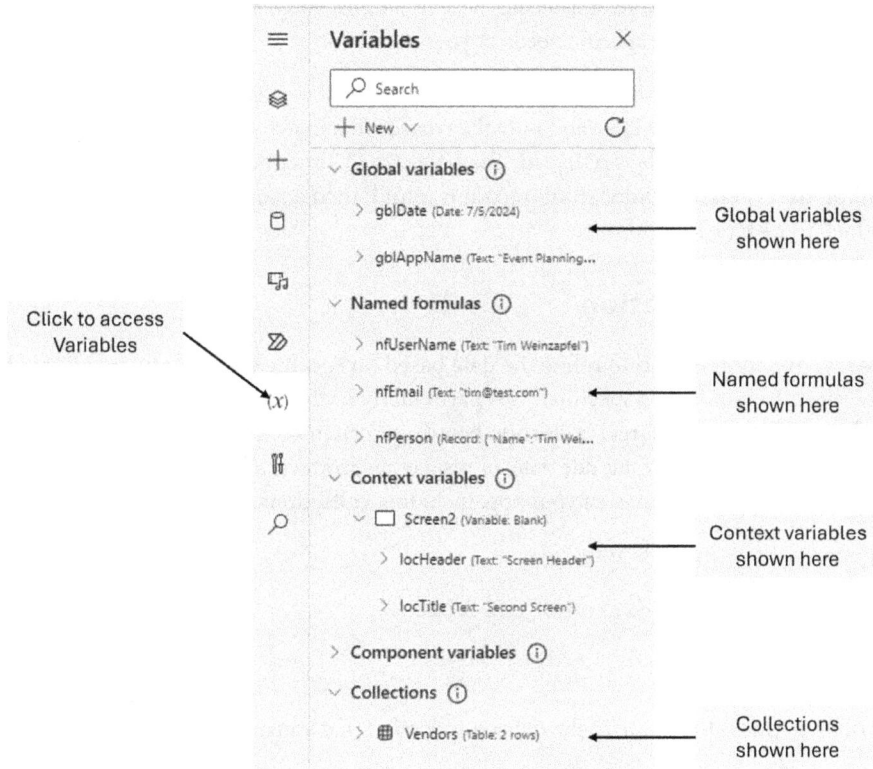

Figure 6.18 – Viewing all variables within the app

Each area contains additional information, which you can see by selecting the > arrows. For example, you can see where the variable is defined as well as where the variable is used. This makes it very easy to manage your variables.

Now that we have covered using variables, in the next section we will provide techniques for filtering your data.

Filtering data

Data filtering is critically important in Power Apps given the vast amount of data that may be available. Not only can large amounts of data negatively impact the performance of your app, presenting users with only the information that they need can significantly enhance the overall user experience. Within Power Apps, there are several functions that help in this process, including `Filter()`, `Search()`, and `LookUp()`. As a developer, it is critical to understand how these functions work in order to build efficient, responsive, and user-friendly applications.

In the following sections, we'll provide detailed instructions on how to use each of these functions, including syntax examples and practical scenarios where they can be applied to enhance data presentation in your Power Apps applications. By mastering these functions, you'll be able to create dynamic and responsive apps that cater to the specific needs of your users.

An important distinction is that both `Filter()` and `Search()` will return a table that contains the same columns as the original table and only the records that match the filter/search criteria. If no records exist, then an empty table is returned. The `LookUp()` function only returns the first record found based on the criteria provided. If no record is found, then a *blank* is returned. Let's begin by looking at the `Filter()` function.

Using the Filter() function

The `Filter()` function is used to refine the data based on specific criteria, returning only those records that meet the conditions specified. It is particularly useful in scenarios where you need to display a subset of data from a larger collection based on certain conditions, such as showing only active projects, filtering tasks by their due date, or displaying products available in a specific category. It can be applied to any data sources in your app, including collections and SharePoint lists.

The basic syntax of `Filter()` is as follows:

```
Filter(Datasource, Condition1, Condition2, …)
```

The parameters are as follows:

- *Datasource*: This is the name of the data or collection that you are looking to filter.
- *Condition*: These are the criteria that are applied to determine which records to return. You can have one or more conditions. This function can accept a variety of operators:
  - And (or &&)
  - Or (or ||)
  - Not (or !)
  - =, <, >, <=, >=, <>,
  - StartsWith, EndsWith
  - IsBlank

Here are some examples:

Example	Description
`Filter(Vendors, Vendor-Type = "Catering")`	This filters the table to records where the `VendorType` column equals `Catering`
`Filter(Vendors, Vendor-Type = "Catering" && VendorRating >= 3)`	This filters the table to records where the `VendorType` column equals `Catering` and the `VendorRating` column is greater than or equal to 3. Note the use of `&&` to mean AND.
`Filter(Vendors, Vendor-Type = "Catering" || VendorRating >= 3)`	This filters the table to records where the `VendorType` column equals `Catering` or the `VendorRating` column is greater than or equal to 3. Note the use of `||` to mean OR.

Table 6.2 – Examples of using the Filter() function

It is important to note that the output of a `Filter()` function is a table. This means that more than one record can be returned. Let's look at a specific example of `Filter()` in use. *Figure 6.19* shows how the `Filter()` is used to display items in a gallery:

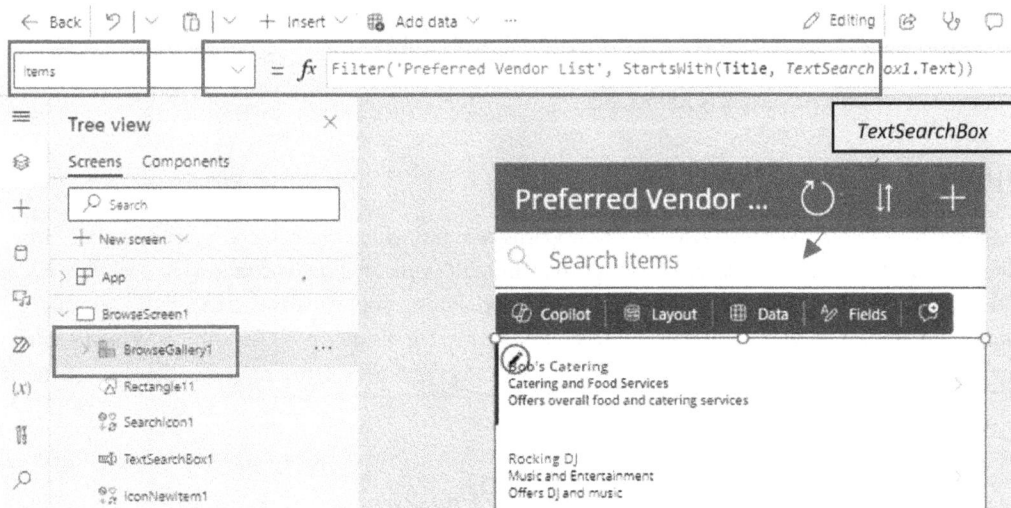

Figure 6.19 – Use of Filter() in a gallery

Let's break down how the function is used:

```
Filter('Preferred Vendor List', StartsWith(Title, TextSearchBox1.
Text))
```

Because `Filter()` is being used within the Gallery's `Items` property, the output of this function determines what records are displayed. Remember that the output of `Filter()` is a table of records. In this example, the data source `Preferred Vendor List` will be filtered to all records in the `Title` field that start with whatever is typed in the **Search** box indicated in *Figure 6.19*.

If a user were to enter `Rock`, they would get the result shown in *Figure 6.20*:

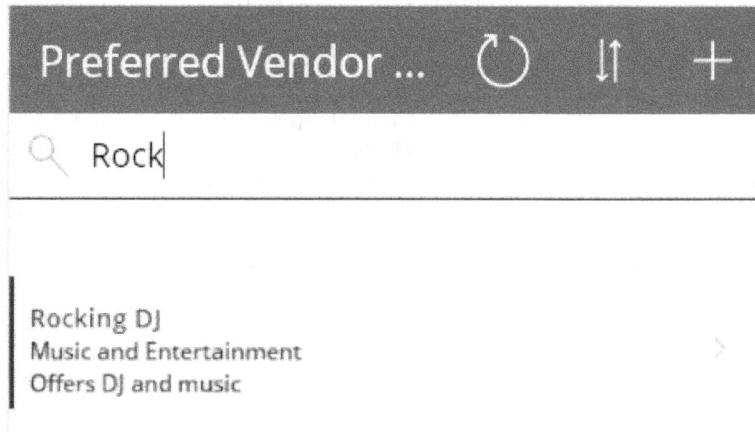

Figure 6.20 – Output of the Filter() function

As you can see, only one record is returned.

Using the Search() Function

The `Search()` function enables users to find records containing text that matches a search term. This function scans through a data source for records where the specified fields contain the search term, making it ideal for creating search features in your app. For example, users can search for products by name, employees by their names or job titles, and more. The `Search()` function is versatile and supports partial matches, helping users find what they need even if they don't remember the exact terms.

The basic syntax of `Search()` is as follows:

```
Search(Datasource, SearchString, Column1, Column2, …)
```

The parameters of each function are as follows:

- *Datasource*: This is the name of the data or collection that you are looking to filter.
- *SearchString*: This is the string to search for. If this is `blank` or empty, then all records will be returned.
- *Column1, Column2, …*: These are the names of the column(s) within the table to search for. One or more columns can be indicated; however, they must contain text.

> **Note**
>
> If the data source is SharePoint or Excel and the column names to be searched contain spaces, then it is necessary to specify each space as _x0020_. For example, the column `Vendor Name` should be referred to as `Vendor_x0020_Name`.

Here are some examples:

Example	Description
`Search(Vendors, "Catering", VendorName)`	This returns a table of all records that have `Catering` in the `VendorName` field
`Search(Vendors, "", VendorName)`	This will return all records

Table 6.3 – Examples of using the Search() function

Let's look at an example of using `Search()`. In *Figure 6.21*, we will use a similar gallery as in *Figure 6.19*, but using the `Search()` instead of `Filter()`.

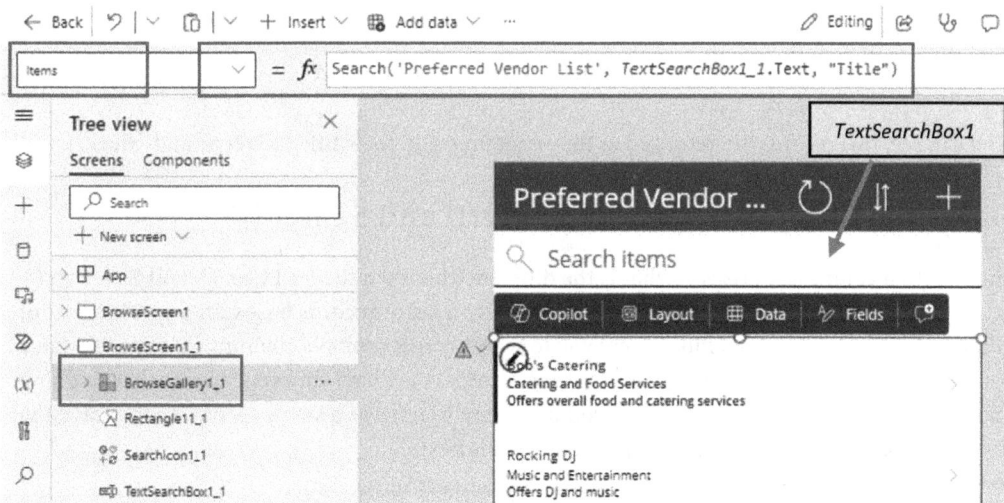

Figure 6.21 – Using Search() in a Gallery

Let's break down how the function is used:

`Search('Preferred Vendor List', TextSearchBox1_1.Text, "Title")`

Because the `Search()` is being used within the Gallery's `Items` property, the output of this function determines what records are displayed. Remember that the output of `Search()` is a table of records. In this example, the data source `Preferred Vendor List` will be filtered to all records in the `Title` field that contain whatever is typed in the **Search** box.

If a user were to enter ee, they would get the result shown in *Figure 6.22*:

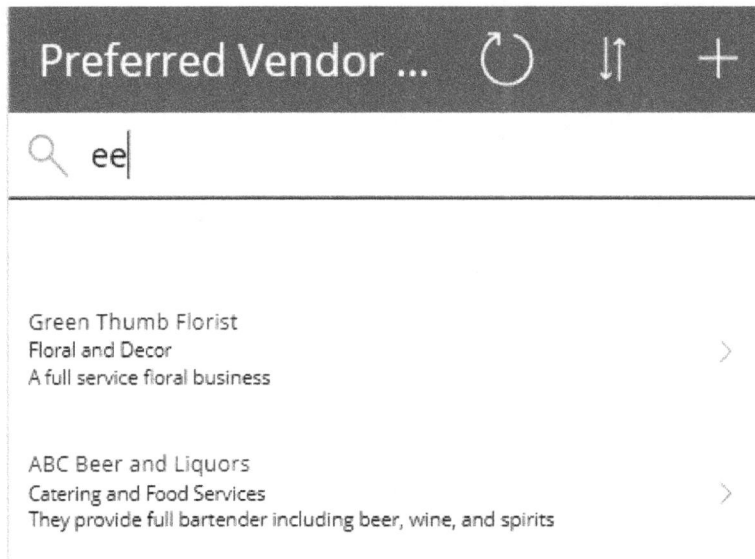

Figure 6.22 – Output of the Search() function

As you can see, two records are returned as they contain ee in their title ("**Green**" and "**Beer**").

What is the difference between Filter() and Search()?

A question that commonly arises is what is the difference between the Filter() and Search() functions? The Filter() function is used to retrieve a set of records based on a specified set of conditions. The Search() function allows you to search across multiple columns for a specific value or set of values. While they both return a table of records, they have different approaches to doing so. But there may also be situations where you only want to retrieve a single record. In this case, the LookUp() function comes in handy. We'll discuss this next.

Using the LookUp() function

The LookUp() function is used to find the first record in a data source that meets a certain condition. It is especially useful when you need to retrieve a single record or a specific value from a record, such as looking up the price of a product, the details of a specific employee, or the status of an order. Unlike Filter(), which can return multiple records, LookUp() always returns a single record or blank if no matching record is found.

The basic syntax of LookUp() is as follows:

```
LookUp(Datasource, Formula, Reductionformula)
```

The parameters are as follows:

- *Datasource*: This is the name of the data or collection that you are looking to filter.

- *Formula*: The formula by which each record of the table is to be evaluated. The first record that meets the formula will be returned, regardless if there are additional records.

- *Reductionformula:* This is an optional parameter that can be a reference to a column in the table. If this is not included, the whole record is returned.

Here are some examples:

Example	Description
LookUp(Vendors, Vendor-Name = "Green Thumb Florist")	This will return the first record where the VendorName equals Green Thumb Florist). The whole record is returned.
LookUp(Vendors, VendorID = 100, VendorName)	This will return the first record where the VendorID equals 100. Only the VendorName field is returned.

Table 6.4 – Examples of using the LookUp() function

Reminder

As noted earlier, it is important to remember the output between Filter(), Search(), and LookUp().

Filter(): Table

Search(): Table

LookUp(): Single Record

Summary

In this chapter, we covered key components of Power Apps that are essential for creating efficient, interactive, and data-driven applications. First, we explored variables, which serve as the foundation for data manipulation and management within your app. This included the different variable types, global and context, and we discussed how to declare and utilize them effectively to enhance app development efficiency and control the flow of operations within your apps. This also included using the With() function and Named Formulas.

We then discussed collections, which allow the storage of data within your app. We explored how to create, update, and manipulate collections, highlighting their significance in managing large datasets or complex data structures.

We then reviewed a comparison across variables and collections. In addition, we covered how to easily identify all the variables and collections used within the app.

Lastly, we focused on data filtering techniques, which are necessary to refine the data that is relevant to the situation. Through practical examples, we illustrated how to implement filtering using functions such as `Filter()`, `Search()`, and `LookUp()`. This aims to enhance user experience by making app navigation and data interaction more intuitive and efficient, thereby making your apps more interactive and user-friendly.

In the next chapter, we are going to continue looking at additional formulas and functions to further enhance your app's functionality.

7

Canvas App Formulas

In this chapter, we're diving into the powerful world of **canvas app formulas**, where we'll unlock the potential to breathe life into your applications. We'll explore essential formulas such as **Search**, **Refresh**, **Filter**, **Sort**, **New form**, **Screen**, and more, equipping you with the tools to enhance user interaction and data manipulation within your apps.

Throughout this chapter, you'll gain practical skills that will empower you to navigate between screens with ease, seamlessly guiding users through your app's functionalities. Additionally, we'll clarify the process of connecting forms to Dataverse tables, enabling efficient data input and management within your canvas apps.

An important best practice we'll emphasize is the consistent renaming of elements within your apps. Naming conventions for forms, galleries, screens, and visual elements are crucial for clarity, maintenance, and debugging. Following these conventions ensures that your app is understandable and maintainable, both for yourself and for any other developers who may work on it in the future.

By the end of this chapter, you'll be proficient in leveraging canvas app formulas to build dynamic and responsive applications tailored to your specific business needs. With newfound expertise in formula usage, screen navigation, data connectivity, and element naming best practices, you'll be well equipped to create engaging and functional canvas apps that drive productivity and efficiency.

In this chapter, we're going to cover the following main topics:

- Submitting a form, **New Screen**, a back button, **New form**, **Refresh**, and **Search**
- Navigating the user between screens when something is clicked
- Connecting a form to a Dataverse table for user input

Technical requirements

Before exploring canvas app formulas, ensure you have the following:

- **Power Apps environment**: Access to a Power Apps environment

- **Canvas app**: A canvas app created within your environment

- **Basic understanding of Power Apps**: Familiarity with the platform

Having these in place will enhance your learning experience and enable you to follow along with the exercises effectively.

Exploring formulas in canvas apps

In this section, we'll delve into the essential formulas that drive canvas apps, including **Search**, **Refresh**, **Filter**, **Sort**, **New form**, **Screen**, and more. These formulas serve as the building blocks for creating dynamic and responsive applications, allowing you to manipulate data, control navigation, and enhance user interaction. Let's explore how each formula works and how you can harness their power to craft intuitive and functional canvas apps:

1. Ensure you are in the **Event Planning Project** environment.

2. Click on **Solutions** on the left navigation.

3. Click on the **Event Planning Project** solution.

4. Find **Event Planning Canvas App**, click on the three dots located to the right of the record, and then click **Edit**.

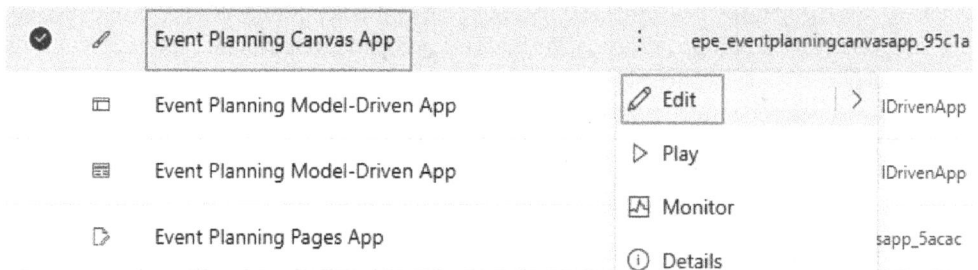

Figure 7.1 – The Edit button for Event Planning Canvas App

5. After clicking edit, the canvas app should look like *Figure 7.2*.

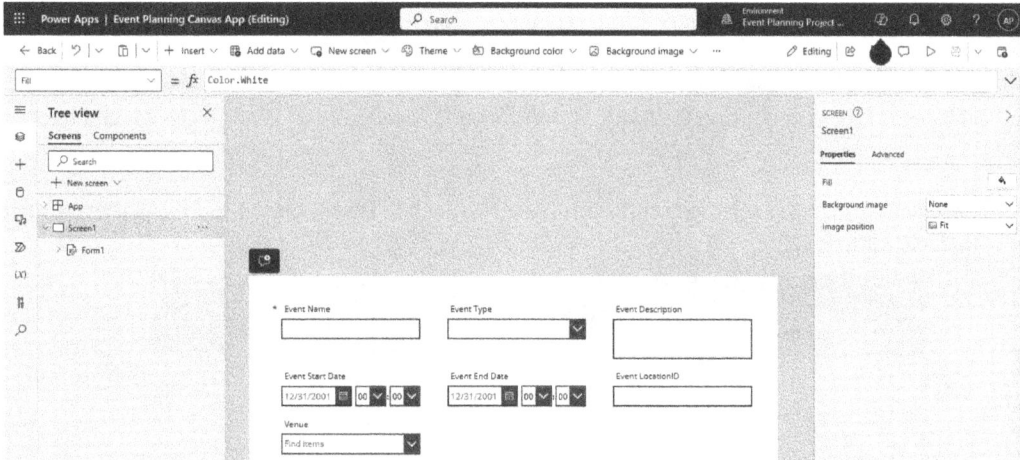

Figure 7.2 – Canvas app with the form created in Chapter 4

6. This form is connected to the **Events** table; click on `Form1` from the left navigation pane and ensure the **Data source** value on the right panel is set to **Events**:

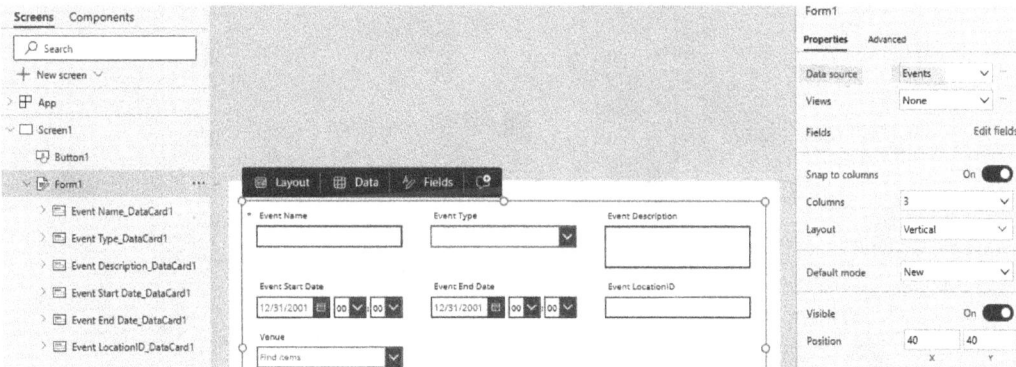

Figure 7.3 – Form1 and Data source to the Events table

In the *Exploring Formulas in Canvas Apps* section, we delve into the essential formulas driving canvas apps, including **Search**, **Refresh**, **Filter**, **Sort**, **New form**, **Screen**, and more. These formulas serve as the building blocks for creating dynamic and responsive applications, allowing you to manipulate data, control navigation, and enhance user interaction. We guide you through understanding how each formula works and harnessing its power to craft intuitive and functional canvas apps. Following this, in the *Creating a Submit Button Formula for Canvas Apps* section, we build upon our foundational knowledge from *Chapter 4*, focusing on crafting a submit button formula to seamlessly send form data back to Dataverse. This step represents a pivotal moment in app development, bridging the gap between user input and backend data processing. Through practical demonstrations and step-by-step instructions, we equip you with the skills to implement a robust submit button formula effectively, enhancing your canvas app's functionality and usability.

Creating a submit button formula for canvas apps

Let's build upon our knowledge from *Chapter 4*, where we learned how to create a new form in canvas apps. Now, we'll focus on crafting a submit button formula to seamlessly send the form data back to Dataverse:

1. Ensure `Screen1` is selected on the left navigation, click **Insert** on the top navigation, and then click **Button**.

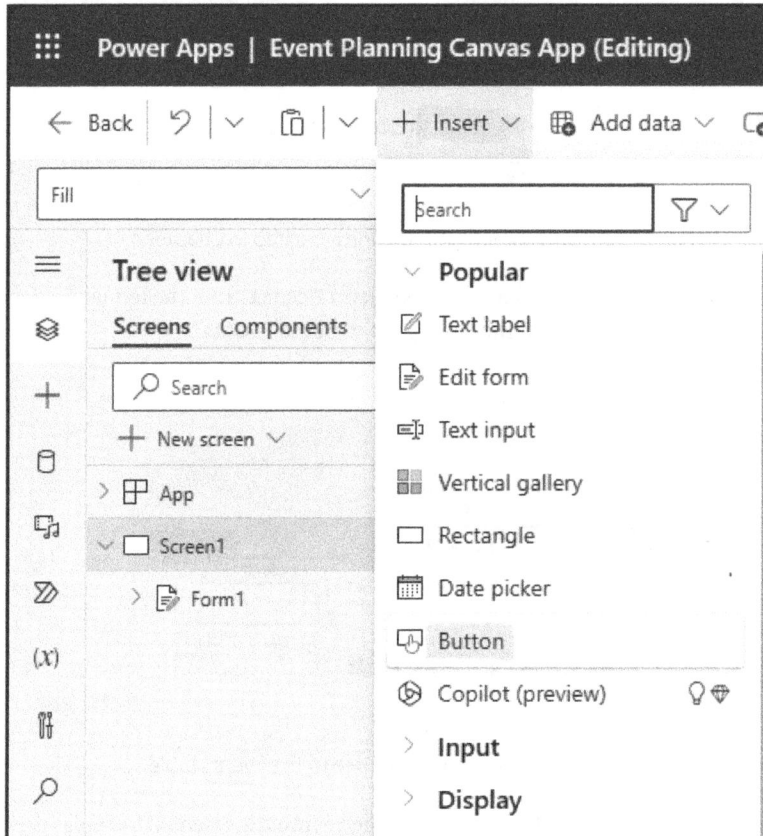

Figure 7.4 – Where to find Screen1, Insert, and Button

2. If `Button1` does not appear within the `Screen1` accordion, please try *Step 1* again and ensure it looks like *Figure 7.4*.

3. Click and drag the button from the top-left corner and place it under the form fields, as you can see in *Figure 7.5*.

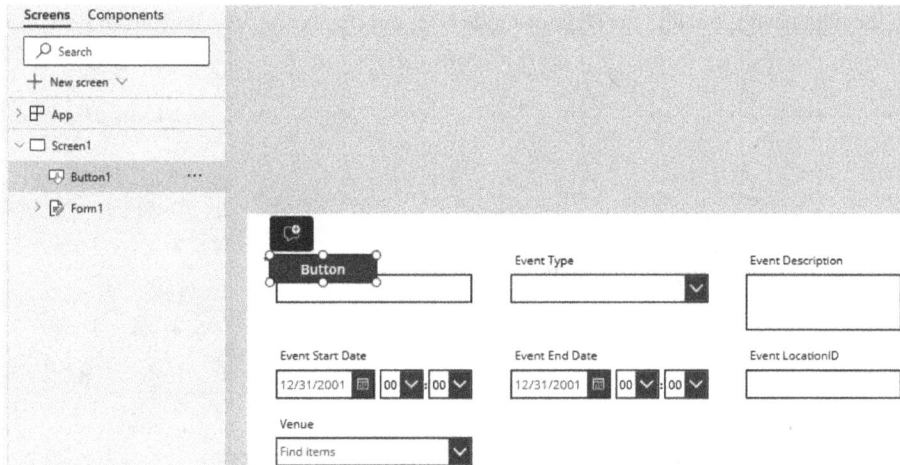

Figure 7.5 – Example of how Button1 should appear under Screen1

4. Double-click the text inside of the button and change it to `Submit`.

   Alternatively, click the button once and change the text inside the **Text** label on the right navigation (*Figure 7.5*).

5. While the **Submit** button is still selected, navigate to the formula bar at the top, ensure `OnSelect` is selected on the left dropdown, remove the *false* text in the formula (**fx**) input box, and replace it with `SubmitForm(Form1);`.

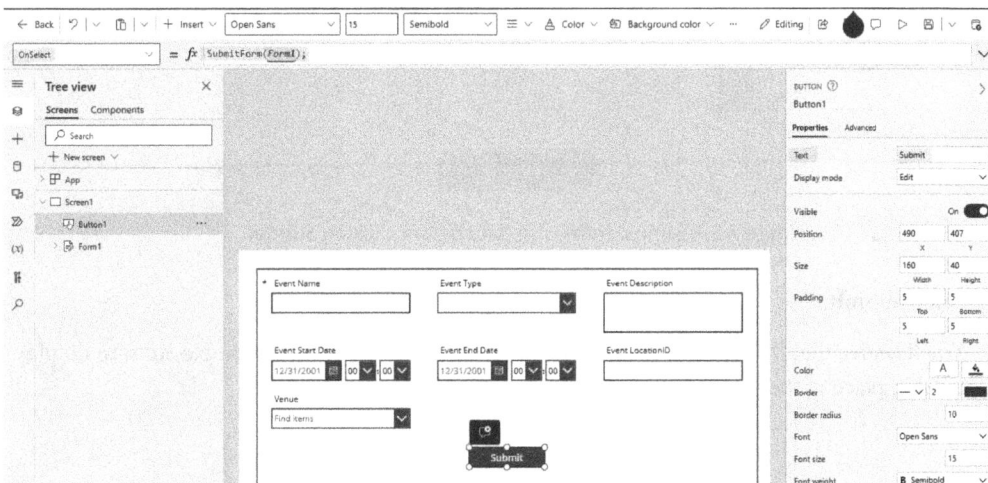

Figure 7.6 – Submit text, OnSelect, and formula to submit the form

6. One last important step is to ensure your form's default mode is set to **New**.

7. Select `Form1` from the left navigation pane.

8. On the right navigation, find the **Default mode** label and change the dropdown from **Edit** to **New**.

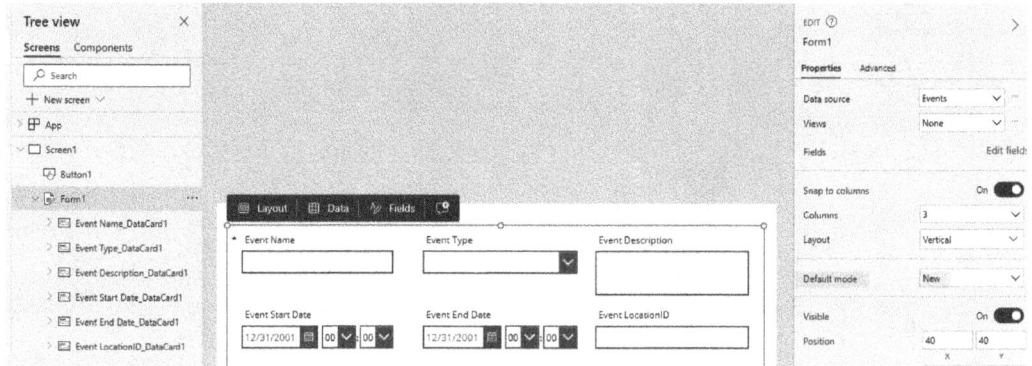

Figure 7.7 – Default mode set to New after selecting Form1

9. Click the play button on the top-right corner of the canvas app to test the form submission

   Fill in the form with any test data you prefer; my version is in *Figure 7.8*.

Figure 7.8 – Sample form with data before clicking Submit

10. Click **Submit**.

    You'll notice that the **Submit** button will still be there and there will be a **No item to display** label in place of the form:

No item to display

Submit

Figure 7.9 – The No item to display screen

11. Click on the **X** button on the top-right corner to go back to edit mode:

> **Note**
>
> This is an incomplete version of the form; in the next steps, we will enhance the **user experience** (**UX**) by hiding the **Submit** button and providing a confirmation message upon successful submission.

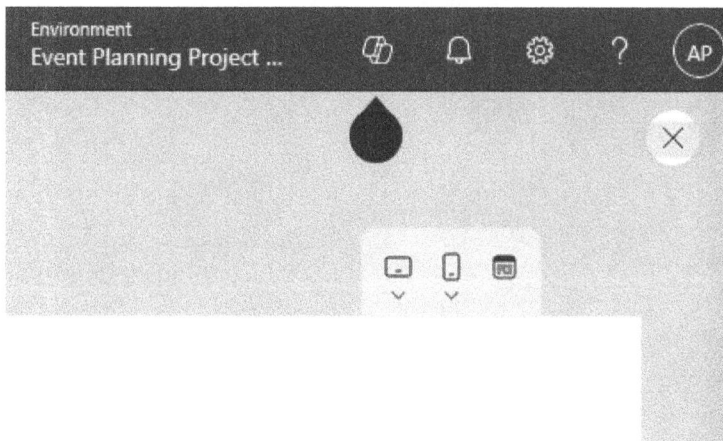

Figure 7.10 – The X button in the top-right corner of the canvas app

In this section, we build on our knowledge from *Chapter 4*, where we learned to create new forms in canvas apps. Now, we focus on crafting a submit button formula to seamlessly send form data back to Dataverse. We guide you through the process, starting with adding a button to the canvas, customizing its appearance and text, and setting up the formula to trigger form submission. Ensuring the form's default mode is correctly configured is emphasized. Practical demonstrations and step-by-step instructions accompany each stage, culminating in testing the form submission. Through this section, you'll gain the skills to integrate user input with backend data effectively, enhancing the functionality of your canvas apps. In the next section, we'll seamlessly integrate a success screen into our canvas app workflow after the submit button is clicked.

Adding a success screen after a Submit button click

Let's explore how to incorporate a success screen into our canvas app workflow after the submit button is clicked. We'll guide you through the process, ensuring that your app provides clear feedback to users upon successful form submission:

1.   Click **New screen**, then scroll down and click the **Success** screen:

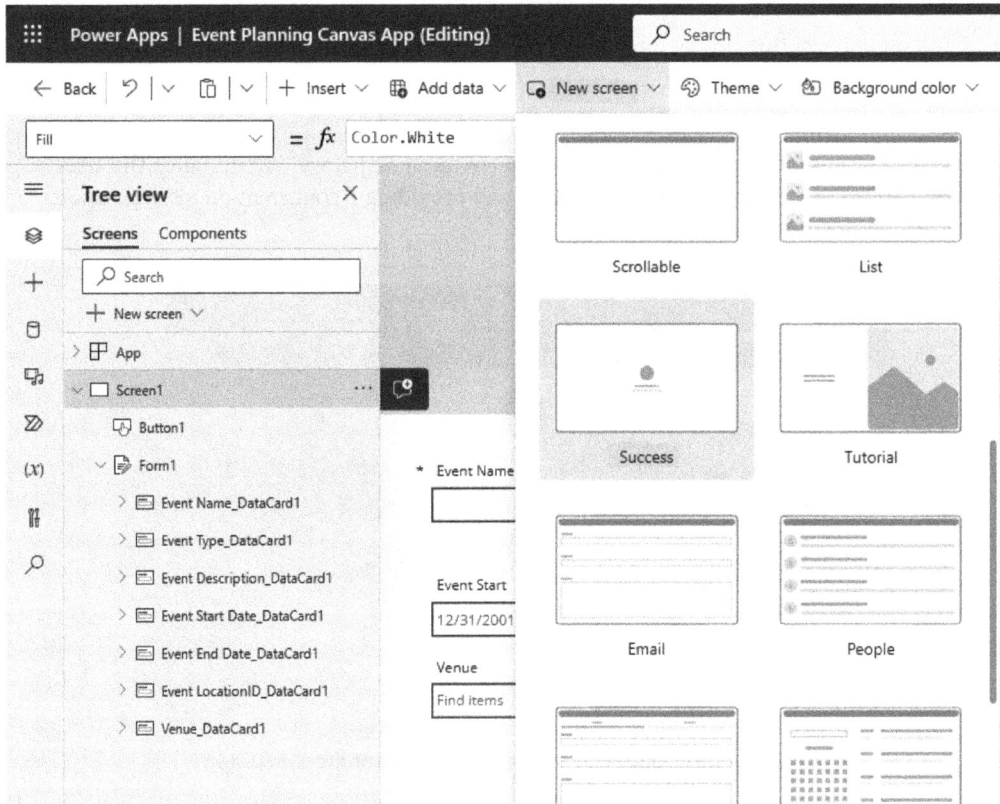

Figure 7.11 – The Success screen option under New screen

2.   Screen2 should appear on the left navigation under Screen1 and a checkmark with the **This was successfully completed** text should appear in the middle of the screen.

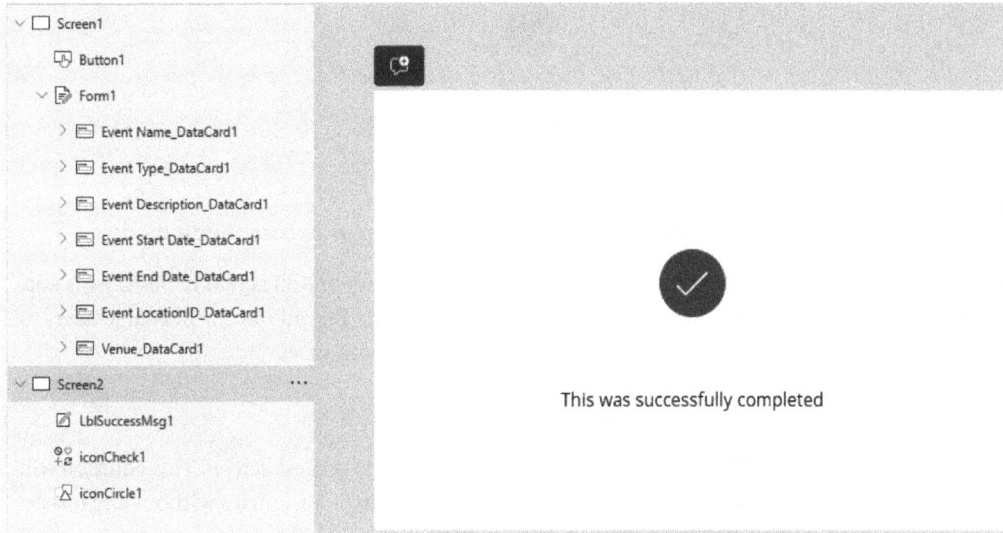

Figure 7.12 – Screen2 and the out-of-the-box success screen page

> **Important note**
>
> We want the **Submit** button created in the previous section to submit the form and take us to the success screen.

3. Click on Button1 on the left navigation.

4. In the formula bar, you should see SubmitForm(Form1);.

5. Update the formula bar to include Navigate(Screen2); after the first formula. The formula bar should now read SubmitForm(Form1); Navigate(Screen2);.

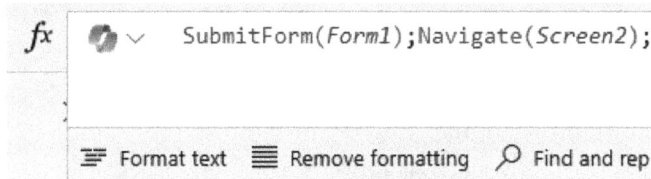

Figure 7.13 – Updated formula for Button1

6. Click the play button in the top-right corner of the canvas app.

7. Add test data to the form again and click **Submit**.

This time, the form should submit, and it should take you to the success screen.

This was successfully completed

Figure 7.14 – Success screen

8.   Click the **X** button in the top-right corner of the canvas app to go back to edit mode.

After detailing the process of integrating a success screen post-submit button action in our canvas app, we'll proceed by incorporating a back button on the success screen. This addition will enable users to seamlessly navigate back to the form for submitting new entries with ease.

Adding a back button to the success screen

Let's now ensure a smooth UX by incorporating a back button on the success screen. This addition will allow users to seamlessly return to the form interface for submitting new entries without any hassle:

1.   Select Screen2 from the left navigation.

2.   Click the **Insert** button on the top navigation.

3.   Open the **Icons** accordion and scroll down to find the **Back arrow** option.

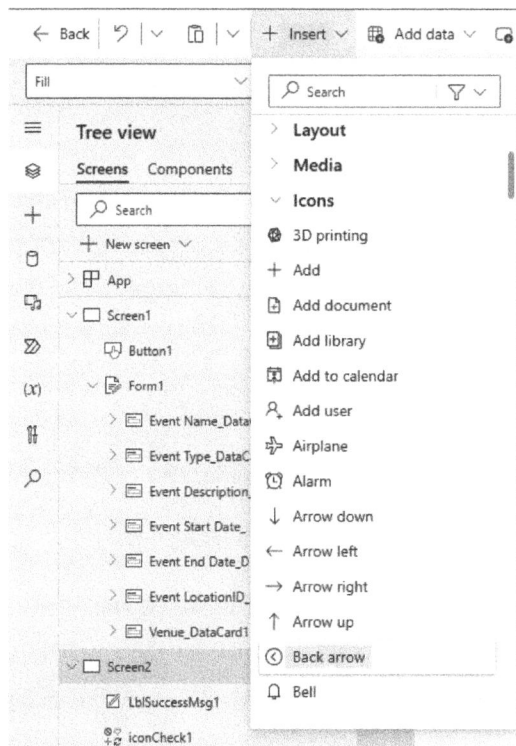

Figure 7.15 – Screen2, Insert | Icons | Back arrow

4. Change the label of the back arrow from `Icon1` to `Back arrow` by clicking the three dots to the right of `Icon1` and clicking **Rename**.

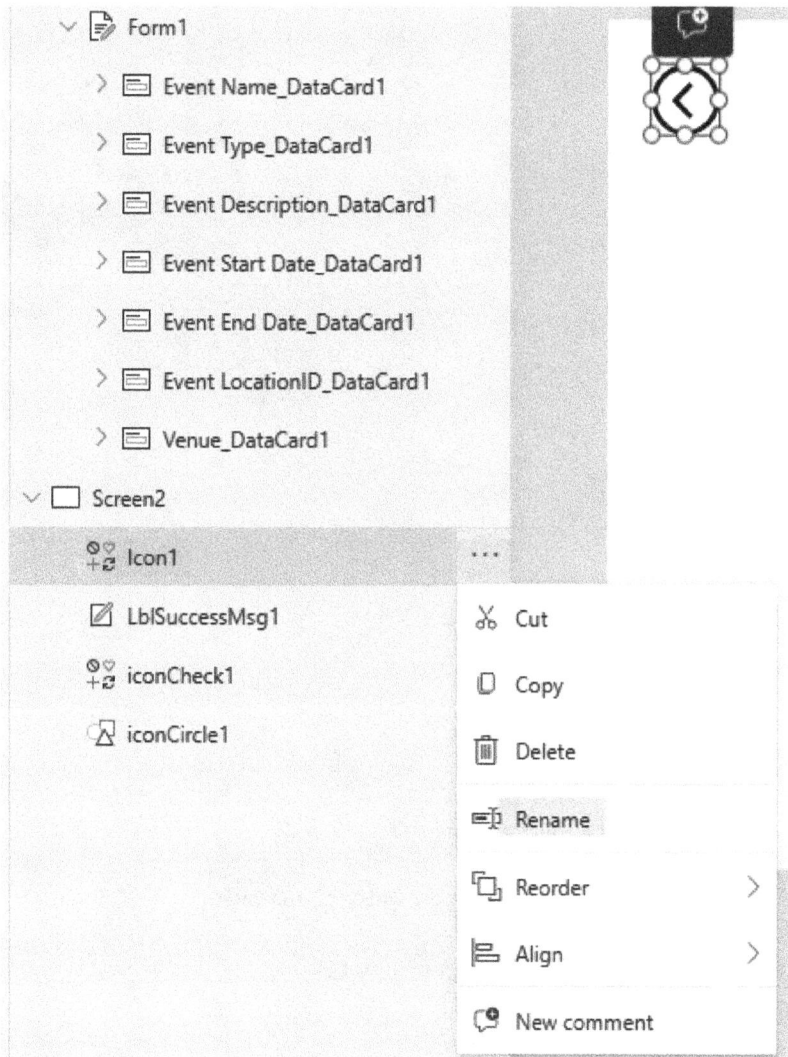

Figure 7.16 – Icon1 and Rename

5. Ensure `Back arrow` is still selected and navigate to the formula bar.

6. Ensure `OnSelect` is selected from the left dropdown and change the formula (**fx**) to `Back();.`

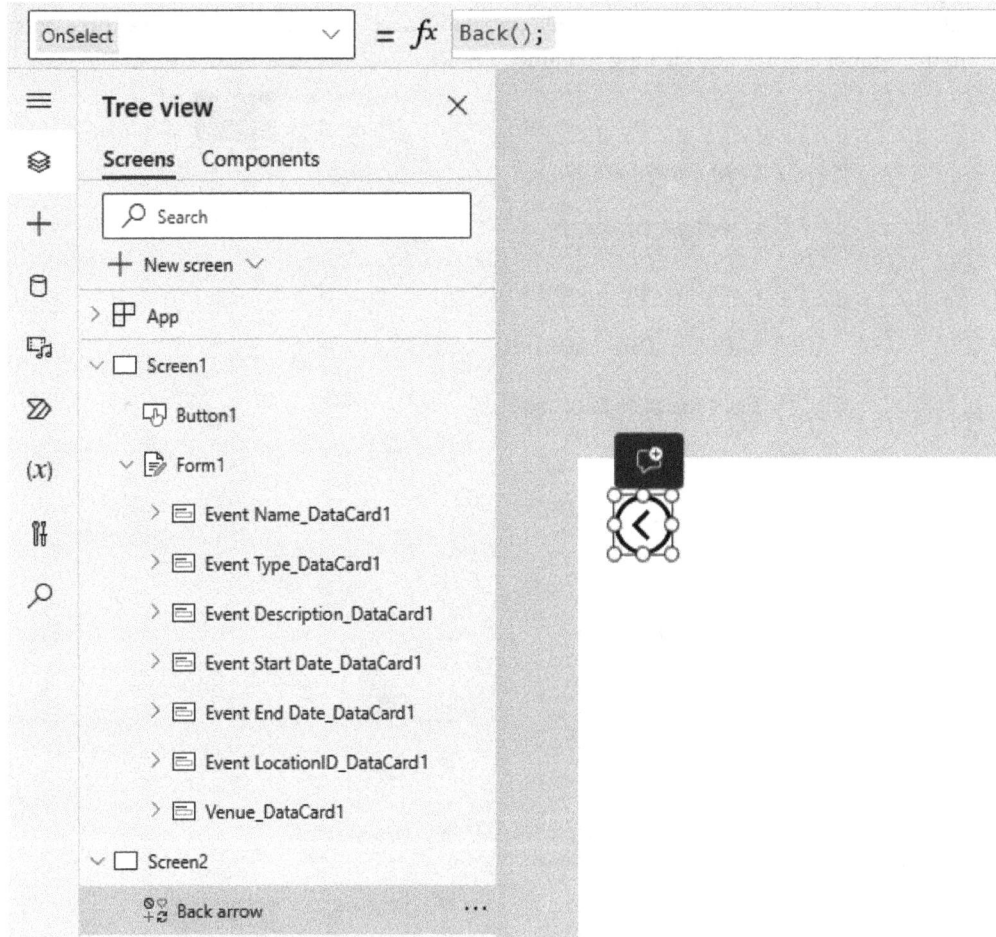

Figure 7.17 – Back arrow, OnSelect, and Back();

7.   Click the play button in the top-right corner of the canvas app and click on the back arrow button.

This will take you back to the form with the **Submit** button.

In conclusion, while the back button successfully navigates users back to the form, it currently displays **No item to display** alongside the **Submit** button, indicating a need for further refinement. Nevertheless, with the integration of the back button into the success screen, users can effortlessly return to the form to submit new entries, enhancing the app's usability. In the next section, we'll address this issue and explore solutions to ensure a seamless UX.

Adding a new form after the back button is clicked

Let's now address the issue of the form not displaying correctly after clicking the back button.

1. Select Button1 from the left **Tree view**.

2. In the formula bar, add a new formula after SubmitForm(Form1);Navigate(Screen2);: please add NewForm(Form1);. See *Figure 7.18* for the full formula.

    The updated formula should now be SubmitForm(Form1);Navigate(Screen2);New-Form(Form1);.

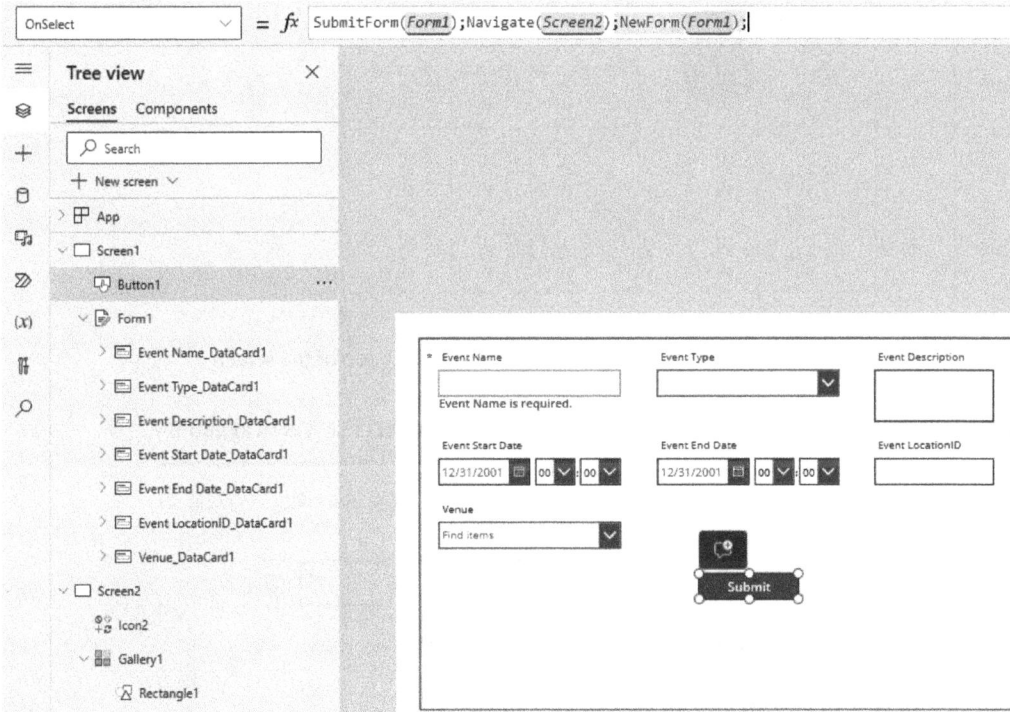

Figure 7.18 – Update to the Button1 formula

3. Test this out by clicking the play button in the top-right corner of the screen, filling out the form, clicking **Submit**, and then clicking the back button to see a new form.

In the next section, we'll explore how to create a list view that showcases all events for enhanced visibility and management.

Displaying all events in a gallery view

Let's now shift our focus to showcasing all the events created in our canvas app by setting up a gallery view. This will allow users to conveniently browse through and interact with the events in a visually appealing format:

1. Click on `Screen2` on the left navigation.

2. Move the success checkbox and the textbox to the top of the screen to make space for the gallery view by clicking and dragging the items.

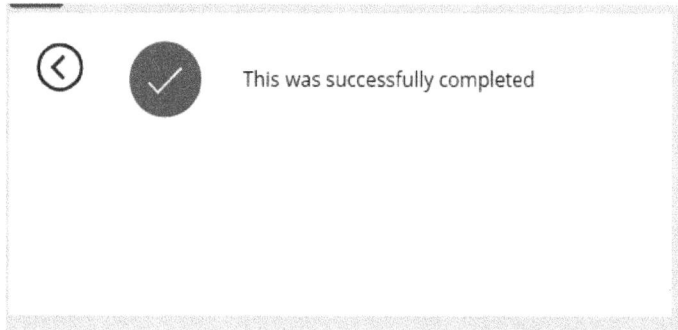

Figure 7.19 – Success screen objects moved to the top of the screen

3. Click the **Insert** button on the top navigation bar and then click **Vertical gallery**.

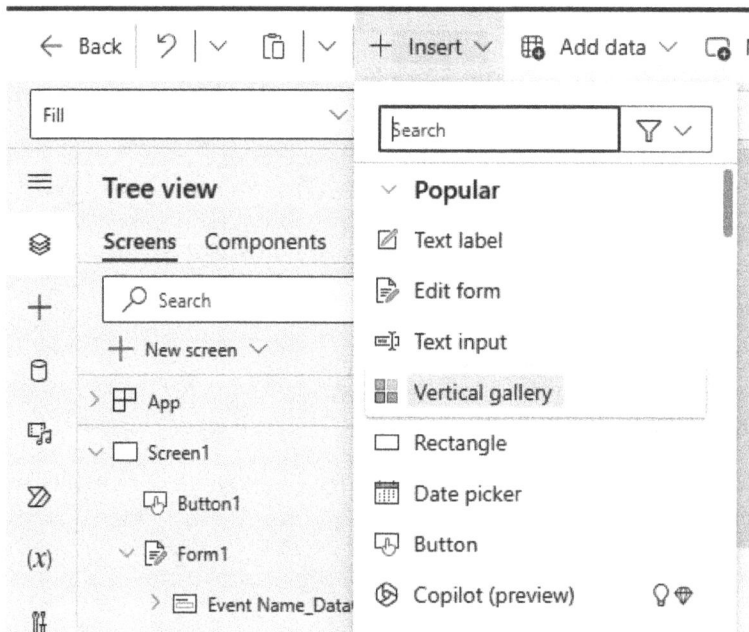

Figure 7.20 – The Insert button and the Vertical gallery button

4. When **Select a data source** appears on the canvas app, select **Events** from **Microsoft dataverse – current environment**.

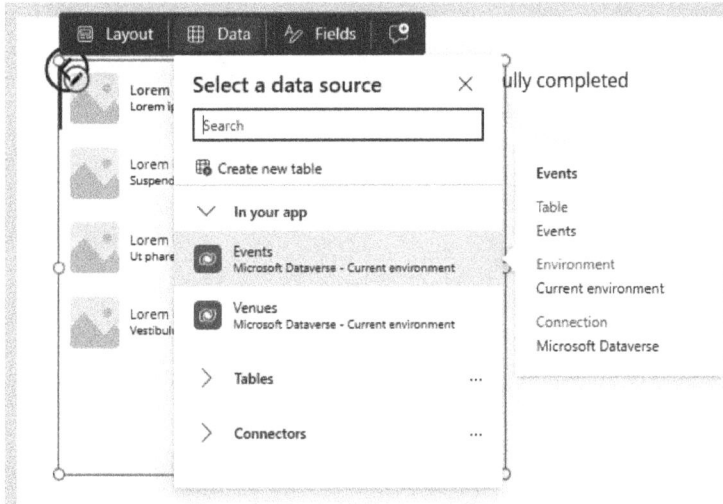

Figure 7.21 – The Events option in Select a data source

5. Drag and drop the vertical display box and place it under the back button and success message.

Notice the list of test events that have been added since we started this project.

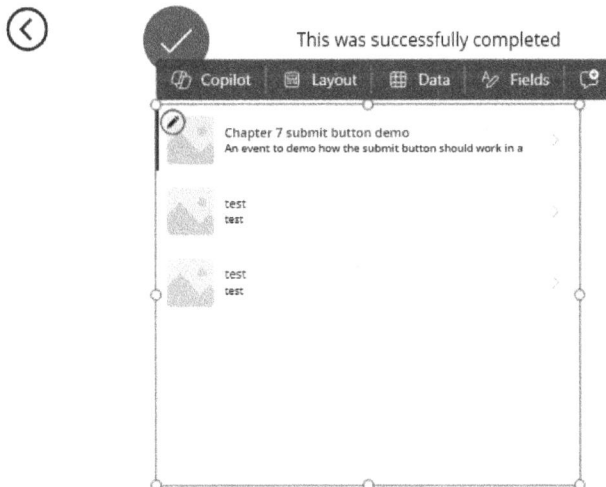

Figure 7.22 – Vertical display list of events

Now that we have added the list of events, let's add a refresh button to refresh the list of events.

Adding a refresh button to the event gallery

Let's enhance the UX by incorporating a refresh button into the vertical display gallery of events, ensuring real-time updates and improved data visibility:

1. Ensure `Gallery1` is selected on the left **Tree view**, select **Insert** on the top navigation bar, and then scroll down and click **Reload**.

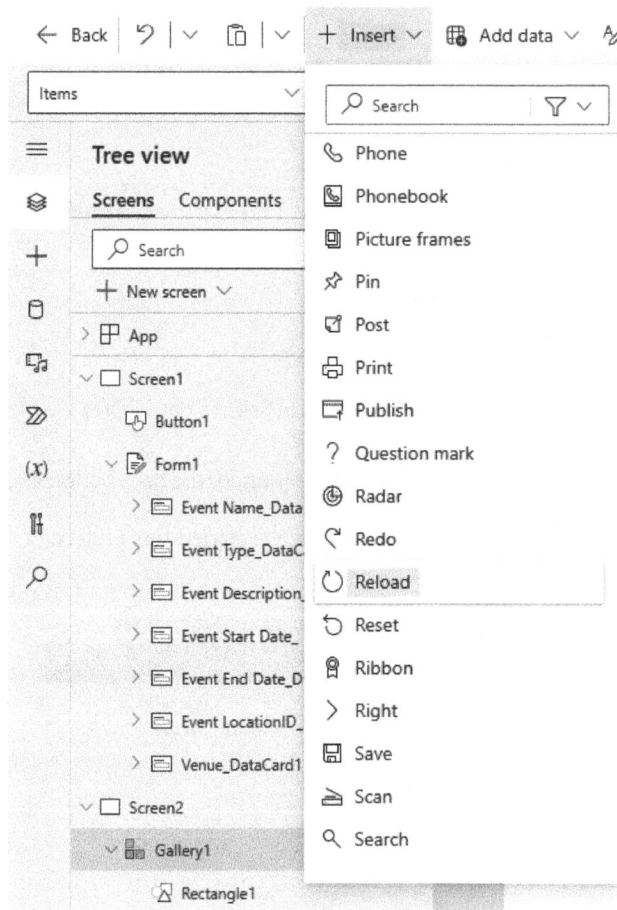

Figure 7.23 – Gallery1, Insert, and Reload

2. Click and drag the reload button to the top-left corner of the gallery list view.

> **Disclaimer**
> I am not a professional UX designer and would rather show you the functionality than try to make the design of this canvas app look good.

3. While having the reload button selected, ensure the `OnSelect` option is selected from the dropdown in the top-left corner, and update the formula (**fx**) to `Refresh(Events);`.

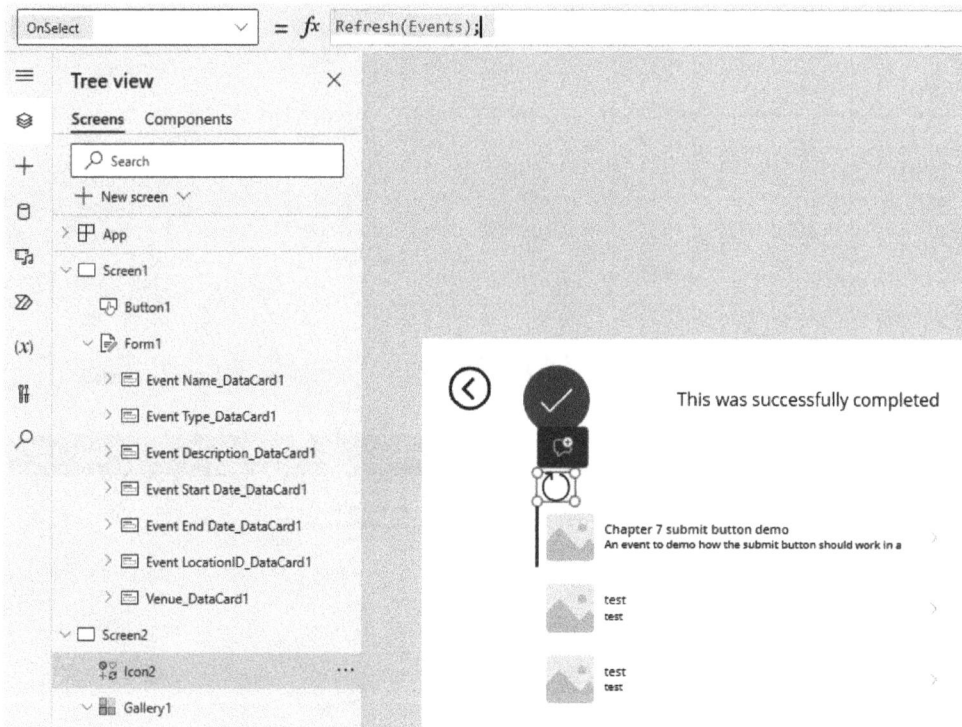

Figure 7.24 – Icon2 (Reload) and the Refresh(Events); formula

4. Click the play button on the top right corner of the screen and then click the refresh button.

    You should see a loading screen but nothing should change since we have not added another event.

With the refresh button implemented to maintain our event data's freshness, let's enhance functionality by introducing a sorting feature. As our event list expands, sorting will facilitate easy navigation and quicker access to specific events.

Adding sorting to the event gallery

In this section, we'll enhance the UX by implementing a sorting feature in the event gallery list. By sorting the events based on date, we'll ensure that the newest events are prominently displayed at the top of the list:

1. Select `Gallery1` from the left **Tree view**.

2. Select **Items** from the top-left dropdown.

3. Update the formula bar to Sort(Events, 'Created'On', SortOrder.Descending).

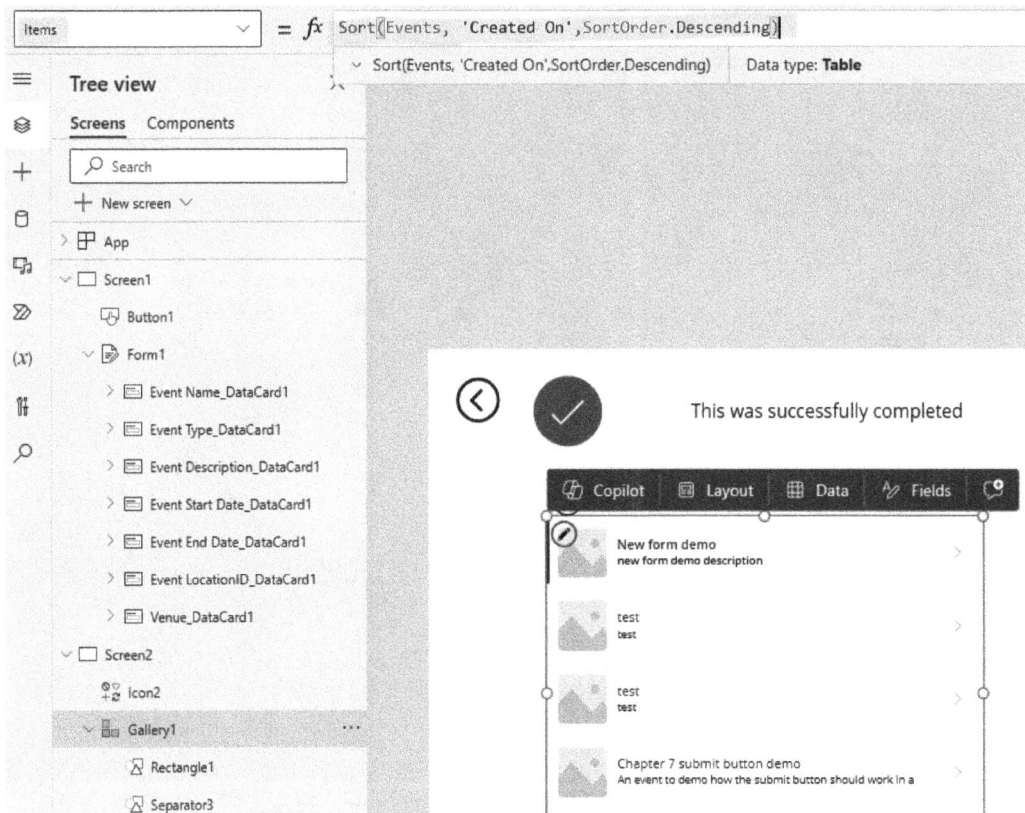

Figure 7.25 – Updated formula for Items in Gallery1

With the sorting feature implemented, users can effortlessly locate the latest events at the top of the list, enhancing their browsing experience and keeping them informed of the most recent additions. In the next section, we'll introduce the **Search** functionality to further optimize event discovery.

Adding search to the event gallery

In this section, we'll enhance the functionality of our event gallery by implementing the **Search** feature. This feature will allow users to easily find specific events from a potentially lengthy list, improving the overall usability of our canvas app. Let's dive into the steps to add this valuable functionality:

1. Add a text input by clicking the **Insert** button on the top navigation and selecting the **Text input** option. Drag that box to the top-right corner of the event gallery list. Update the **Text input** text to Search from either the formula bar, double-clicking the text input box, or by updating the Default label on the right panel.

2. Click the **Insert** button on the top navigation, open the **Icons** accordion, scroll down to **Search**, and select it. Drag the search icon to the inside of the **Text input** box and adjust the size to fit inside the box.

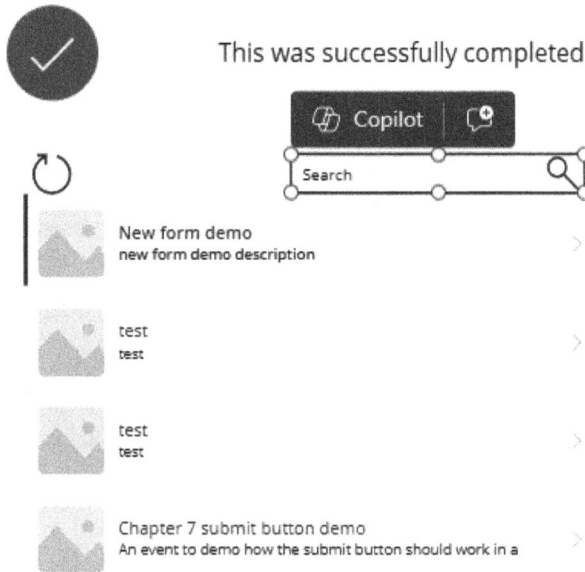

Figure 7.26 – Example of Search box

3. Click on `Gallery1` from the left **Tree view** and ensure **Items** is selected from the top-left dropdown.

4. Update the formula (**fx**) to the following: `Sort (Search (Events, TextInput1.Text, "epe_eventname"), 'Created On', SortOrder.Descending)`.

5. Click the play button and start typing one of the test event names in the search bar to see it search.

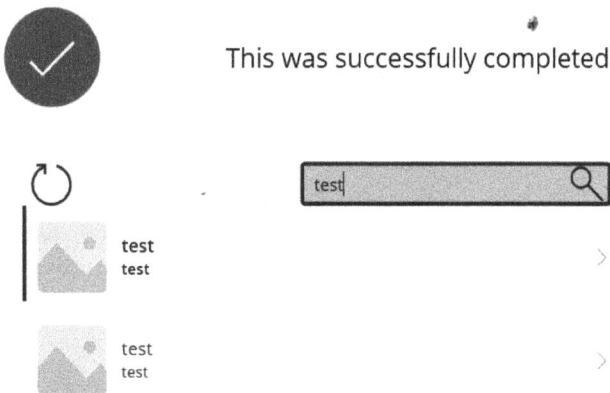

Figure 7.27 – Testing the search bar

The provided formula utilizes the sort function within a search function in canvas apps. It first searches for records in the **Events** data source where the epe_eventname column matches the text inputted into TextInput1. Then, it sorts the results based on the Created On column in descending order, ensuring that the most recently created events appear at the top of the list.

> **Note**
> While this example uses default element names such as TextInput1 and Gallery1, it is best practice in canvas app development to rename elements to more descriptive names. For instance, renaming TextInput1 to txtSearchEvent or Gallery1 to galEvents helps maintain clarity, eases maintenance, and aids in debugging. This practice ensures that other developers can easily understand the purpose of each component, facilitating better collaboration and code management.

Filtering active events

In this section, we'll focus on filtering the events for display only those that are currently active. This filtering ensures that users see relevant and up-to-date information when interacting with the canvas app.

To implement this functionality, we'll utilize the **Filter** function in Power Apps. This function allows us to narrow down the records based on specific criteria.

The following is the step-by-step guide for it:

1. Navigate to the screen where the event gallery is located within your canvas app.
2. Take a copy of Gallery1 and place it next to the original Gallery1.
3. Locate the formula that defines the Items property of the copied Gallery1.
4. Replace the existing formula with the following: Filter (Events, Status = "Active.
5. Save your changes and test the app to verify that only active events are displayed in the copied gallery.

By applying this filter, users will have a streamlined experience, focusing solely on the events that are currently active, enhancing the usability and efficiency of the app.

Summary

In this chapter, we explored various formulas and techniques to enhance canvas apps, including adding a submit button and a success screen, incorporating navigation buttons for improved UX, implementing sorting and search functionalities for better data organization, and more. By understanding and applying these techniques, readers can create more dynamic and user-friendly apps that meet their specific needs. Understanding how to leverage formulas and features effectively is crucial for building robust applications that drive productivity and engagement.

In the next chapter, we will explore advanced customization options including conditional formatting, dynamic data display, and URL deep linking. These techniques will enable readers to create even more dynamic and interactive apps, enhancing UX and functionality.

8

Conditional Formatting and URL Deep Linking

In the previous chapter, we covered several canvas app formulas that are extremely useful in building user-friendly applications. This chapter continues with the approach of enabling you to expand the capabilities of your Power Apps for an improved user experience. This includes working with two advanced features of Power Apps that are essential for enhancing the interactivity and functionality of your applications: **conditional formatting** and **URL deep linking**. Each of these features not only enriches user interaction but also streamlines processes and improves application efficiency. We will then provide an example that combines these to add functionality to an app.

By the end of this chapter, you'll be proficient in applying conditional formatting to dynamically change the appearance of app components based on specific conditions. In addition, we will explore the concept of URL deep linking and how this can also be used to add additional functionality. Lastly, we will put these concepts to use to further enhance the overall user experience and user interface, commonly referred to as the **user experience/user interface (UX/UI)**.

In this chapter, we're going to cover the following main topics:

- Applying conditional formatting
- Implementing URL deep linking
- Enhancing the overall UX/UI

Technical requirements

Before exploring canvas app formulas, ensure you have the following:

- **A Power Apps environment**: Access to a Power Apps environment
- **A Canvas app**: A canvas app created within your environment
- **A basic understanding of Power Apps**: Familiarity with the platform

Having these in place will enhance your learning experience and enable you to follow along with the exercises effectively.

Applying conditional formatting

Conditional formatting in Power Apps is a technique used to dynamically change the appearance of app components based on specific conditions. This feature allows you to highlight, emphasize, or conceal elements within an app according to the data they display or the user's interaction. By applying conditional formatting, you can create more intuitive and visually appealing interfaces that respond to context, enhance data readability, and provide a more personalized UX. By adjusting colors, visibility, and other properties of controls based on the conditions, developers can highlight important information, alert users to critical issues, or simply make the app easier and more engaging to use.

Importance of conditional formatting

The strategic use of conditional formatting can transform a standard application into a dynamic interface that reacts to its environment. This responsiveness not only improves aesthetics but also contributes to a more functional and user-friendly application. For instance, changing the color of a text box based on the value it contains can immediately alert users to areas that require attention, such as approaching deadlines or thresholds being exceeded. Another example is showing or hiding various input controls (input boxes, dropdowns, etc.) based on data that has been completed in the app so far.

Conditional formatting extends beyond visual tweaks. It can control the flow of processes and ensures users are only presented with relevant options and information at appropriate times. This customization makes apps more efficient, reducing the cognitive load on users and preventing user errors by hiding or displaying elements as needed.

Core benefits

There are a number of benefits to using conditional formatting. These include the following:

- **Enhanced user engagement**: Visually distinct elements grab attention where necessary, improving user interaction with the application

- **Improved data visualization**: Data representation becomes more intuitive using color codes, visibility adjustments, and other formatting changes

- **Increased app efficiency**: Conditional formatting can streamline user workflows by showing or hiding elements and enabling actions based on specific conditions

- **Error reduction**: By dynamically changing options and the visibility of elements, apps can guide users more effectively, reducing mistakes and enhancing data integrity

In summary, conditional formatting is not just about making your app look attractive; it's about making it work smarter. It enables applications to adapt to their environment and user inputs, presenting a seamless, interactive experience that boosts productivity and satisfaction. In the following sections, we'll explore how to implement and maximize conditional formatting in your Power Apps projects.

Concepts within conditional formatting

Understanding the foundational concepts and principles of conditional formatting in Power Apps is crucial for effectively applying it to enhance app functionality and the user interface. Next, we delve into the essential aspects that form the basis of conditional formatting, including controls and properties, data types, and the logical constructs used to define conditions.

Understanding controls and properties

Controls and properties are essential to all apps as they are the building blocks of your apps. Let's dig into each of these specifically.

- **Controls**: Controls are the components of your Power Apps application. Examples include labels, buttons, text inputs, and shapes. Conditional formatting typically involves changing the properties of these controls in response to user interactions or data changes. See *Figure 8.1* for an example of various controls.

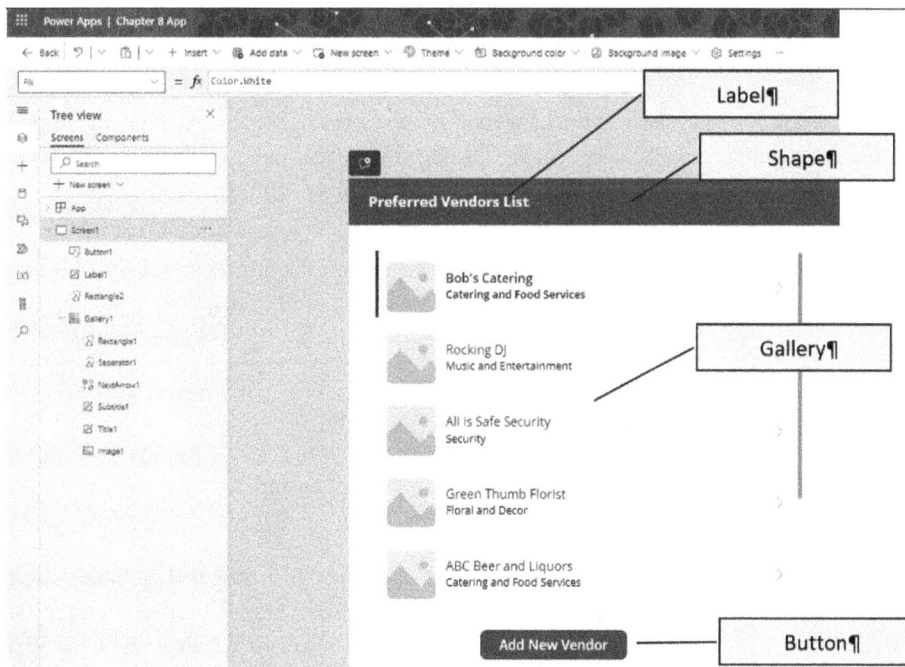

Figure 8.1 – Example of different controls

Figure 8.1 shows four different controls that have been added to the app. This includes a label displaying an overall title for the page, a blue rectangle shape, a gallery, and a button.

- **Properties**: Properties are attributes of controls that determine their appearance or behavior. For conditional formatting, properties such as color, visibility, and size can be dynamically altered based on specific conditions. Each control has a set of customizable properties, which can be manipulated using expressions. See *Figure 8.2* for an example.

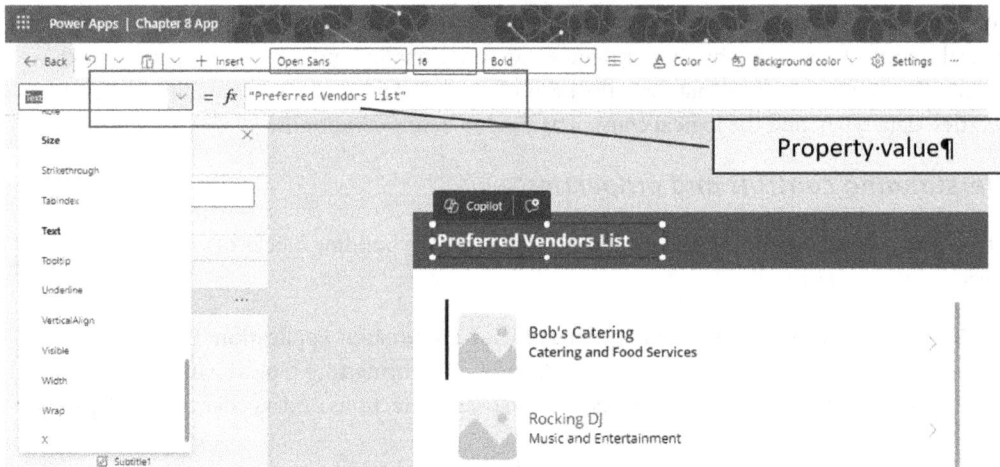

Figure 8.2 – Example of a property for a label

If we examine *Figure 8.2*, we can see how the **Label** control is selected. Then, by selecting the dropdown in the **Properties** area, all the available properties for that control are made available. Specifically, the **Text** property has the value **Preferred Vendors List**, which displays this text.

> **Quick tip**
>
> Any properties that have been modified will be shown in bold in the **Properties** dropdown. This makes it easy to identify those properties that have been edited from their default.

Property data types and expressions

Before diving too deep into conditional formatting, it is important to understand how data types and expressions work within properties:

- **Data types**: Power Apps properties support various data types, including text, number, date, and Boolean (true/false). Understanding the data type associated with each property is essential, as it affects the type of operations or comparisons that can be performed. For example, the **Visible** property of a control accepts either `true`, meaning the control should be visible, or `false`, meaning the control should not be visible. However, the **Height** property requires numerical inputs to indicate the height of the app. Let's provide another example using the **Fill** property for a rectangle shape, as shown in *Figure 8.3*.

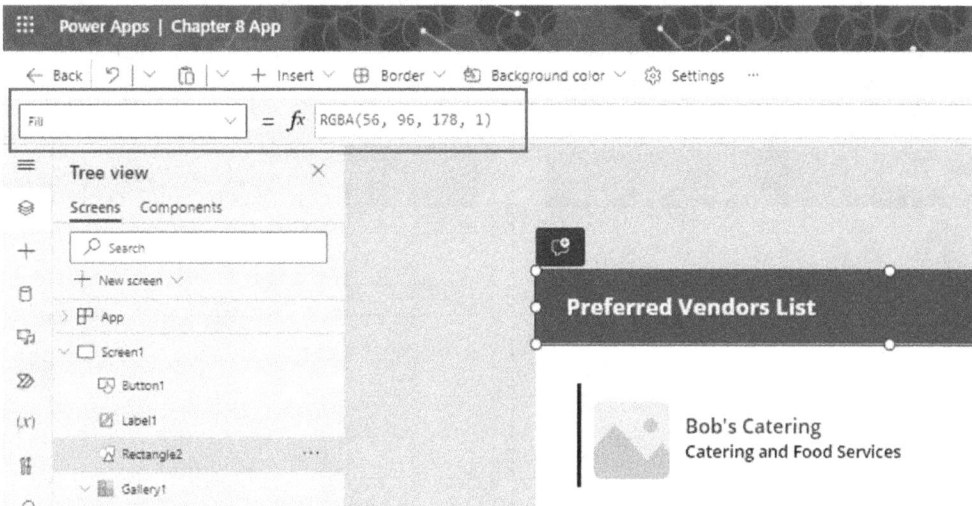

Figure 8.3 – Example of the Fill property data type

In this example, a shape is selected, and the **Fill** property uses numerical RGBA values to show the blue color. Changing any of these values will also change the color.

- **Expressions**: Expressions are used to define the logic for conditional formatting. They can compare values, perform calculations, or evaluate data states. Expressions are written using syntax similar to Excel formulas, making it intuitive for those familiar with spreadsheet applications. To provide an example, let's say that you have an edit form on your page with a corresponding button to submit the form. However, you want the text on the blue button to change depending on whether the form is adding a new record or editing an existing record (as shown in the screenshot). This can easily be done by updating the **Text** property of the button, as shown in *Figure 8.4*.

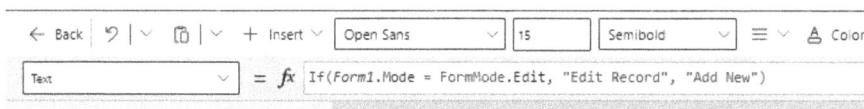

Figure 8.4 – Using an expression to dynamically change button text

In this example, an If() function is used to determine the mode of the form. If it is in edit mode, the text **Edit Record** should be displayed. Otherwise, it should display **Add New**. By using an expression within the **Text** property, this allows you to dynamically update the text. Let's continue with this example and walk through a way to specifically set this up.

Example of basic conditional formatting

Now that we've covered the concepts of Power Apps controls, properties, and expressions, let's dig into a specific example of how conditional formatting can be used. We'll use the example that is shown in *Figure 8.4*.

Consider a scenario where you have an edit form. When a record is being edited, the button will display **Edit Record**, but if a new record is being added the button will display **Add New**. *Figure 8.5* provides a side-by-side comparison of this.

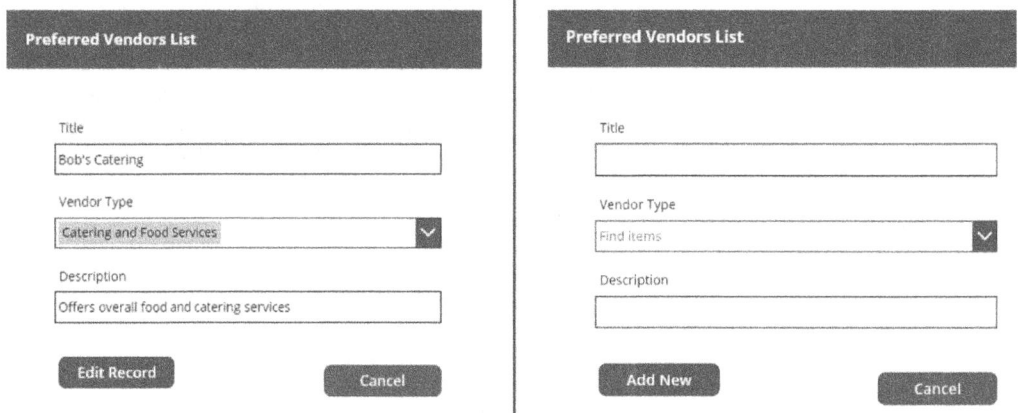

Figure 8.5 – Example of an edit form in edit mode versus New mode

The screenshot on the left shows the form with an existing record, and the user is allowed to make edits to it. However, the screenshot on the right is what is shown when a new record is needed. The text on the blue button, however, will change depending on this situation. This is easily addressed through a conditional expression in the **Text** property of the button control. This is shown in *Figure 8.6*.

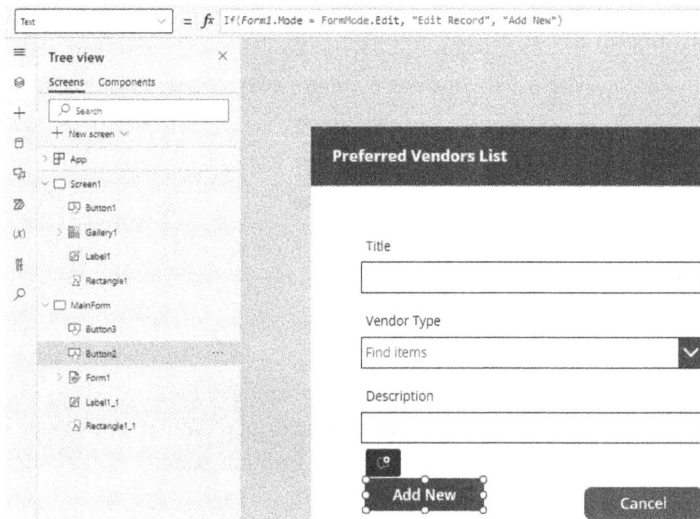

Figure 8.6 – Using an expression to dynamically change button text

By mastering these basic concepts and principles, you can begin to implement more complex conditional formatting rules in your Power Apps apps, significantly enhancing the functionality and user experience of your applications. In subsequent sections, we'll explore how to apply these fundamentals through detailed examples and best practices.

Examples of conditional formatting

To illustrate the power and versatility of conditional formatting in Microsoft Power Apps, let's explore several practical examples. These examples will demonstrate how to use conditional formatting to enhance app functionality, improve user experience, and make your applications more intuitive and visually appealing.

Example 1 – Changing text color based on values

Scenario: you have a gallery that displays a list of projects and their current status. There is a status column with values including **Not Started**, **In Progress**, **Complete**, and **Overdue**. You want the status to be highlighted in red for projects that have a status of **Overdue**. *Table 8.1* provides the expression for the **Color** property.

Control	A label displaying the project status (`StatusLabel`)
Property to modify	Color
Condition	When the status equals Overdue
Expression	`If(StatusLabel.Text) = "Overdue", Color.Red, Color.Black)`

Table 8.1 – Updating the Color property

In this example, the `If()` statement is used to evaluate the text value of the **Label** control named `StatusLabel`. If the value of this label is `Overdue`, the color of the text will be red. Otherwise, the color will be black.

Example 2 – Altering the background color of a form based on input validity

Scenario: in a user registration form, you want the background of the email input box to turn yellow if the email entered is invalid. *Table 8.2* provides the expression for the **Fill** property.

Control	A text input for email (`Email`)
Property to modify	Fill
Condition	When the format of the text entered into the text input does match an email format
Expression	`If(!IsMatch(Email.Text, Match.Email), Color.Yellow, Color.White)`

Table 8.2 – Updating the Fill property

In this example, the `IsMatch()` function is used to evaluate the format of the text within the text input control. This function has a number of predetermined patterns that can be used. In addition, custom patterns can also be created using regular expressions. **Regular expressions** are sequences of characters that can be used to identify patterns. Further discussion of regular expressions is beyond the scope of this book.

Example 3 – Toggling the visibility of admin controls based on one's role

Scenario: certain controls should only be visible to users with administrative privileges within the app, such as a button that takes the user to an admin page.

Control	A button that navigates the user to an admin screen (`AdminButton`)
Property to modify	Visible
Condition	When the user's role is defined as `admin`
Expression	`User().Email in colAdminList`

Table 8.3 – Updating the Visible property

In this example, the `User().Email in colAdminList` expression is used to see whether the user's email address is contained within a variable or collection called `colAdminList`. If so, it returns `true`, which makes the button appear.

Each of these examples showcases how conditional formatting can be applied in Power Apps to create dynamic, responsive, and effective UIs. By implementing these types of formatting, you can significantly enhance the functionality and aesthetics of your applications.

Best practices

Implementing conditional formatting in Microsoft Power Apps not only enhances the visual appeal and interactivity of your applications but also improves usability. To ensure that your applications remain efficient and user friendly, here are some best practices for design and usability when applying conditional formatting.

Maintain clarity and consistency

Ensure that the formatting choices you make do not confuse users. Use colors, fonts, and visibility changes that clearly communicate the intended message or status without ambiguity. Apply similar formatting rules for similar types of data or conditions across your app. Consistent visual cues help users learn the interface faster and navigate the app more intuitively.

Prioritize accessibility

Conditional formatting should enhance the accessibility of your application, not hinder it. Consider the following:

- **Color contrast**: Make sure there is sufficient contrast between text and background colors to aid users with visual impairments.

- **Color blindness**: Avoid using color as the only means of conveying information, as this can be ineffective for color-blind users. Combine colors with icons or labels.

Avoid over-formatting

While conditional formatting can be very useful, overusing it can make your app look cluttered and confusing. Use conditional formatting sparingly and only where it adds clear value to the UX.

Test on multiple devices

Power Apps apps are often used across a range of devices with different screen sizes and capabilities:

- **Responsive design**: Ensure that your conditional formatting looks good and functions well on all device types, from smartphones to tablets to desktops

- **Performance**: Test the performance impacts of your conditional formatting, particularly in data-heavy applications or on older devices

Keep performance in mind

Conditional formatting can impact app performance, especially if complex expressions or data bindings are involved. Optimize expressions and limit real-time data processing where possible to maintain a smooth UX.

Documentation and comments

As conditional formatting rules can become complex, documenting your logic and adding comments to your expressions will help you and others understand the setup better. This is especially important for maintenance or future modifications.

Iterate based on feedback

Gather user feedback on your conditional formatting implementations to understand their impact. Users might provide insights into how the formatting affects their interaction with the app and suggest improvements.

By expanding your use of conditional formatting, you can create Power Apps apps that are not only more interactive and engaging but also capable of handling complex scenarios and data-intensive operations efficiently. Now that we have reviewed conditional formatting, let's move to another approach that can add functionality to your app.

Implementing URL deep linking

URL deep linking in Power Apps enhances navigation and usability by allowing users to jump directly to specific screens or states within an app. This is particularly useful in complex applications with multiple UI layers. By using deep links, developers can construct a more structured and direct path to information, improving the user journey within the app. Deep linking facilitates easier sharing of app locations, simplifies navigation, and supports more intricate application designs.

What is URL deep linking?

In the context of Power Apps, URL deep linking refers to the ability to develop a URL that opens an app and navigates to a particular screen or initializes certain states. This URL can be shared via email, social media, or a Power Automate flow that sends emails based on scheduled or other triggers, or it can be embedded in other applications, allowing for a seamless transition directly to specific app content without navigating through the app's initial screens. In addition, the URL can include additional parameters that can be used within the app itself.

Using URL deep linking offers several benefits:

- **Improved navigation**: Users can jump directly to the content they need, which is especially beneficial in complex apps with multiple layers of navigation

- **Increased engagement**: By reducing the number of steps needed to reach desired content, deep links can help increase user engagement and satisfaction

- **Streamlined processes**: Deep links can be used to streamline processes by automating navigation and prefilling data, thus reducing manual input and errors

Now that we have provided a brief introduction and discussed the benefits, let's delve further into URL deep linking.

Understanding URL structures in Power Apps

Before going deep into creating the URL itself, it's important to break down the components of a Power Apps URL. First, to obtain the URL of your app, simply click on the three dots and select **Details**, as shown in *Figure 8.7*:

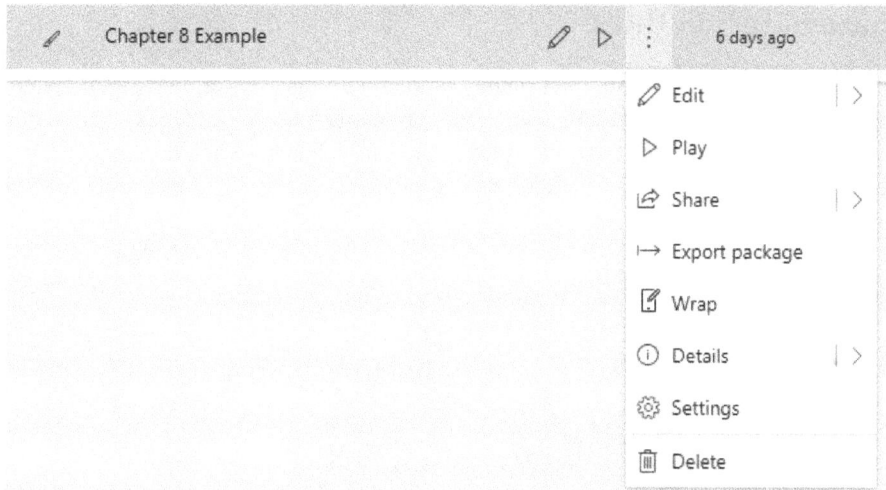

Figure 8.7 – Finding the Details of a Power Apps app

This will bring up the details of the app, including the URL along with other pertinent information. This is shown in *Figure 8.8*.

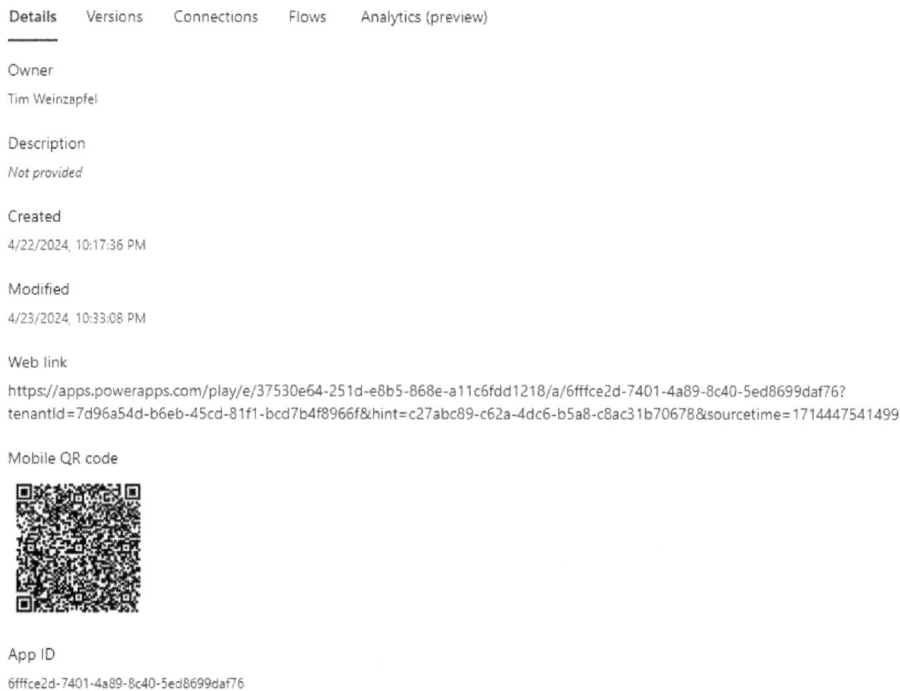

Details   Versions   Connections   Flows   Analytics (preview)

Owner
Tim Weinzapfel

Description
Not provided

Created
4/22/2024, 10:17:36 PM

Modified
4/23/2024, 10:33:08 PM

Web link
https://apps.powerapps.com/play/e/37530e64-251d-e8b5-868e-a11c6fdd1218/a/6fffce2d-7401-4a89-8c40-5ed8699daf76?
tenantId=7d96a54d-b6eb-45cd-81f1-bcd7b4f8966f&hint=c27abc89-c62a-4dc6-b5a8-c8ac31b70678&sourcetime=1714447541499

Mobile QR code

App ID
6fffce2d-7401-4a89-8c40-5ed8699daf76

Figure 8.8 – Details of the Power Apps app, including the web link

Using parameters in URLs

Parameters play a crucial role in the functionality of deep linking. They allow developers to pass information directly to an app via the URL, which the app can then use to perform specific actions or display certain data. Parameters are typically added to the URL following a question mark ("?") using a key-value pair. In addition, multiple parameters can be added and are separated by an ampersand (&). *Figure 8.9* provides an example:

https://apps.powerapps.com/play/[screenid]?*VendorID=1&Screen=EditScreen*

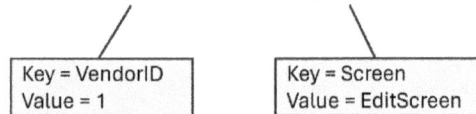

| Key = VendorID |
| Value = 1 |

| Key = Screen |
| Value = EditScreen |

Figure 8.9 – Example of parameters embedded in a URL

> **Note**
>
> The URL provided is condensed for demonstration purposes. What is important to note is that all parameters are added to the end of your specific app's URL using the format shown in *Figure 8.9*.

In this example, there are two URL parameters after the question mark (?). These parameters are VendorID, with a value of 1, and Screen, with a value of EditScreen. When the Power Apps app is opened up, these two parameters and their values will now be available for use in the app.

Retrieving incoming URL parameters in your Power Apps app

Once a URL link is set up with parameters, as shown in the example in *Figure 8.9*, the next step is accessing these in your Power Apps app. This is done using the Param() function, as follows:

```
Param("parameterName")
```

Using the URL example from *Figure 8.9*, the functions for each would be as follows:

```
Param("VendorID")
Param("Screen")
```

Using parameters in app logic

Now that you understand the basics of URL parameters, let's delve into how these can be utilized. First, let's provide an example of a very simple app, which is shown in *Figure 8.10*.

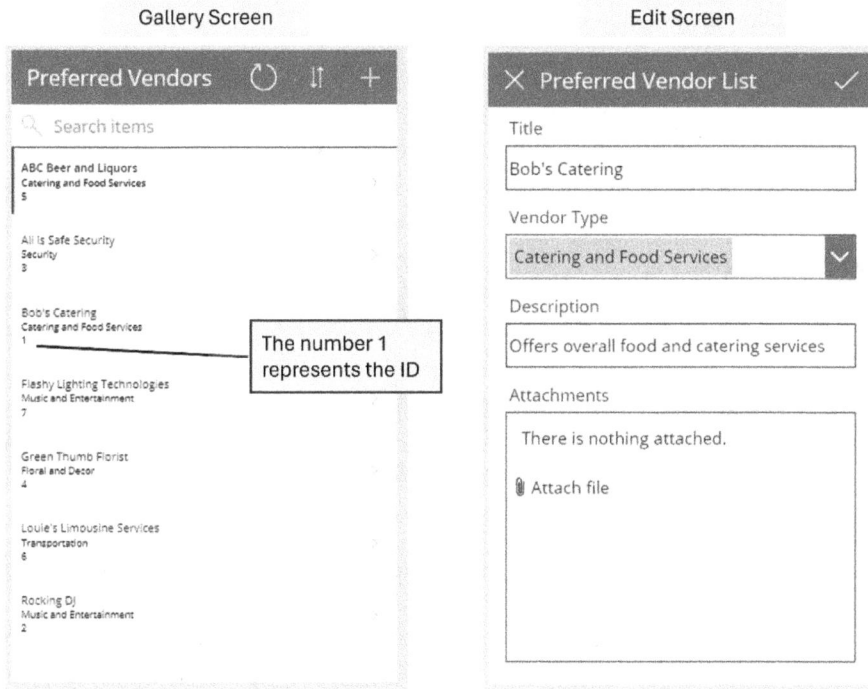

Figure 8.10 – Layout of a simple app

This app has two screens. The one on the left is a simple gallery that displays all the vendors. There is also an edit screen on the right, where the user can add new records or edit existing records. *Figure 8.10* displays a record for **Bob's Catering**.

When the app is opened normally, the gallery screen appears by default. However, our goal is to create a web link that, when clicked on, will take the user directly to the edit screen, already filtered to a specific record. We'll use **Bob's Catering** as an example.

To do this, we are going to take advantage of the following URL parameters, which will be attached to our applicable Power Apps web link:

https://apps.powerapps.com/play/[screenid]?*VendorID=1&Screen=EditScreen*

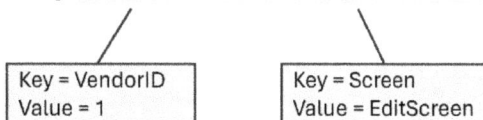

Key = VendorID
Value = 1

Key = Screen
Value = EditScreen

Figure 8.11 – URL parameters that will be used

The steps we then follow are as follows:

1.   Initialize our variables in the **OnStart** property of the overall app. See *Figure 8.12*.

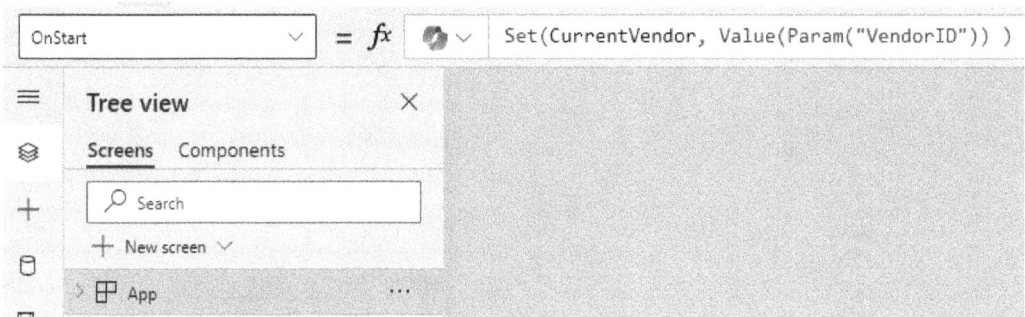

OnStart   ∨   = *fx*   ● ∨   Set(CurrentVendor, Value(Param("VendorID")))

≡   **Tree view**   ✕

⊗   **Screens**   Components

+   🔍 Search

        + New screen ∨

🗄

⌐   > ▱ App   ⋯

Figure 8.12 – Initializing variables in the OnStart property of the app

This creates a variable called `CurrentVendor` and sets the value equal to the value passed from the URL. Looking at *Figure 8.11*, this would be the value of `1`. As you will see in a subsequent step, we will use this ID in a `LookUp()` function. However, because parameter values are always passed in string format, we use the `Value()` function to convert this to a number.

> **Note**
>
> All parameter values are always in string format. This will cause errors if you compare this to a numerical field. An easy way to address this is to use the `Value()` function, as shown in *Figure 8.12*.

2.   Our next step is to use the **StartScreen** property of the overall app to determine which screen should be displayed. When no URL parameters exist, the gallery screen should be displayed. However, when we have a URL parameter as shown in *Figure 8.11*, we want to direct the user to the edit screen. *Figure 8.13* shows this screen:

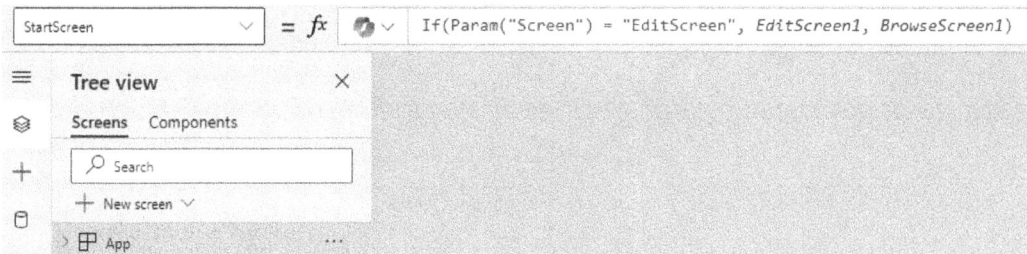

StartScreen   ∨   = *fx*   ● ∨   If(Param("Screen") = "EditScreen", EditScreen1, BrowseScreen1)

≡   **Tree view**   ✕

⊗   **Screens**   Components

+   🔍 Search

        + New screen ∨

🗄

        > ▱ App   ⋯

Figure 8.13 – Setting the StartScreen property for the app

We use an `If()` statement to check whether the `Screen` URL parameter equals `EditScreen`. If so, proceed to the edit screen (*EditScreen1*). Otherwise, send the user to the gallery screen (*BrowseScreen1*).

3. Our last step is to update the form in the edit screen to account for this new functionality, as shown in *Figure 8.14*.

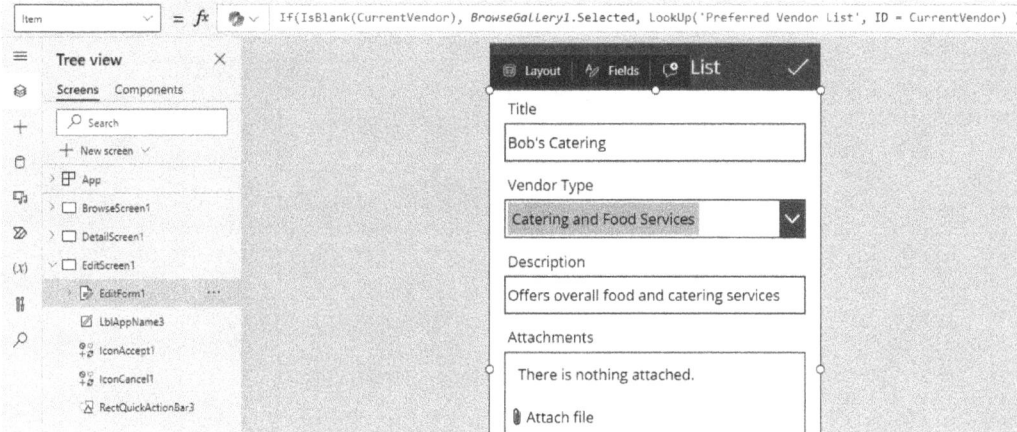

Figure 8.14 – Updating the Item property in the edit form

Here, we update the `Item` property of the edit form and use the following code:

```
If(IsBlank(CurrentVendor), BrowseGallery1.Selected,
LookUp('Preferred Vendor List', ID = CurrentVendor) )
```

We use an `If()` statement that first checks to see whether the `CurrentVendor` variable is blank or has a value. Remember that, in *Step 1*, when a URL parameter is used, this variable is set to a specific ID value. If no value is present, which means no URL parameter was used, then the item from the gallery screen is displayed. However, when the `CurrentVendor` variable does have a value that was set from the URL parameter, then a `LookUp()` function is used to find the specific entry.

Figure 8.15 provides a summary of which screen the user is sent to based on whether URL parameters are used are not.

Standard URL *Gallery Screen Displayed*	URL with ?VendorID=1&Screen=EditScreen *Edit Screen Displayed*
Preferred Vendors ↻ ⇅ + 🔍 Search items ABC Beer and Liquors Catering and Food Services 5 > All Is Safe Security Security 3 > Bob's Catering Catering and Food Services 1 > Flashy Lighting Technologies Music and Entertainment 7 > Green Thumb Florist Floral and Decor 4 > Louie's Limousine Services Transportation 6 > Rocking DJ Music and Entertainment 2 >	✕ Preferred Vendor List ✓ Title Bob's Catering Vendor Type Catering and Food Services ⌄ Description Offers overall food and catering services Attachments There is nothing attached. 📎 Attach file

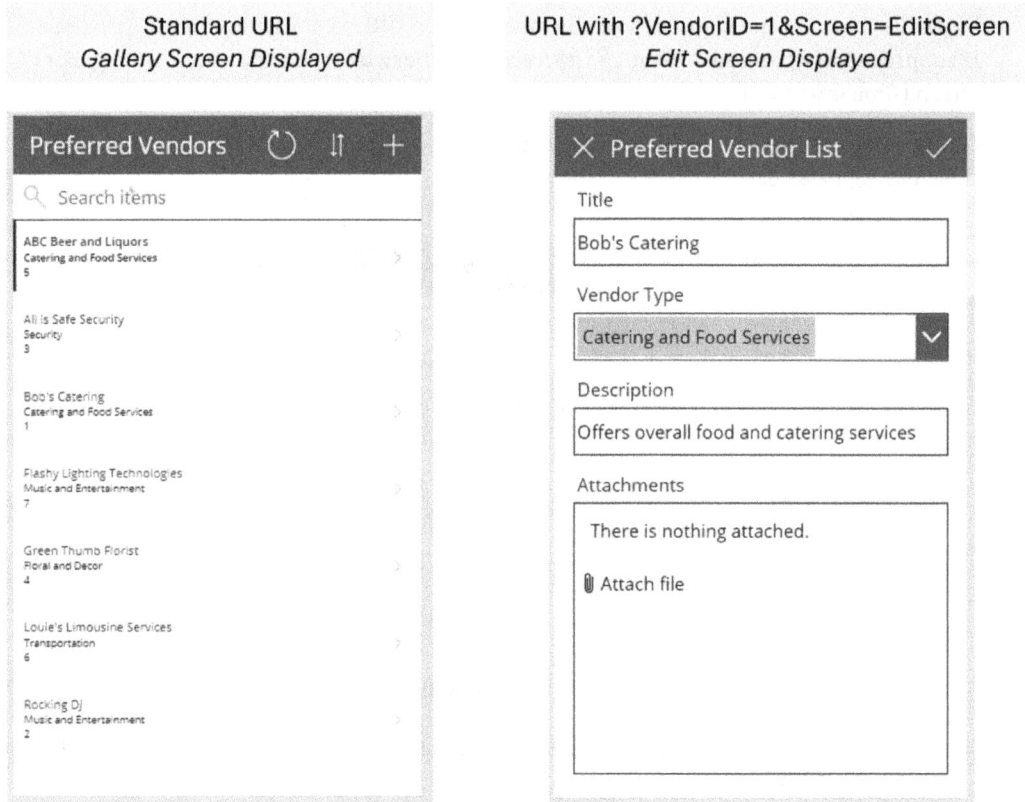

Figure 8.15 – Summary of the initial screen based on the URL

This is just one example of how URL parameters can be used. Handling incoming URL parameters effectively is key to leveraging the full potential of deep linking in Power Apps. By retrieving, using, and validating these parameters, you can create more dynamic, responsive, and user-friendly applications.

Now that we have covered two topics that can add functionality to your app, let's dive into a specific example and show how these concepts can be utilized to improve the overall use and experience of the app for the end user.

Enhancing the overall UI/UX

Now that we have explored examples of functionality, we want to put these to use in a sample use case. In this example, we have a simple Power Apps app that tracks the list of preferred vendors for our catering company. For this example, we want to have the following functionality:

- Display an **Admin** button, but only for individuals who are considered administrators. This button will take the user to an admin screen, which we will cover shortly.

- When new vendors are added, use conditional formatting to ensure that the email address for the contact person is in a valid format.

- Show a URL that, when a user clicks on it, takes the user directly to the app and allows the user to edit a specific record.

Figure 8.16 provides a sample view of the current app.

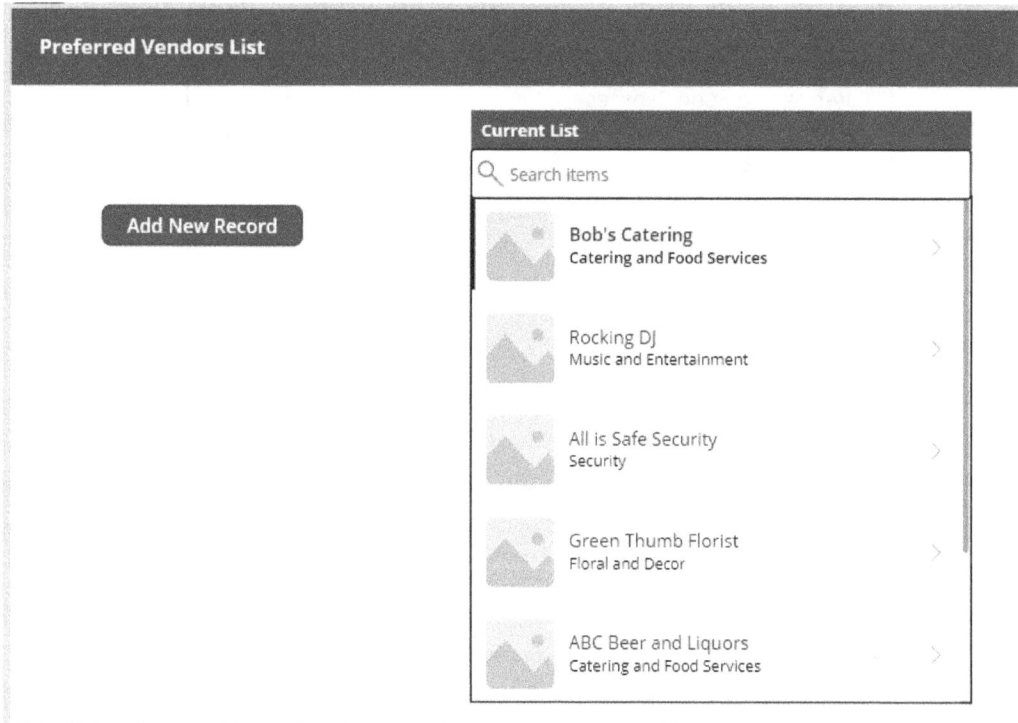

Preferred Vendors List

Add New Record

Current List

Q Search items

Bob's Catering
Catering and Food Services

Rocking DJ
Music and Entertainment

All is Safe Security
Security

Green Thumb Florist
Floral and Decor

ABC Beer and Liquors
Catering and Food Services

Figure 8.16 – Layout of our sample app

Currently, this app consists of a button to allow new records to be added along with a gallery on the right that displays existing vendors.

Let's dig into each of these requirements and add some more functionality.

Creating an Admin button

For this requirement, we want to display a button that will take the user to an administration screen. In our case, we want to be able to update the list of available categories that our preferred vendors can be in. *Figure 8.17* provides a list of the current categories:

Vendor Type

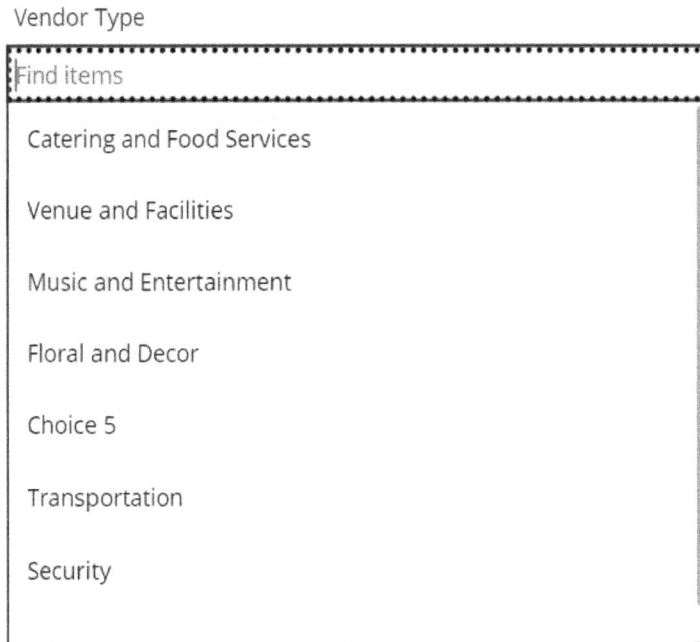

Figure 8.17 – Current list of available categories

In this example, we quickly identify that the list has **Choice 5**, which we would like to edit. We want to have an admin screen that will allow us to do this rather than having to update the underlying data source. Let's walk through the steps on how this can be achieved:

1. Add a button control to the main page. For additional clarity, we have added a label to the right that displays the current username. This is shown in *Figure 8.18*.

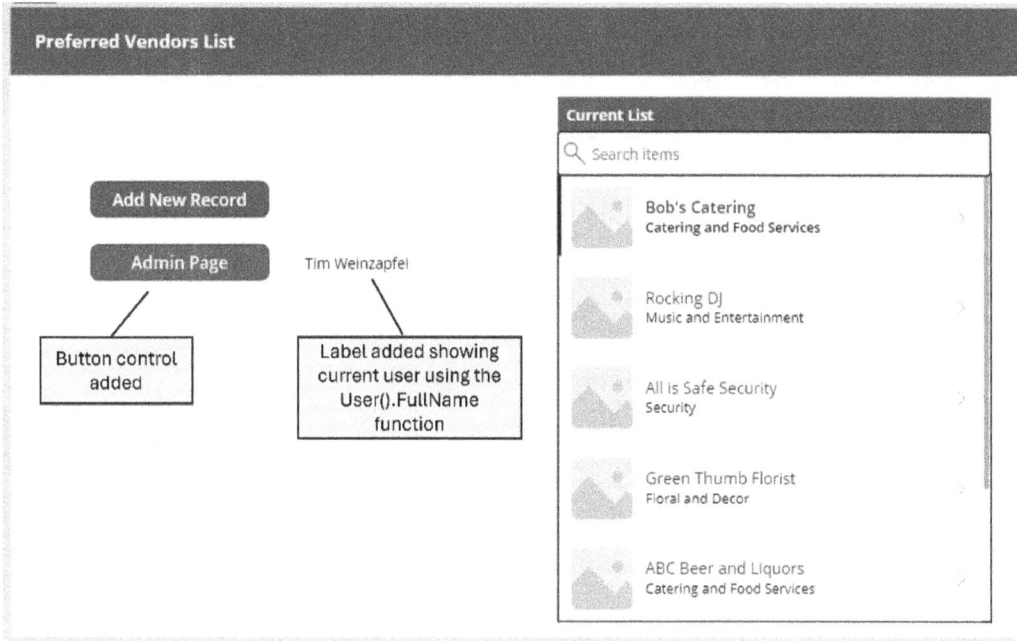

Figure 8.18 – Button control added

2. We then will update the **Visible** property of the button to determine whether the current user is an administrator. If they are, the button will appear. If the user is not an admin, the button will not appear. In this example, we'll just presume that the only admins are Tim and Andrea.

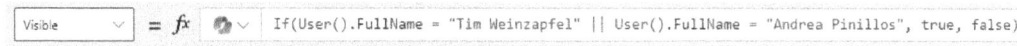

Figure 8.19 – Visible property of the button control

3. For the **Admin** button, we'll simply navigate the user to an admin screen, called `AdminScreen`, using the `Navigate()` function of the **OnSelect** property. This is shown in *Figure 8.20*.

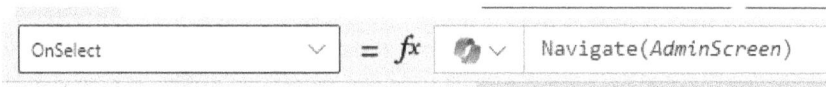

Figure 8.20 – Updating the OnSelect property of the button control

This functionality is now set up and will enable administrators to access an admin screen. Now that we have this completed, let's move on to the next requirement, which is to use conditional formatting to validate email address formats.

Using conditional formatting to validate email address formats

In this example, we want to use conditional formatting to alert the user when email addresses are not entered using the correct email format. The concept for this was discussed previously in the section on conditional formatting. Now we will put this to use. The following steps will address this:

1. First, we create a simple edit form that includes the fields we want to add. This form is shown in *Figure 8.21*.

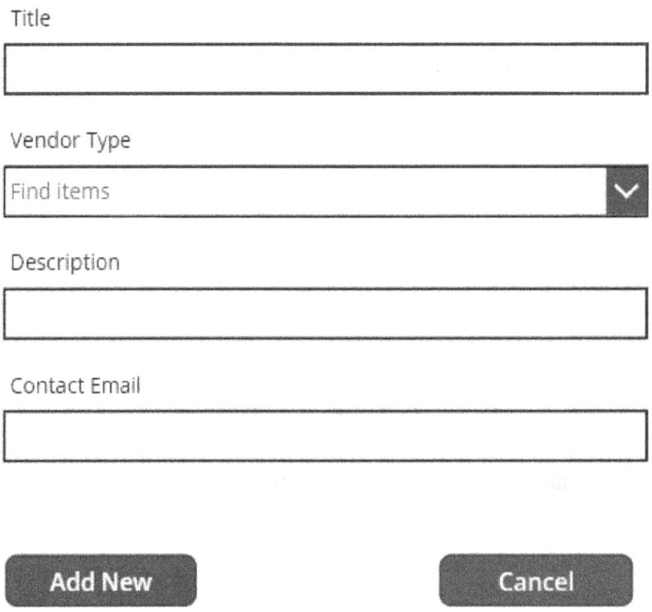

Figure 8.21 – Layout of the edit form

2. You will note that we have an input for `Contact Email`. It is this field that we want to highlight when an incorrect email address is entered. For reference, this field has the name `DataCardValue4`, as this will be needed for our conditional formatting. The **Fill** property of this control is then updated with our conditional formatting code, as follows:

```
If(!IsMatch(DataCardValue4.Text, Match.Email), Color.Yellow,
Color.White)
```

A complete view of this is shown in *Figure 8.22*.

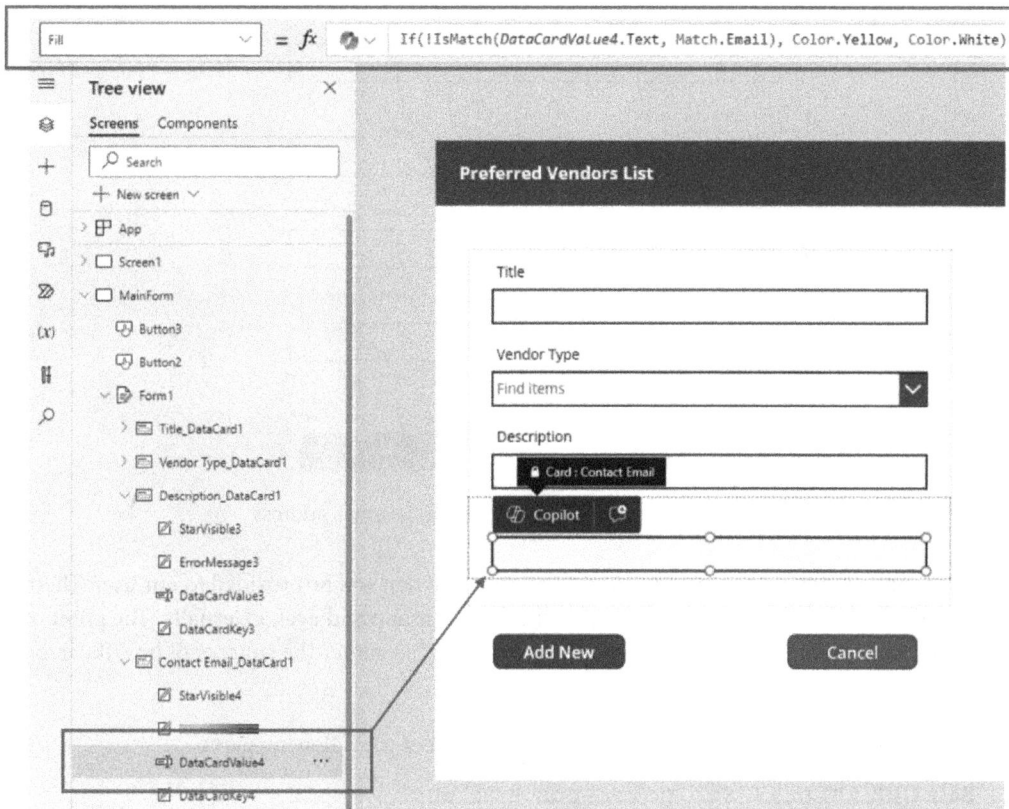

Figure 8.22 Fill property of the email input field set with conditional formatting

As shown, the box currently has a yellow color. However, entering in a valid email format will convert this back to a white background. Let's move on to our final example by using URL deep linking to send a user to a specific record.

URL deep linking for editing a record

In this last example, we want to send a URL to a user that will automatically send them to a vendor record to be edited. A simple example in our case would be if we had missing information on a vendor and wanted someone to update the record. In this case, let's presume that we have a vendor who is missing an email address, and we want one of our staff members to update this. This is shown in *Figure 8.23*.

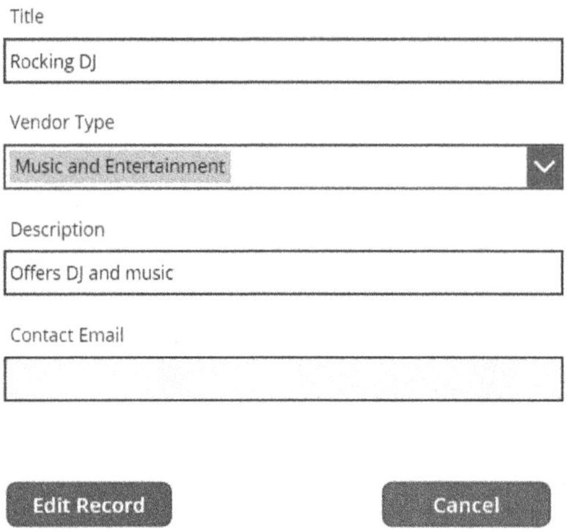

Figure 8.23 – Example of a record missing an email address

Let's walk through the steps of enabling this. We need a URL that will be provided to our user. There are different ways that URLs can be created, both within this app and even externally. The point of this example is not to dig into creating the URL specifically, but rather the concept of how it can be structured to enable our functionality:

1. We first need to create the URL. The URL will consist of the following parts:

    Base URL: `https://web.powerapps.com/apps/<AppID>`

    First parameter: `VendorID`

    Value: 2

    Second parameter: `Screen`

    Value: `EditScreen`

    As a reminder, the `<AppID>` value can be found in the details of the app. Please refer to *Figure 8.8* as an example. In addition, although the vendor ID is not shown in *Figure 8.23*, because we have access to the underlying data, this is the correct ID for that vendor. Your parameters will be different and applicable to your situation.

    The full URL thus becomes this: `https://web.powerapps.com/apps/<AppID>?VendorID=2&Screen=EditScreen`

2. We then set up the app as described in the section that we covered entitled *Implementing URL deep linking*. By doing this, once the user clicks on the URL provided, it will take them to the specific form, displaying the record that we want to be displayed, and the form is then put into edit mode.

And that's it! We have now added some additional functionality to the app, which should provide a better UI and UX.

Summary

In this chapter, we covered how to apply conditional formatting to various elements within a Power Apps app. We showed how this can improve the overall functionality of your app.

We then discussed the concept of URL deep linking and how this can be used to direct users to specific areas within your app. First, we covered the overall concept of URL parameters. We then showed how URL parameters can be used to affect the app itself.

Lastly, we combined both areas into a single use case. We showed how conditional formatting can be used to provide two key functionalities for our app. The first one was displaying an **Admin** button, but only for individuals designated as administrators. The second was to use conditional formatting to provide immediate feedback to the user on the validity of an email address format. We then showed how a URL link could be created to direct a user to a specific record that would be automatically opened to the edit screen, thus avoiding the user having to perform multiple clicks.

In the next chapter, we begin our journey toward integrating Power BI with other applications, starting with Power Automate, Microsoft Teams, and Microsoft Outlook.

Part 3:
Power Platform and
Other Integrations

This part focuses on the integration capabilities of Power Apps with other Microsoft tools, enhancing functionality and user experience. You'll learn how to integrate Power Apps with Power Automate, Teams, and Outlook, where you'll explore how to add additional functionality to your Power Apps by automating tasks such as sending Outlook emails, calendar invites, and Teams notifications using Adaptive Cards.

Next, you'll discover how to Integrate Power BI into your Power Apps, enabling you to embed Power BI dashboards and reports directly into your apps, as well as incorporating a Power App within a Power BI report for seamless data interaction. Moving forward, you'll delve into Integrating Power Apps with SharePoint, where you'll learn how to embed Power Apps into SharePoint sites, create custom SharePoint list forms, and utilize the `SharePointIntegration` component to enhance collaboration. Finally, you'll be introduced to Integration with Power Virtual Agents and Copilot, where you'll explore how to leverage AI through Microsoft Copilot Studio to create virtual chatbots and use Copilot to assist in app creation. These integrations provide a powerful way to extend the capabilities of your Power Apps, making them more dynamic and interactive.

This part has the following chapters:

- *Chapter 9, Integration with Power Automate/Teams/Outlook*
- *Chapter 10, Integration with Power BI*
- *Chapter 11, Integrating Power Apps with SharePoint*
- *Chapter 12, Integration with Power Virtual Agents/CoPilot*

9

Integration with Power Automate/Teams/Outlook

Welcome to the integration powerhouse of Power Apps, where the synergy between Power Automate, Teams, and Outlook opens up a world of possibilities. In this chapter, we'll embark on a journey to master the art of integrating these powerful tools seamlessly, enhancing your capabilities to automate tasks, streamline communication, and optimize workflows.

Throughout the upcoming lessons, you'll not only grasp the fundamental concepts but also roll up your sleeves and immerse yourself in practical applications. By this chapter's end, you'll possess the skills to orchestrate the flow of data and actions between Power Apps, Teams, and Outlook with finesse, revolutionizing how you work and collaborate.

In this chapter, we're going to cover the following main topics:

- Sending an Outlook email using Power Automate
- Sending an Outlook calendar invite using Power Automate
- Sending a Teams notification using Power Automate using adaptive cards

Through these essential topics, you'll gain the proficiency to automate processes, facilitate communication, and leverage the full potential of Microsoft Power Platform. These lessons aren't just useful; they're transformative, empowering you to drive efficiency, productivity, and innovation in your organization.

Get ready to unlock the power of integration and propel your productivity to new heights. Let's dive in and embark on this journey together, equipping you with the skills to thrive in the digital landscape of modern workplaces.

Technical requirements

To successfully engage with the materials in this chapter, you'll need the following:

- **Microsoft Power Apps**: Access to Power Apps for creating and managing your applications
- **Microsoft Power Automate**: Access to Power Automate for creating automated workflows
- **Microsoft Teams**: Access to Teams for collaboration and communication
- **Microsoft Outlook**: Access to Outlook for email and calendar management

Ensure that you have the necessary permissions and licenses to utilize these Microsoft tools effectively. Without the appropriate permissions, you may encounter limitations or be unable to access certain features or integrations. Verify with your organization's IT department or Microsoft account administrator to confirm that your accounts are properly set up and licensed. You can find the files used in this chapter on GitHub: `https://github.com/PacktPublishing/Power-Apps-Tips-Tricks-and-Best-Practices/tree/main/Chapter09`

Sending an Outlook email using Power Automate

In this section, you'll embark on a journey to streamline your communication workflows by sending Outlook emails directly from Power Apps using Power Automate. By mastering this capability, you'll empower yourself to automate email notifications, alerts, and updates, enhancing your efficiency and responsiveness in various scenarios.

Sending Outlook emails programmatically through Power Automate offers a seamless way to integrate your Power Apps with your email communication. Whether you need to notify stakeholders about updates, send automated responses, or trigger actions based on specific events, this integration provides a powerful solution for streamlining your processes.

Before you start: Ensure that you have your Power Automate environment ready and authenticated with your Microsoft account. Verify that you have the necessary permissions and licenses to utilize Power Automate and Outlook effectively. Without proper setup, you may encounter limitations or issues with email integration. Once everything is set up, we can dive into the configuration process to enable smooth email integration within our Power Apps ecosystem. Follow these steps:

1. **Access Power Automate**: Begin by navigating to the Power Automate portal (`https://flow.microsoft.com/`) and log in with your Microsoft account credentials.

2. **Create a new flow**: Once logged in, ensure you are in the **Event Planning Project Development** environment, click on the **My flows** tab in the sidebar, and then select **New Flow** to create a new flow.

3. Select **Automated cloud flow** from the dropdown. This will allow us to create a trigger from one of our Power Apps:

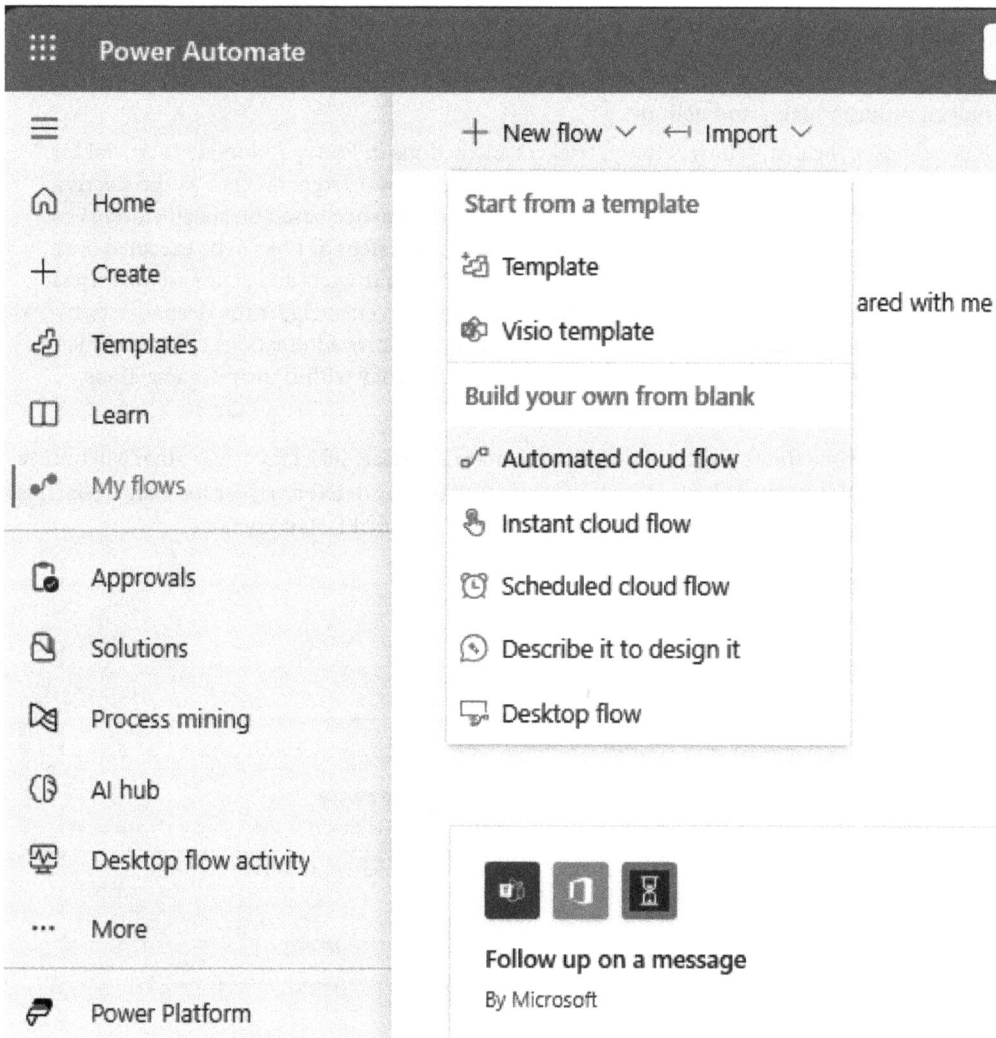

Figure 9.1 – The Automated cloud flow option under the New flow button

4. **Give your flow a name**: For this demo, we will be sending an email to each attendee that is added with event details:

- **Flow name**: `Attendee Email`

Understanding triggers and actions

Understanding the distinction between triggers and actions in Power Automate is crucial for efficiently designing and orchestrating automated workflows. Triggers serve as the starting points of flows, determining when a workflow should commence based on specific events or conditions. On the other hand, actions define the subsequent steps or tasks to be executed once the trigger initiates the flow. Knowing the difference between triggers and actions allows users to accurately configure their flows, ensuring that the right events trigger the desired actions. This understanding empowers users to create robust and effective automations tailored to their unique business needs, maximizing productivity and efficiency within their organizations.

- **Choose your flow's trigger**: In the flow creation interface, select the trigger that will initiate the flow. Select the **When a row is created, updated or deleted** trigger for Dataverse. This trigger monitors for changes in records within a specified Dataverse table.

5. After selecting the trigger, click **Create**:

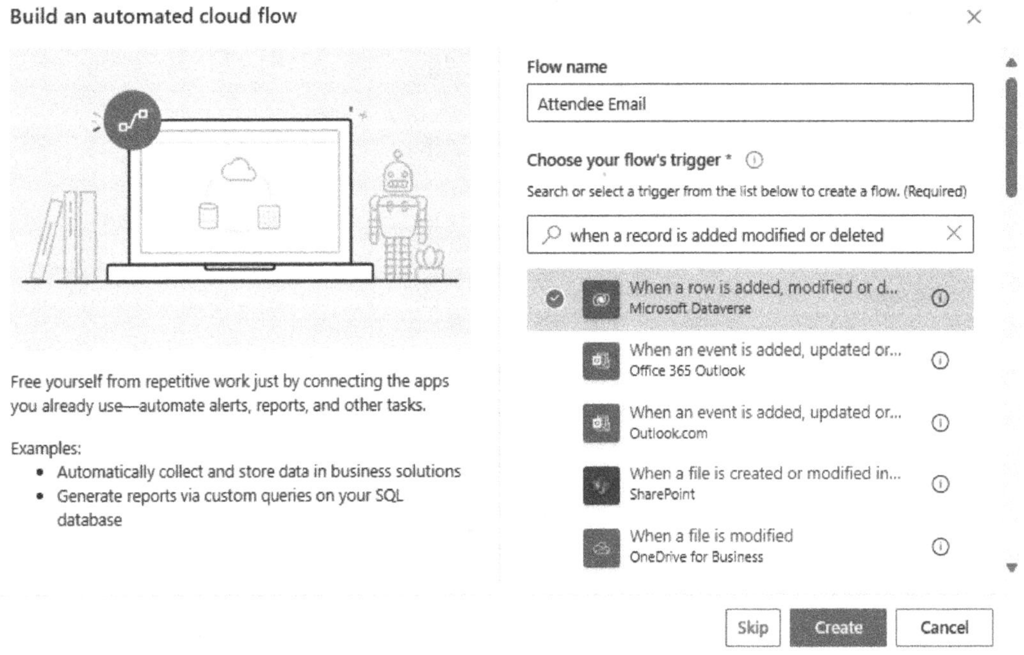

Figure 9.2 – The When a row is added, modified or deleted trigger option

6. **Connect to Dataverse**: Authenticate and connect to your Dataverse environment. Choose the table from which you want to monitor record changes:

- **Change Type**: `Added`
- **Table Name**: `Attendees`
- **Scope**: `Organization`

Scope explanation

Organization: Monitors changes across the entire organization's Dataverse environment.

Business Unit: Focuses on changes within a specific business unit in Dataverse.

Parent: Child Business Unit: Tracks changes in hierarchical relationships between parent and child business units.

User: Filters trigger events based on specific users initiating the actions.

---

7. **Configure your trigger settings**: Set up the trigger by specifying any additional conditions or filters, such as specific record types or fields to monitor:

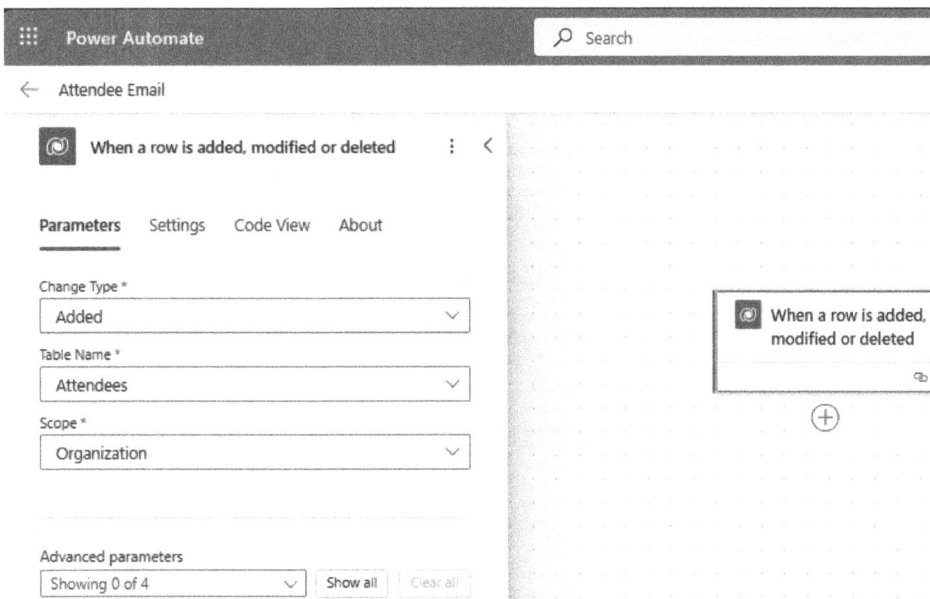

Figure 9.3 – Parameters for the When a row is added, modified or deleted trigger

8. **Add actions**: Once the trigger has been set up, you can add actions to the flow.

9. Click the **plus** icon under the trigger and select **Add an action**.

10. **Select the email action**: To send an email, search for and select the **Send an email (V2)** action.

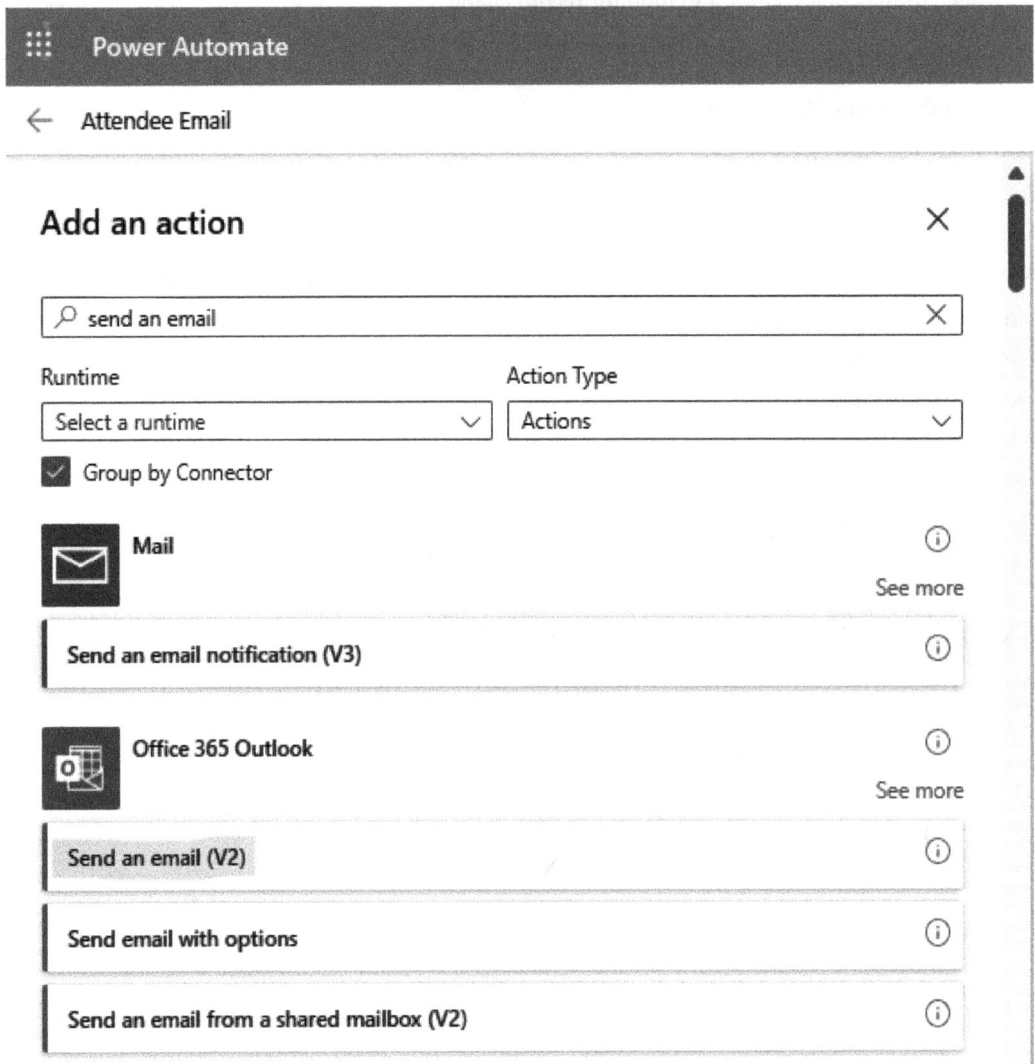

Figure 9.4 – Send an email (V2) in the Add an action form

11. **Configure email settings**: Click the email action box to open the settings. Here, specify the recipient's email address, subject, body, and any other relevant details. You can also dynamically populate these fields using data from the trigger or other sources.

12. **Add dynamic content**: Click in the **To** box and select the **lightning bolt** icon for dynamic content. Select **Attendee Email** from the box:

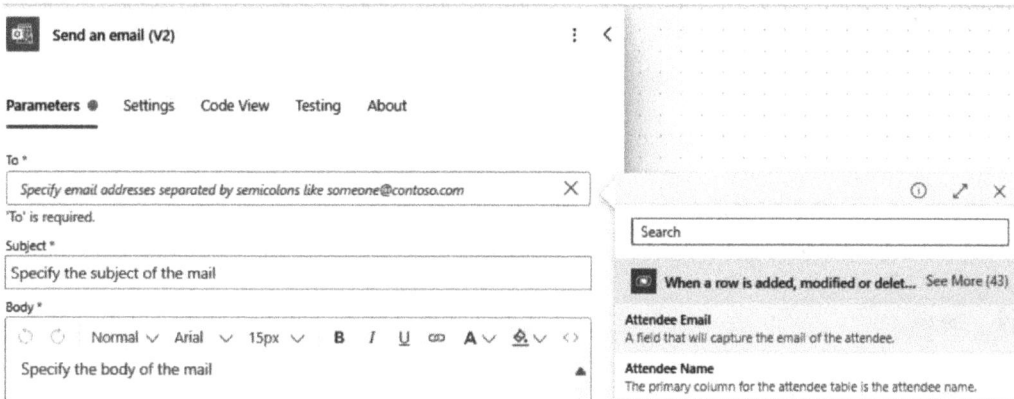

Figure 9.5 – Attendee Email in the When a row is added, modified or deleted dynamic box

13. **Set the email subject**: Enter You're attending a new event! in the **Subject** line.

14. **Configure the email body**: For the **Body** area of the email, our goal is to dynamically include the event name from the event table. However, directly adding the **Event** value from the **Attendee** table poses a challenge as it retrieves the GUID of the event rather than its name.

In the following subsection, we will address how to extract and utilize string values from lookup fields to properly display the event name in the email body.

Extracting a string value from a lookup field

When dynamically adding data from a related table, such as extracting the event name from the event table to include it in the body of the email, a common challenge arises. If we directly insert the **Event** value from the **Attendee** table, we'll retrieve the GUID of the event instead of its name. To overcome this, let's explore how to extract a string value from a lookup field:

1. Insert a step between When a row is added, modified, or deleted and Send an email (V2) by clicking the + button and then clicking **Add an action**.

2. In the **Add an action** slide-out, start typing Get a row by ID and select it under **Microsoft Dataverse**.

3. In the **Get a row by ID** settings, choose **Events** under **Table Name**, type event in the search bar, and select **Event (Value)** for **Row ID**:

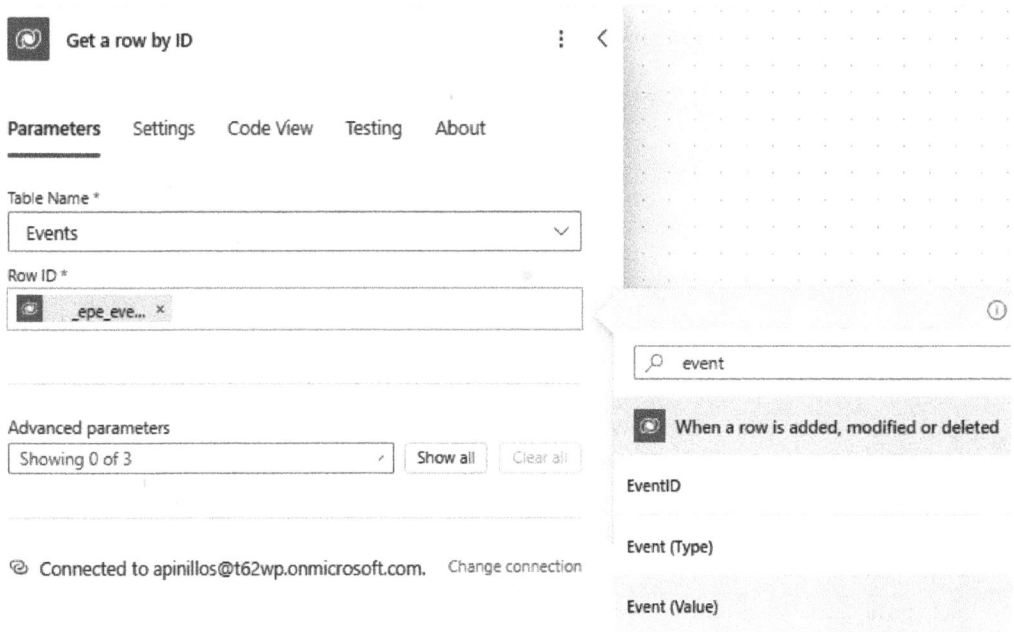

Figure 9.6 – Event (Value) in the dynamic box

4. Insert a new step between Get a row by ID and Send an email (V2) by clicking the + button and then **Add an action**.

5. Type Compose in the **Add an action** slide-out and select **Compose** under **Data Operation**.

6. In the **Compose** settings, click inside the **Inputs** box, then click the **lightning bolt** icon to add a dynamic field. Select Event Name under **Get row by ID**.

After completing these steps, your flow will seamlessly retrieve the event name from the Dataverse table and dynamically include it in the email body, overcoming the challenge of extracting string values from lookup fields.

Dynamically including the event's name in the attendee's email body

When sending emails to attendees, it's essential to provide relevant information about the event they're attending. One critical piece of information is the event name, which adds context and clarity to the email content. Let's utilize the steps outlined in the previous section to dynamically add the event name from the event table to the body of the attendee email:

1. Click the **Send an email (V2)** action box.

2. Click inside the **Body** section of the **Send an email (V2)** settings, click the **lightning bolt** icon, and select **Attendee Name** under **When a row is added, modified or deleted**:

   A. Add the word `Hello` before that dynamic field.

3. Click *Enter* and add the words `You have been invited to attend`. Then, click the **lightning bolt** icon and select **Outputs** under **Compose**. Finish the sentence with an exclamation point.

4. Double-check that the **Send an email (V2)** settings look similar to what's shown in *Figure 9.7*:

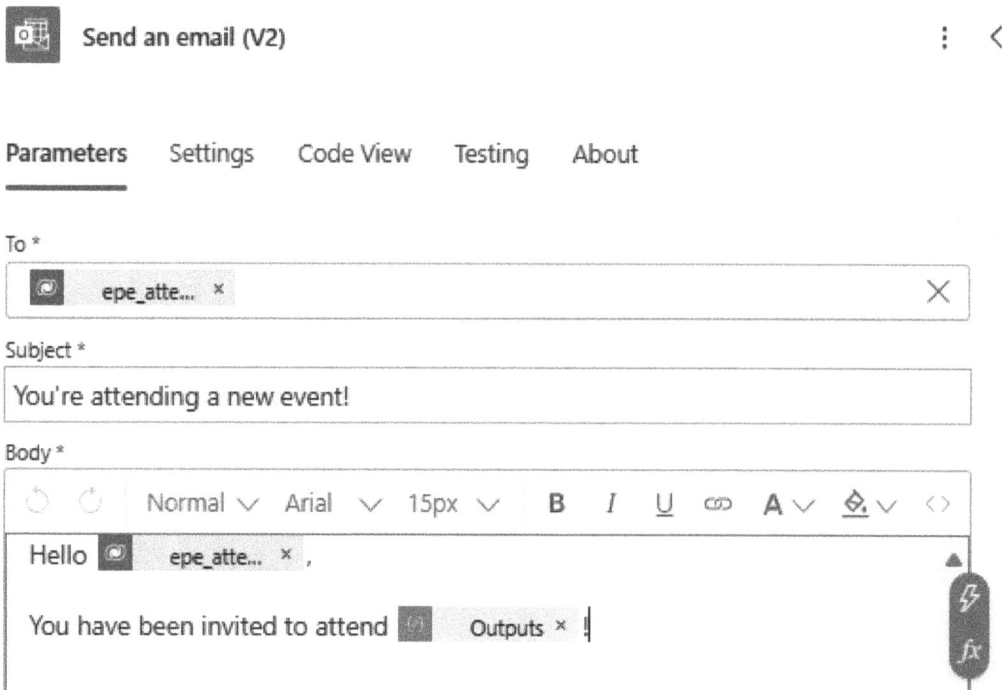

Send an email (V2)		⋮ ‹

Parameters   Settings   Code View   Testing   About

To *

epe_atte... ×	×

Subject *

You're attending a new event!

Body *

↺ ↻   Normal ∨   Arial ∨   15px ∨   **B** *I* U ∞ A ∨ ✎ ∨ ‹›

Hello epe_atte... × ,

You have been invited to attend Outputs × !

Figure 9.7 – Send an email (V2) sample

Incorporating the event name dynamically enriches the attendee email experience, ensuring recipients have clear and pertinent details about the event they're participating in.

Testing the flow

Before activating the flow, it's crucial to conduct thorough testing to ensure that it functions correctly and meets the desired objectives. You can use sample data or trigger the flow manually to test its functionality. Follow these steps to manually test the flow using the model-driven app:

1. After configuring the flow, ensure that it is saved. Then, click on the **Test** button located at the top-right corner of the Power Automate designer. Select **Manually** to initiate manual testing.

    In *Chapter 4*, we created a model-driven app and exposed the **Event** table. We will use the previously created model-driven app to trigger the flow.

2. Navigate to the Power Apps maker portal, select the **Event Planning Project Development** environment, click **Solutions**, click **Event Planning Project**, click **Apps**, click the three dots next to **Event Planning Model-Driven App**, and click **Play**.

    In the **Event Planning Model-Driven App** area, find the **Event** table. Here, we will find the related Attendee table in the **Related** tab:

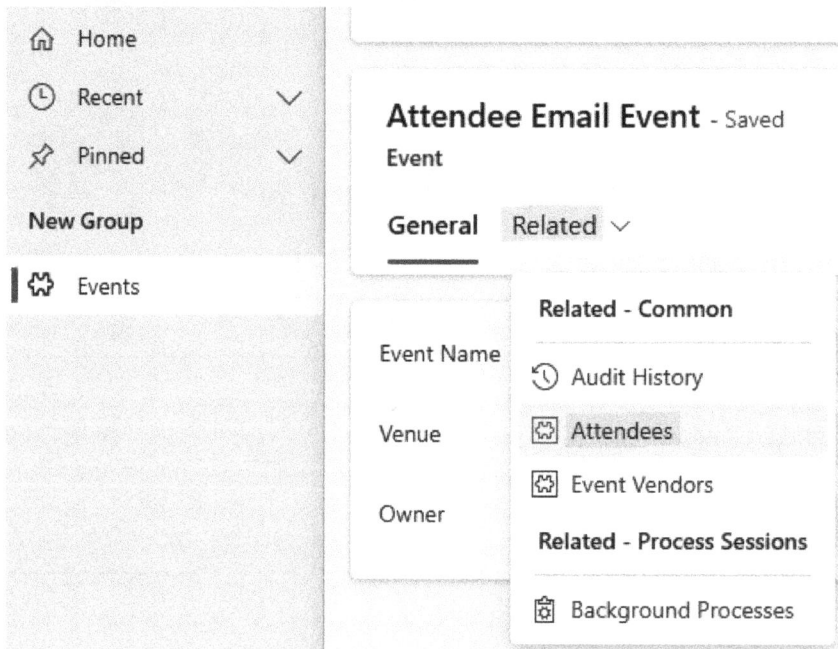

Figure 9.8 – Attendees under the Related tab

3. Click **New Attendee**. Then, in the form, add an attendee name and your email and click **Save & Close**:

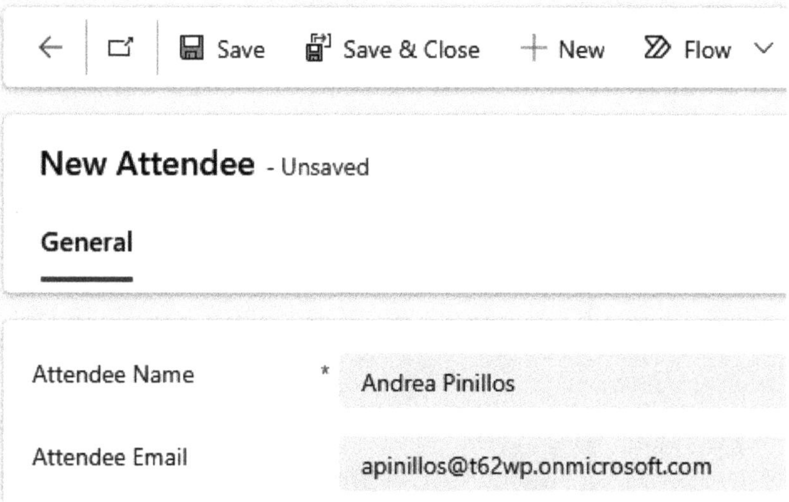

Figure 9.9 – New Attendee sample form

4. **Monitor flow execution**: After making the changes, monitor the execution of the flow. You can do this by navigating to the Power Automate designer and checking the flow's run history in the Power Automate portal. If your flow ran successfully, your Power Automate designer will look like what's shown in *Figure 9.10*:

> **Tip**
>
> As a best practice, during the development phase, ensure that any email notifications sent by the flow are directed only at yourself or a few colleagues. This approach prevents real users from being spammed with "test" notifications while you are still refining your flow.

←   Attendee Email   🖉

✓ Your flow ran successfully.   ✕

>

0s ✓
When a row is added, modified or deleted

0.1s ✓
Get a row by ID

0s ✓
Compose

0.9s ✓
Send an email (V2)

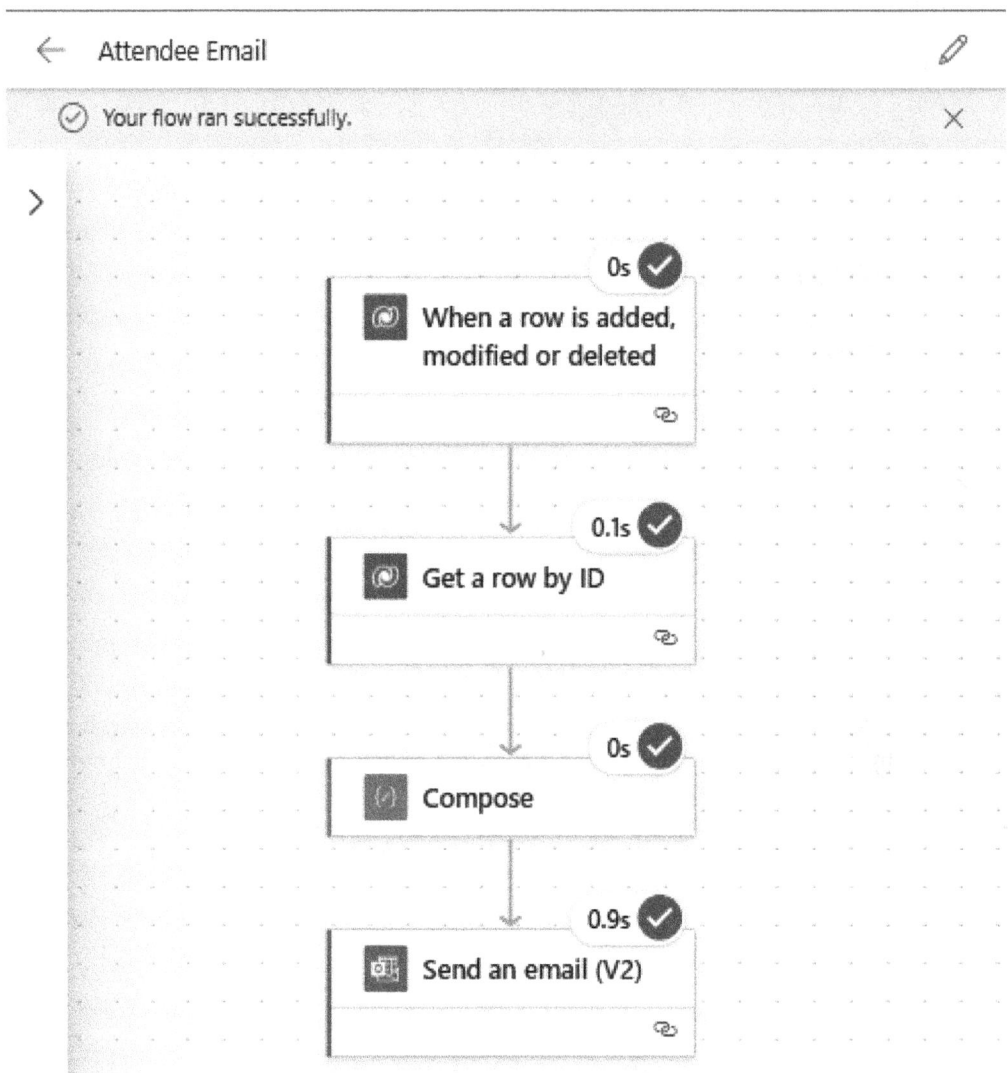

Figure 9.10 – The steps that passed during testing

5. **Verify and review email generation**: Once the flow has been triggered, verify that the email is generated successfully. Check the recipient's inbox to ensure that the email containing the event name is delivered as expected:

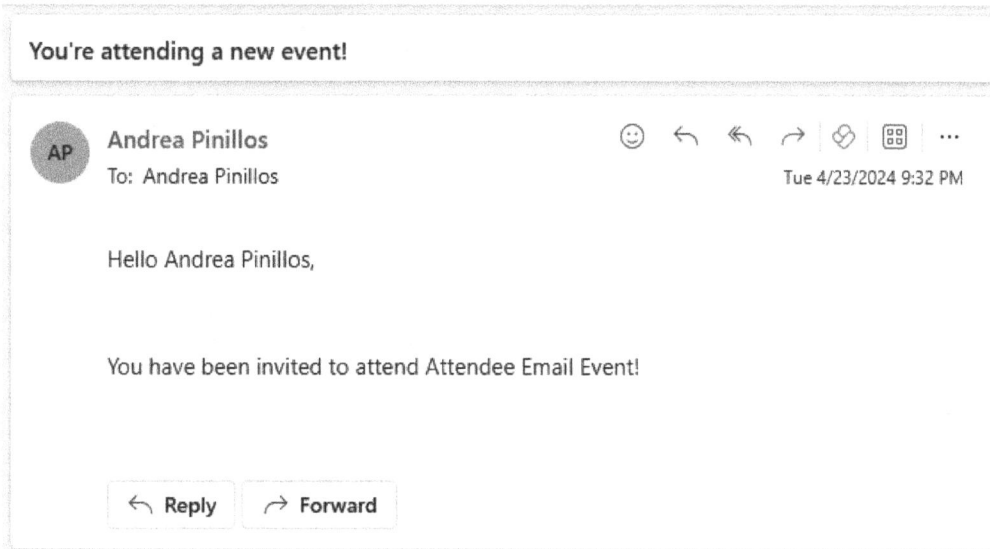

Figure 9.11 – Email example

6. **Check for errors**: If any errors occur during flow execution or if the email is not generated as expected, review the flow's configuration and troubleshoot any issues. Check the flow's run history and error messages for insights into potential problems.

7. **Repeat testing as needed**: Repeat the testing process as needed to ensure the flow operates reliably under various scenarios. Make adjustments to the flow configuration as necessary to address any issues or improve performance.

8. **Save and activate the flow**: Once tested successfully, save your flow and activate it. Click "Turn on" on the top ribbon to activate the flow. This will enable the flow to run automatically whenever the trigger conditions are met.

So far, we've established a connection between Power Apps and Power Automate by creating a flow that listens for changes in a specific Dataverse table. This flow is triggered whenever data in that table is modified or added, with the table being updated through interactions with a model-driven app. This approach is a common pattern – using Power Apps to write or modify data in tables and then leveraging Power Automate to trigger notifications, validations, or other processes. Despite being two distinct platforms, Power Apps and Power Automate integrate seamlessly, offering a smooth and cohesive experience for the end user.

As we've seen, the capability to send Outlook emails via Power Automate has transformed communication workflows, enhancing efficiency and responsiveness. In the next section, we'll dive into sending Outlook calendar invites and develop logic to automate this process based on Power Apps triggers. This will further optimize your productivity, allowing you to seamlessly integrate calendar invites into your automated workflows and efficiently manage responses within Power Apps.

Automating Outlook calendar invitations with Power Automate

Imagine how powerful it would be for companies to leverage data collected in one place to drive specific actions. For example, in this use case, the action is sending targeted Outlook calendar invitations based on the data. This approach can be incredibly valuable for businesses that rely on precise scheduling and efficient communication.

In this section, we'll explore how to automate Outlook calendar invitations triggered by changes in your data source, such as additions, modifications, or deletions. Before diving into the implementation, it's essential to enhance our event table by adding two new fields: one for the event start date and time, and another for the event end date and time. These fields will serve as crucial components in dynamically populating the calendar invitation details.

By framing the process in this way, you can better envision how similar solutions could be applied to your unique needs. Let's proceed with setting up the automation and incorporating these enhancements to streamline your event management processes:

1. **Update the Event table**: Before configuring the trigger, navigate to your event table in Microsoft Dataverse and add two new date and time fields: one for `Event start date and time` and another for `Event end date and time`. These fields will enable you to accurately capture and utilize the event timing information in your calendar invitations. If you need assistance creating new date and time fields in your event table, navigate back to *Chapter 2*:

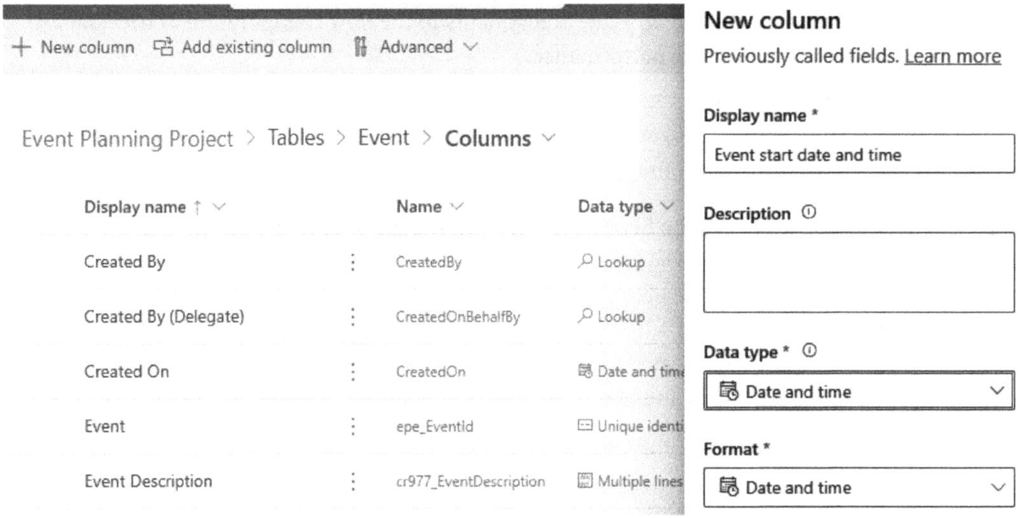

Figure 9.12 – New column form with filled-out fields

2. **Access Power Automate**: Log in to the Power Automate portal (`https://flow.microsoft.com/`) using your Microsoft account credentials.

3. **Locate previous email flow**: Navigate to the **Attendee Email** flow you previously created for sending emails to attendees.

4. **Edit the flow**: Open the flow for editing.

5. **Replace the email action**: Locate the action responsible for sending the email to attendees. Delete this action.

6. **Add a Create event (V2) action**: Add the **Create event (V2)** action instead. This action will be responsible for generating the Outlook calendar invitation.

7. **Configure calendar invitation settings**: In the **Create event (V2)** action, configure the available settings, such as event title, start time, end time, location, and attendees. Ensure that you dynamically populate these fields with data from the trigger. See *Figure 9.12*:

   A. **Calendar Id**: **Calendar**:

       i. **Subject**: Type `You're invited to attend` [select the **lightning bolt** icon and then select **Outputs** under **Compose**] ! :

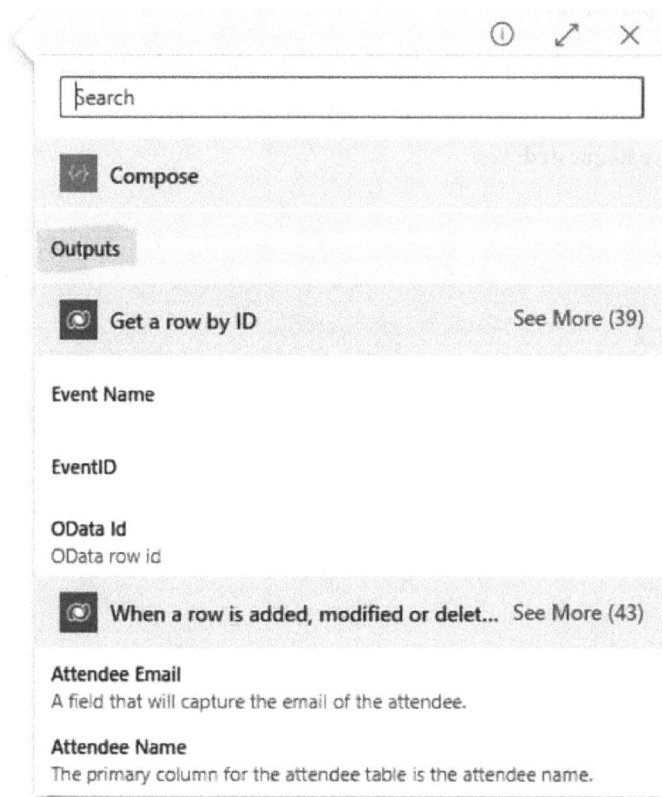

Figure 9.13 – The Outputs option in the dynamic box

B. **Start Time**: Click on **fx** to enter the following expression:

```
formatDateTime(outputs('Get_a_row_by_ID')?['body/epe_
eventstartdateandtime'], 'yyyy-MM-dd HH:mm:ss')
```

C. **End Time**: Click on **fx** to enter the following expression:

```
formatDateTime(outputs('Get_a_row_by_ID')?['body/epe_
eventenddateandtime'], 'yyyy-MM-dd HH:mm:ss')
```

When to use triggerOutputs versus outputs

Use triggerOutputs within a trigger action to access data specific to the trigger event.

Use outputs to access data generated by the previous action in the flow and pass it along to subsequent actions.

D. **Time Zone: (UTC-08:00) Pacific Time (US & Canada)**

E. **Advanced parameters**:

   i. **Required attendees**: Select the **lightning bolt** icon and then select **Attendee Email**

   ii. **Show As: busy**

   iii. **Response Requested: Yes:**

Create event (V4)   ⋮

Parameters   Settings   Code View   Testing   About

Calendar Id *

| Calendar | ∨ |

Subject *

You are invited to Outputs × !

Start Time *

| *fx* formatDat... × |

End Time *

| *fx* formatDat... × |

Time Zone *

| (UTC-08:00) Pacific Time (US & Canada) | ∨ |

Advanced parameters

| Showing 3 of 17 | ∨ | Show all   Clear all

Required Attendees

| epe_atten... × ; | ✕ | ✕

Show As

| busy | ∨ | ✕

Response Requested

| Yes | ∨ | ✕

Figure 9.14 – Create event (V4) parameters

I went ahead and added the **Event start date and time** and **Event end date and time** fields to the main form of the Event table. This will allow us to fully test by adding the event start and end date. Please go back to *Chapter 2* if you need help adding fields to the Event table:

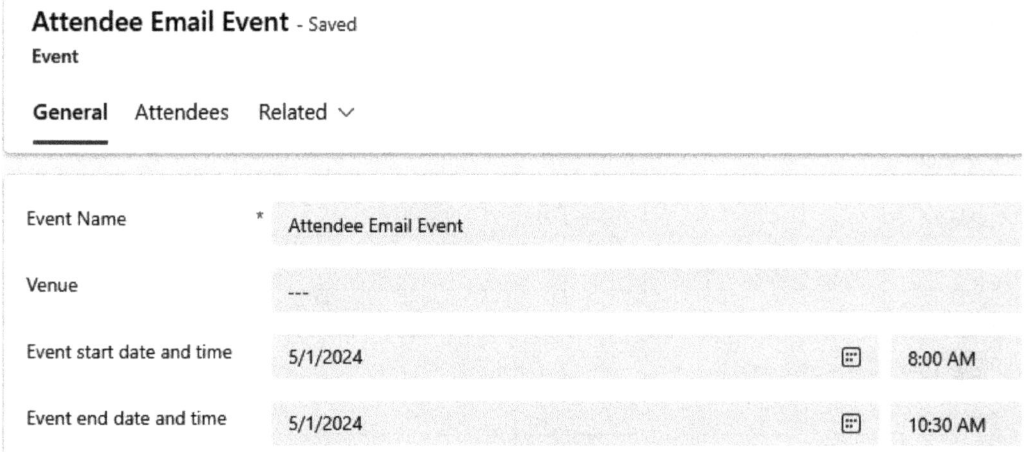

Attendee Email Event - Saved
Event

General	Attendees	Related ∨

Event Name	*	Attendee Email Event		
Venue		---		
Event start date and time		5/1/2024	⊞	8:00 AM
Event end date and time		5/1/2024	⊞	10:30 AM

Figure 9.15 – The Attendee Email Event form in the model-driven app

8. **Save and test**: Save the changes to the flow and test its functionality to ensure that Outlook calendar invitations are generated correctly based on changes in the event table:

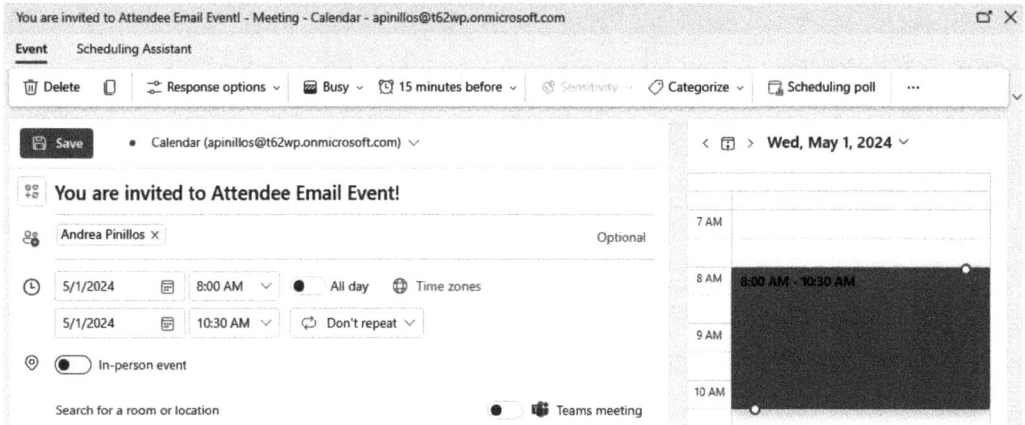

Figure 9.16 – Calendar invite example

By exploring the intricacies of automating Outlook calendar invitations with Power Automate, you've unlocked a powerful tool for efficient event management and coordination. Whether it's scheduling meetings, organizing appointments, or coordinating team activities, the ability to automate calendar invitations streamlines processes and ensures seamless communication. With these newfound skills, you're well-equipped to optimize your workflow, save time, and stay on top of your busy schedule with ease. In the dynamic world of team collaboration, timely communication plays a pivotal role in fostering productivity and cohesion. With Power Automate's robust capabilities, you can now automate the delivery of scheduled Teams notifications, ensuring that important messages reach your team members exactly when they're needed most.

Using adaptive cards in Teams notifications

In today's fast-paced work environment, effective communication is the cornerstone of successful collaboration. Leveraging Microsoft Teams' powerful capabilities, we can enhance communication further by incorporating actionable cards into our notifications. These cards provide interactive elements that enable users to take immediate action directly from the notification itself, streamlining workflows and improving productivity.

In this section, we'll explore how to harness the potential of actionable cards within Teams notifications, empowering you to create dynamic and engaging messages that drive actionable outcomes seamlessly. Let's dive into the world of actionable cards and unlock new possibilities for streamlined collaboration in Teams.

Building an adaptive card

Let's navigate to the **Adaptive cards** interface to get familiar with how to use the JSON code and the designer:

1. Navigate to `https://adaptivecards.io/designer/`.

> **Attention**
> Pay close attention to the **CARD PAYLOAD EDITOR** area to familiarize yourself with the JSON structure that we'll be utilizing in Power Automate. This JSON code will serve as the foundation for building our actionable card in Power Automate, enabling us to create dynamic and interactive experiences within our workflows.

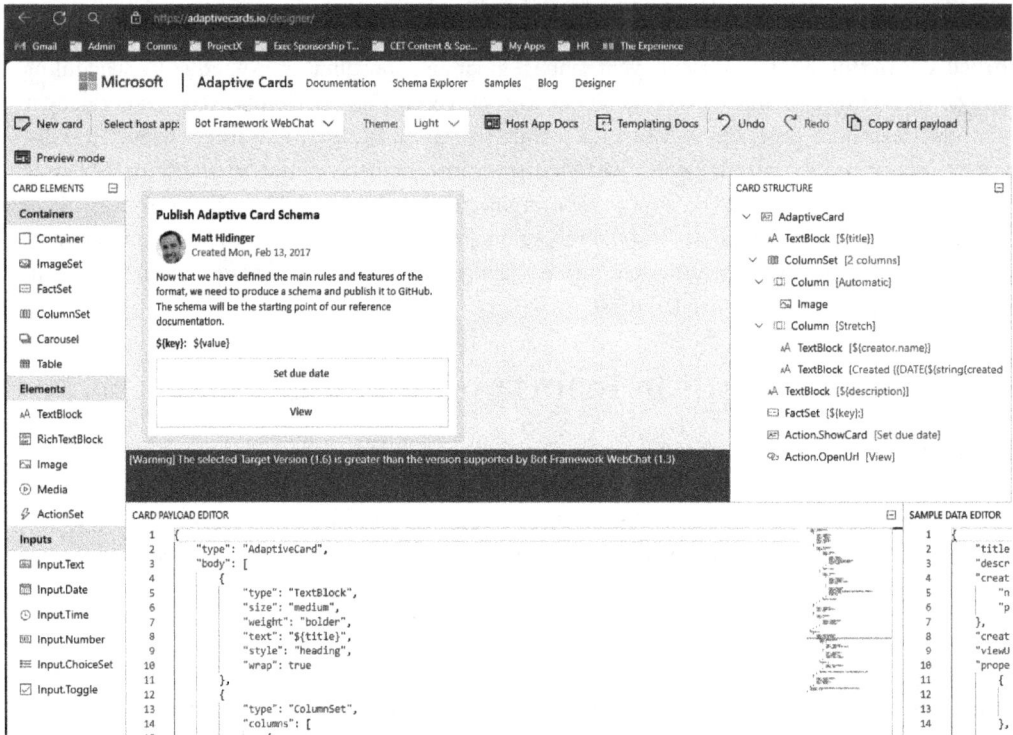

Figure 9.17 – Adaptive card designer

2.   Select **New card** at the top-left corner of the page. This will open different template options and let you create a blank card or a card from a JSON schema.

Explore

Feel free to play around and explore the different templates available. Adaptive cards can be used to gather information from a user and post that data back to Power Apps. We will be starting from a blank card because we will be posting information about an event to an attendee who has just been added to that event.

---

3. Select **Blank Card** under **Create** so that we can build out our card to inform the attendees that they have been added to an event:

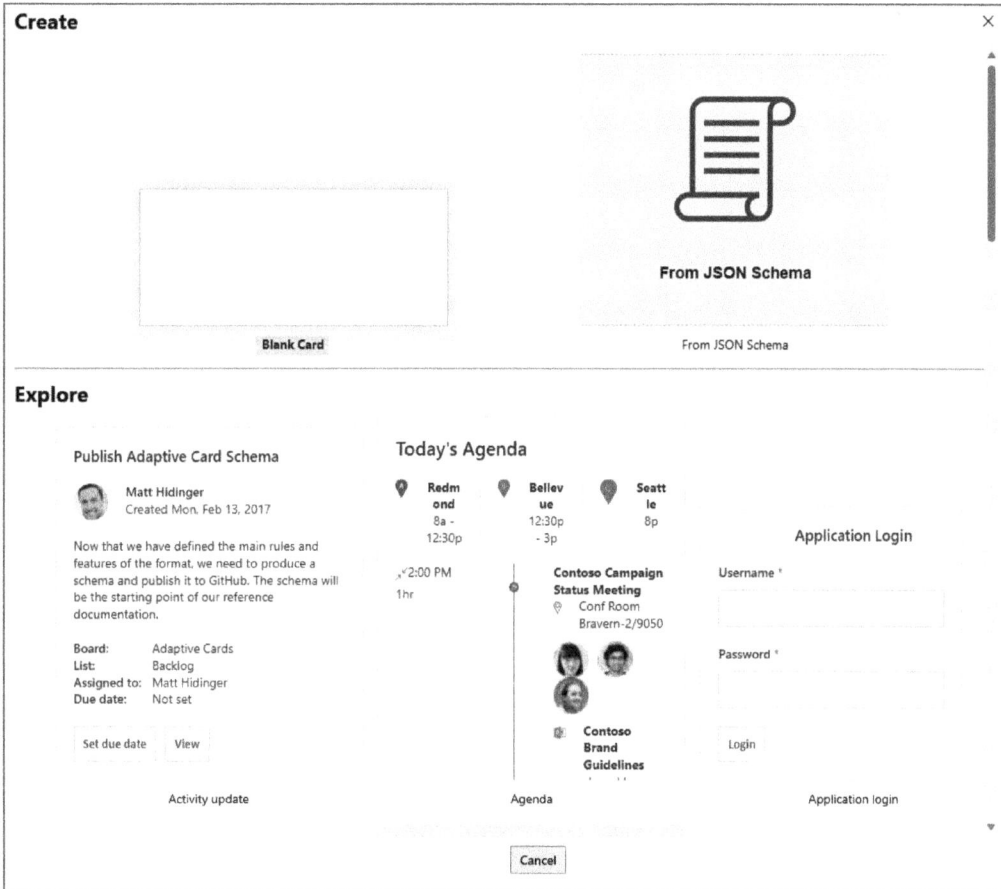

Figure 9.18 – Adaptive card templates

4. After selecting **Blank Card**, your **CARD PAYLOAD EDITOR** JSON should only contain the container code. Let's add to it by using **CARD ELEMENTS** from the left panel:

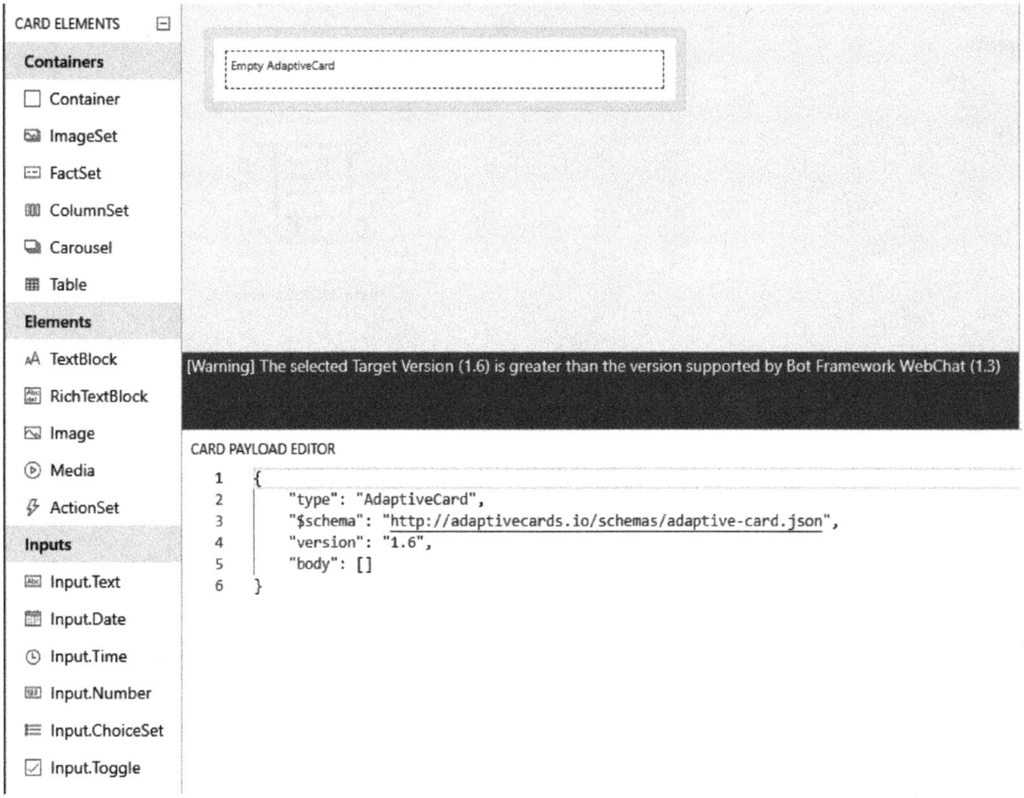

Figure 9.19 – Empty adaptive card in the designer

5. Click and drag three **TextBlock** elements from under the **Elements** section.

    Your adaptive card should look like what's shown in *Figure 9.19*:

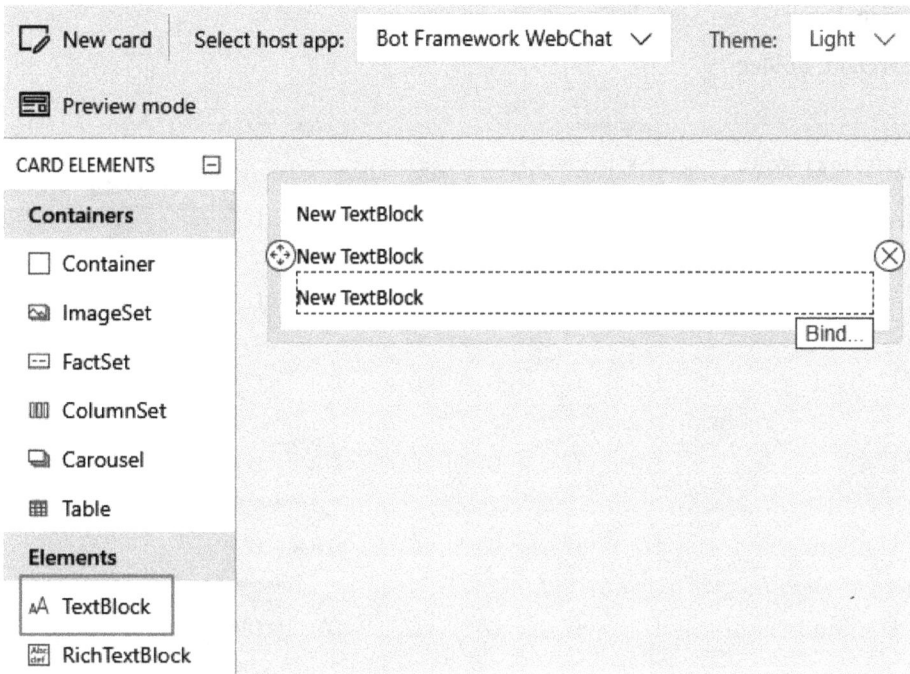

Figure 9.20 – Adding three Textblock elements to an empty adaptive card

6. Select the top **New TextBlock** element so that it is highlighted with dotted lines.

7. Under **Element Properties** on the right panel, change the following variables:

    - **Text**: You're invited to an event!

    - **Size**: **Large**

    - **Weight**: **Bolder**:

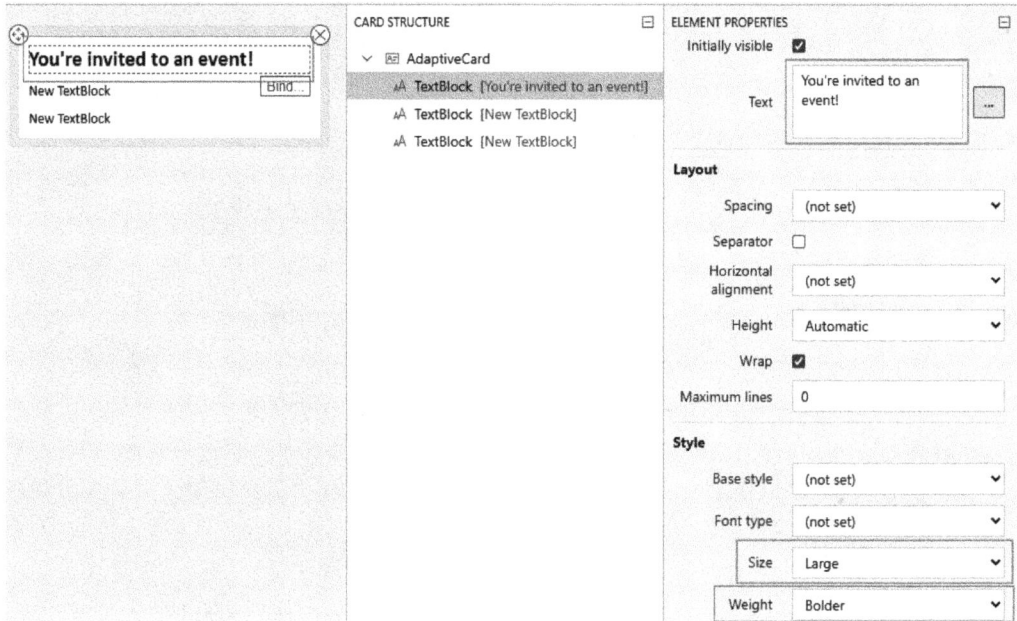

Figure 9.21 – Properties for TextBlock

8. Select the middle **New TextBlock** element and update the following variables:

    - **Text**: Event Name:

    - **Size**: **Medium**

- **Color: Accent:**

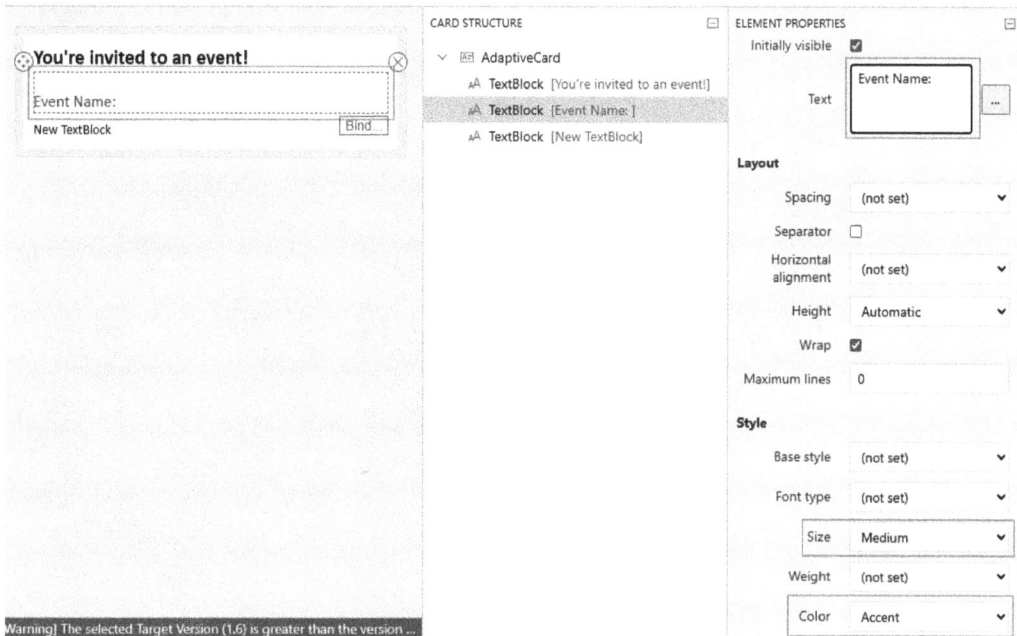

Figure 9.22 – Second TextBlock properties

9. Select the bottom **New TextBlock** element and make the following change:

   - **Text**: `Date(s)::`

Figure 9.23 – Third TextBlock properties

> **Note**
>
> For the middle and bottom text blocks, we will be adding dynamic data from Dataverse in Power Automate. We want to be able to pull the actual event name and dates from Power Apps and display it on this adaptive card when we're sending the Teams notification to the attendee.

10. Now, let's copy that JSON code from the **CARD PAYLOAD EDITOR** area and go back to Power Automate:

```
CARD PAYLOAD EDITOR
 1    {
 2        "type": "AdaptiveCard",
 3        "$schema": "http://adaptivecards.io/schemas/adaptive-card.json",
 4        "version": "1.6",
 5        "body": [
 6            {
 7                "type": "TextBlock",
 8                "text": "You're invited to an event!",
 9                "wrap": true,
10                "weight": "Bolder",
11                "size": "Large"
12            },
13            {
14                "type": "TextBlock",
15                "text": "Event Name: ",
16                "wrap": true,
17                "size": "Medium",
18                "color": "Accent",
19                "spacing": "Medium"
20            },
21            {
22                "type": "TextBlock",
23                "text": "Date(s): ",
24                "wrap": true
25            }
26        ]
27    }
```

Figure 9.24 – JSON code for our adaptive card

Now that we know how to create a JSON payload and have seen the JSON come to life in the designer, we can build our Power Automate flow so that it can include this code. After, we'll be able to post that card to a Teams notification for an attendee that has been added to an event.

Building a Power Automate flow with an adaptive card

Let's dive into Power Automate and construct a flow that incorporates an adaptive card. Through this process, we'll leverage the dynamic capabilities of adaptive cards to create engaging and interactive experiences within our workflows. Follow along as we explore building a Power Automate flow with an adaptive card, empowering you to enhance your automation solutions with user-friendly interfaces.

Creating the flow

Before we start configuring our flow, it's essential to set up a new automated flow with the right parameters. This section will guide you through creating the flow and defining its trigger to suit our event notification needs:

1. Similar to the other two flows, let's create a new automated flow and name it `Attendee Teams Notification with Adaptive Card`.

2. Similar to the other two flows, the trigger will be **When a row is added, modified or deleted** in Dataverse. Ensure you have the following values:

    A. **Change Type**: **Added**

    B. **Table Name**: **Attendees**

    C. **Scope**: **Organization**

3. Click the plus (+) button under this trigger and select **Add an action**.

> **Callout**
>
> Since our trigger is pulling data from the Attendee table, we want to add a step to pull data from the Event table so that we can pull the event name. If we simply pull **Event (Value)** from the Attendee table, we will be pulling the GUID instead of the event's name.

Integrating event data

Since our trigger is pulling data from the Attendee table, we want to add a step to pull data from the Event table to retrieve the event name. If we only pull **Event (Value)** from the Attendee table, we will get the GUID instead of the event's name:

1.   Select **Microsoft Dataverse** from the card options:

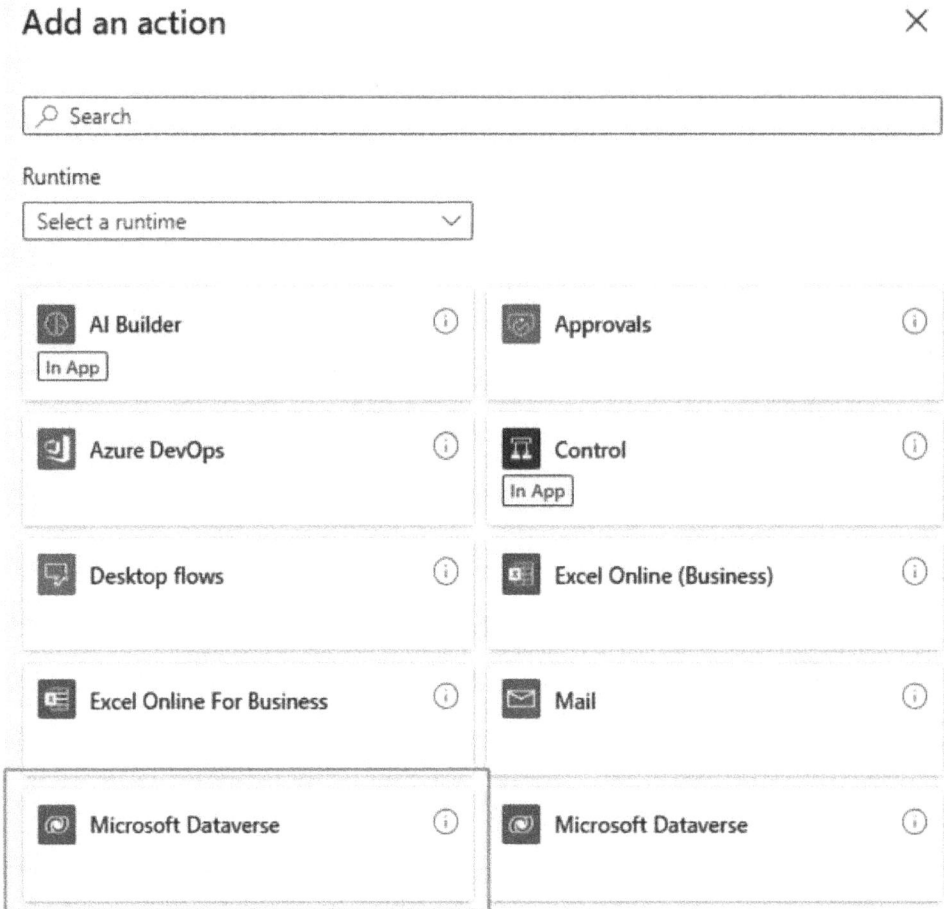

Add an action   ✕

⌕ Search

Runtime

Select a runtime   ⌄

🔲 AI Builder ⓘ In App	🔲 Approvals ⓘ
🔲 Azure DevOps ⓘ	🔲 Control ⓘ In App
🔲 Desktop flows ⓘ	🔲 Excel Online (Business) ⓘ
🔲 Excel Online For Business ⓘ	🔲 Mail ⓘ
🔲 Microsoft Dataverse ⓘ	🔲 Microsoft Dataverse ⓘ

Figure 9.25 – Microsoft Dataverse in Add an action

2. Select **Get a row by ID** from the **Microsoft Dataverse** options:

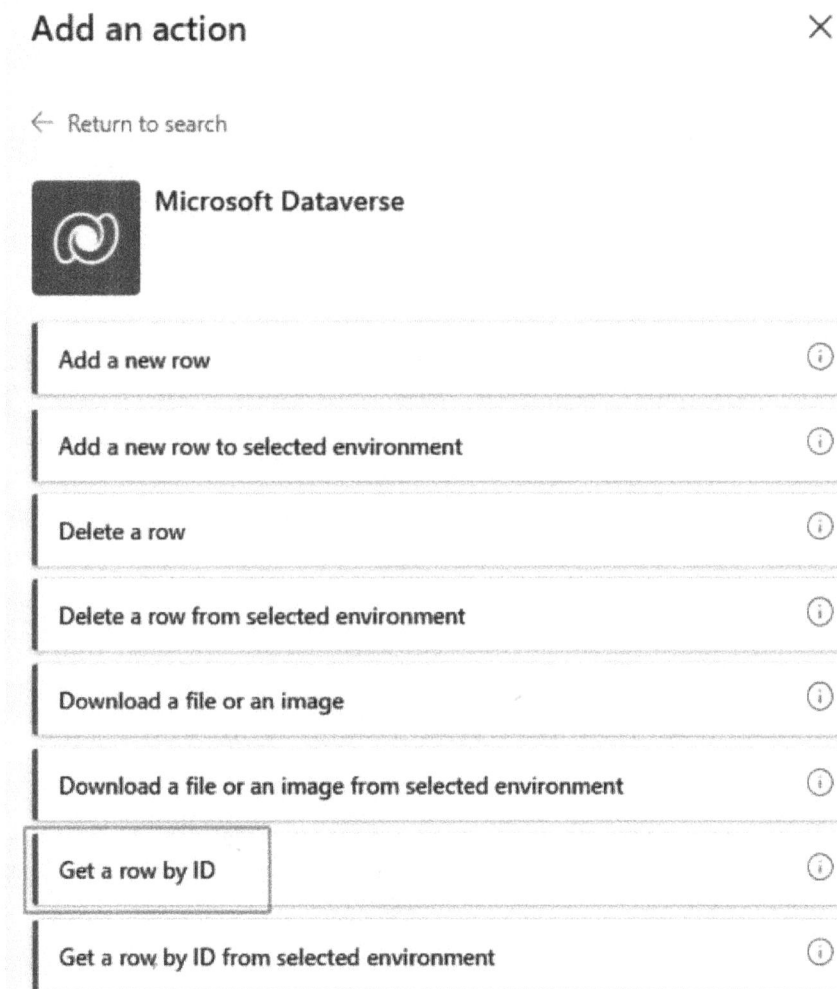

Add an action   ✕

← Return to search

Microsoft Dataverse

| Add a new row | ⓘ |

| Add a new row to selected environment | ⓘ |

| Delete a row | ⓘ |

| Delete a row from selected environment | ⓘ |

| Download a file or an image | ⓘ |

| Download a file or an image from selected environment | ⓘ |

| Get a row by ID | ⓘ |

| Get a row by ID from selected environment | ⓘ |

Figure 9.26 - Get a row by ID in Add an action

3. In the **Get a row by ID** panel, make the following changes:

A. **Table Name: Events**

B. **Row ID**: Dynamically choose the **Event (Value)** option under **When a row is added, modified or deleted**.

C. Change the name of this action to Get event:

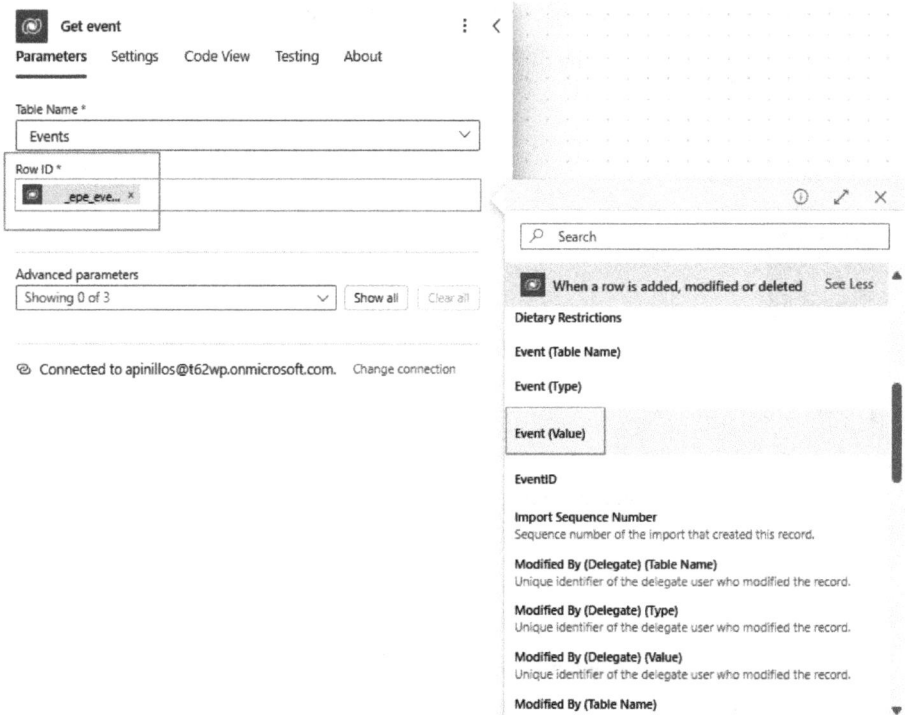

Figure 9.27 – Event (Value) in dynamic box

> **Callout**
>
> It's important to understand that this is only possible if there is a relationship between tables. The Event table is related to the Attendee table, which is how we can add attendees to our event in our model-driven app.

The trigger and action steps should look like what's shown in *Figure 9.27*:

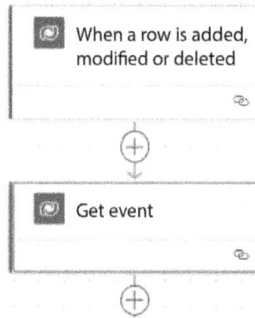

Figure 9.28 – Trigger and action

4. Click on the + sign under **Get event** and select **Add an action**.

5. In the **Add an action** panel, find the **Microsoft Teams** card and select it:

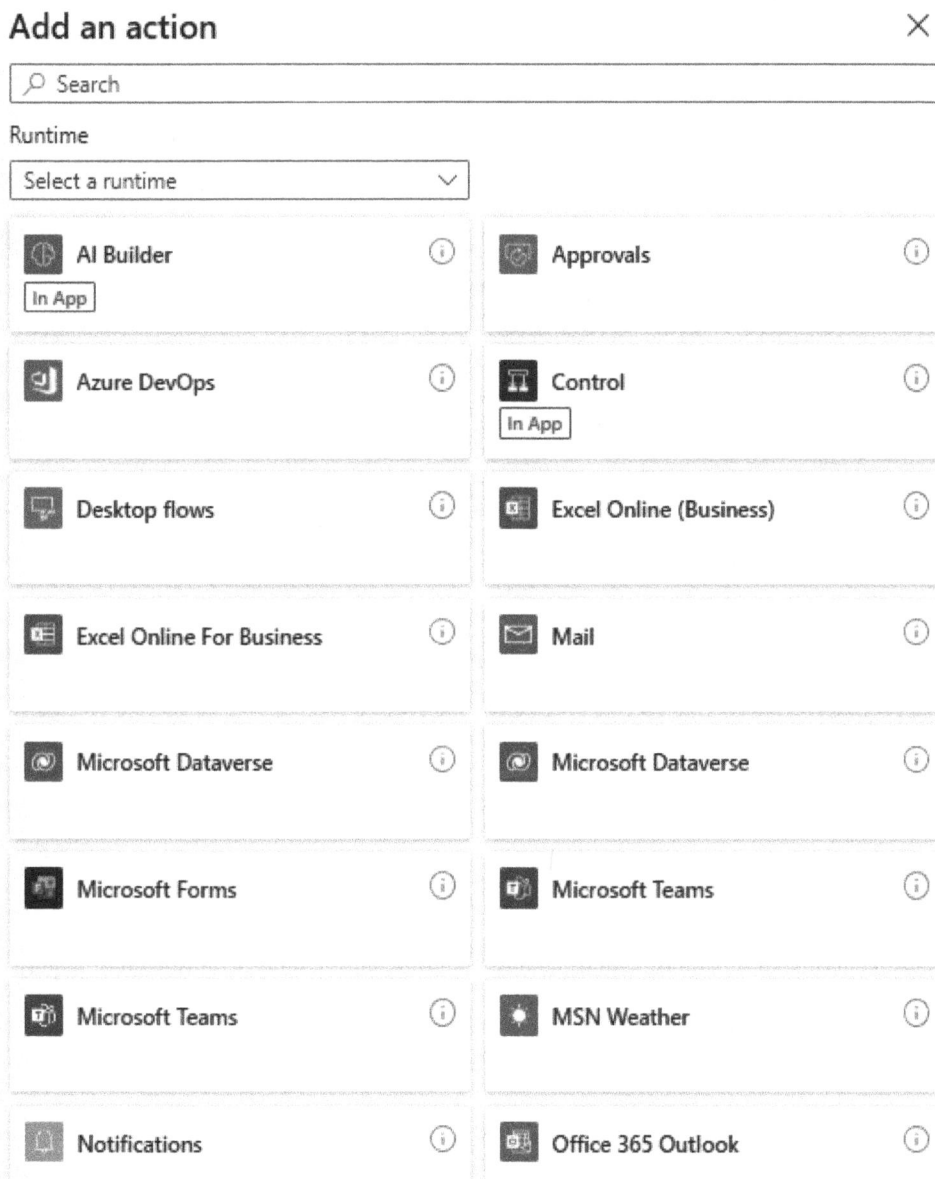

Add an action   ✕

🔍 Search

Runtime

Select a runtime   ⌄

🔘 AI Builder ⓘ	🔘 Approvals ⓘ
In App	

🔷 Azure DevOps ⓘ	ⲏ Control ⓘ
	In App

🖥 Desktop flows ⓘ	📗 Excel Online (Business) ⓘ

📗 Excel Online For Business ⓘ	✉ Mail ⓘ

◉ Microsoft Dataverse ⓘ	◉ Microsoft Dataverse ⓘ

🔲 Microsoft Forms ⓘ	🔷 Microsoft Teams ⓘ

🔷 Microsoft Teams ⓘ	☀ MSN Weather ⓘ

🔔 Notifications ⓘ	📧 Office 365 Outlook ⓘ

Figure 9.29 – Microsoft Teams in Add an action

6.   Under the **Microsoft Teams** action panel, find and select **Post card in a chat or channel**:

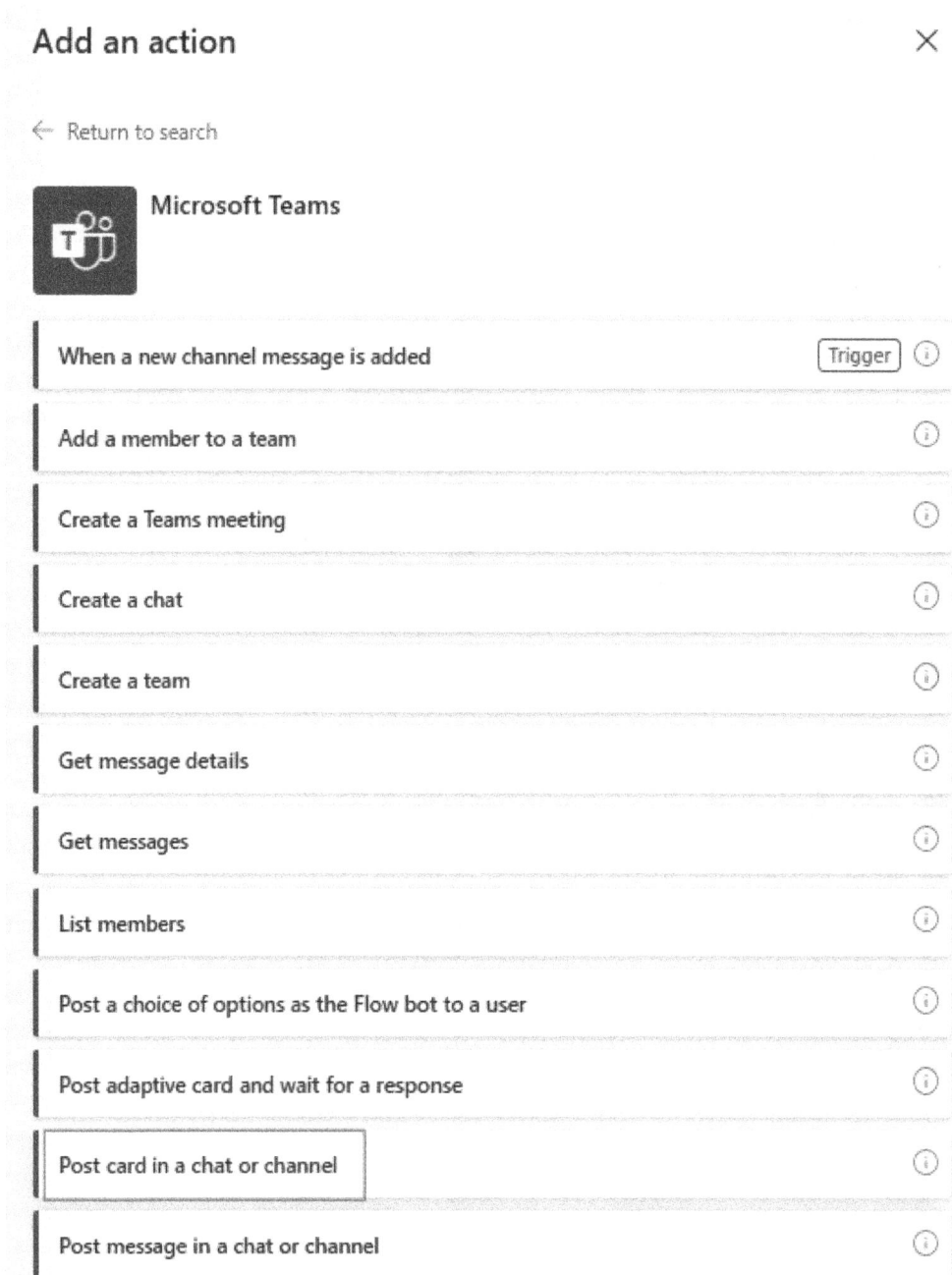

Add an action   ✕

← Return to search

Microsoft Teams

When a new channel message is added	Trigger ⓘ
Add a member to a team	ⓘ
Create a Teams meeting	ⓘ
Create a chat	ⓘ
Create a team	ⓘ
Get message details	ⓘ
Get messages	ⓘ
List members	ⓘ
Post a choice of options as the Flow bot to a user	ⓘ
Post adaptive card and wait for a response	ⓘ
Post card in a chat or channel	ⓘ
Post message in a chat or channel	ⓘ

Figure 9.30 – Post card in a chat or channel under Microsoft Teams in Add an action

7.   Under the **Parameters** tab of this action, make the following changes:

    A.   **Post As**: **Flow bot**

    B.   **Post In**: **Chat with Flow bot**

    C.   **Message**: Paste the JSON code you copied from the adaptive card designer

8.   Regarding the **Message** value, we will make a few adjustments:

Figure 9.31 – Post card in chat or channel parameters

9. Move the schema and version to the bottom of the object:

```
"$schema": "http://adaptivecards.io/schemas/adaptive-card.json",
"version": "1.4"
```

10. Add msteams to the bottom of the object:

```
"msteams": {
"width": "Full"
},
```

11. Dynamically add the attendee email by clicking the **Recipient** box, clicking the **lightning bolt** icon, then selecting **Attendee Email** from the **When a row is added, modified or deleted** section:

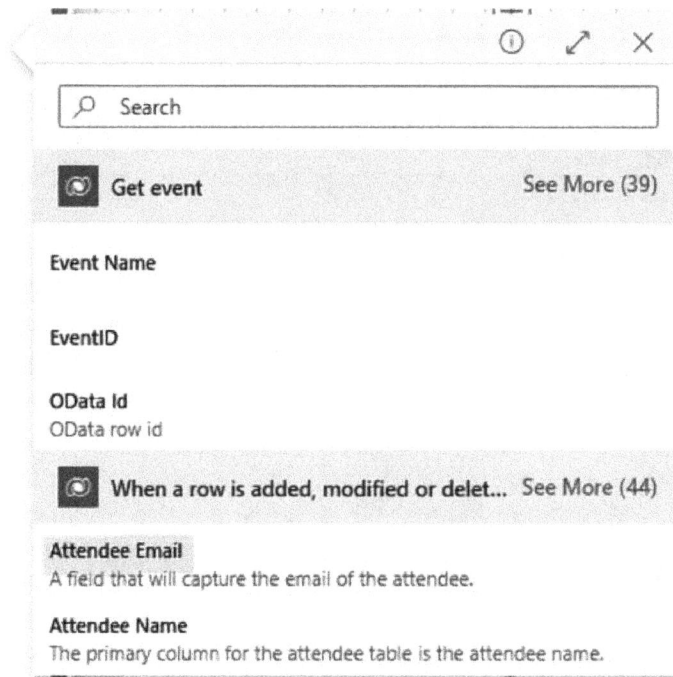

Figure 9.32 – Attendee Email in dynamic box

12. Dynamically add the event name by clicking after the colon in the line where it says `"text"`, `"Event Name: "`, then clicking the **lightning bolt** icon, then selecting **Event Name** under the **Get event** section:

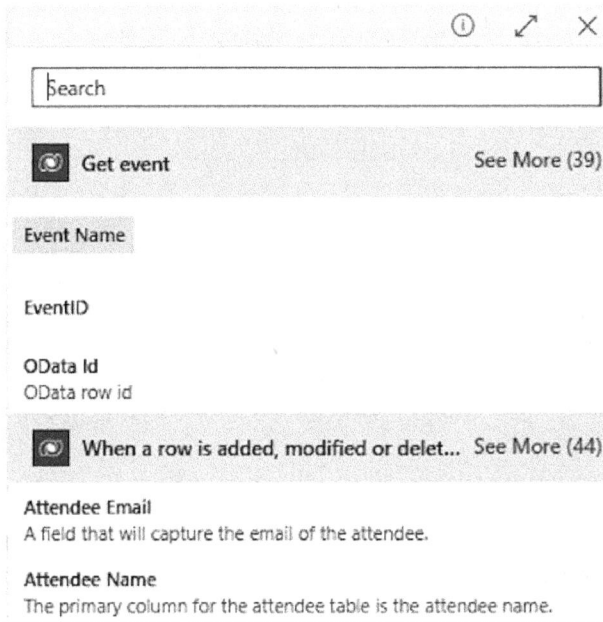

Figure 9.33 – Event Name in dynamic box

13. The full JSON should include the following code:

```json
{
"type": "AdaptiveCard",
"body": [
{
"type": "TextBlock",
"text": "You're invited to an event!",
"wrap": true,
"weight": "Bolder",
"size": "Large"
},
{
"type": "TextBlock",
"text": "Event Name: body('Get_event')?['epe_eventname']",
"wrap": true,
"size": "Medium",
"color": "Accent",
"spacing": "Medium"
},
{
"type": "TextBlock",
```

```
"text": "Date(s): formatDateTime(outputs('Get_event')?['body/
epe_eventstartdateandtime'], 'yyyy-MM-dd HH:mm:ss')
  - formatDateTime(outputs('Get_event')?['body/epe_
eventenddateandtime'], 'yyyy-MM-dd HH:mm:ss')
",
"wrap": true,
"spacing": "None"
},
],
"msteams": {
"width": "Full"
},
"$schema": "http://adaptivecards.io/schemas/adaptive-card.json",
"version": "1.4"
}
```

After defining the JSON code for the adaptive card, which specifies the format and content of the message to be sent, you need to ensure that the trigger, actions, and card configurations have all been set up so that they can work together:

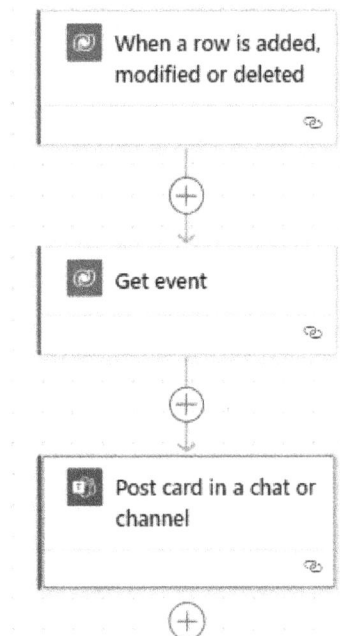

Figure 9.34 – Initial trigger, the Get event action, and the Microsoft Teams action

The finished Power Automate flow will have a trigger for **When a row is added, modified or deleted**. It will also have an action called **Get event** so that you can pull fields from the event table. Finally, it will have an action to **Post card in a chat or channel**.

Testing our adaptive card

With the flow configured and ready, it's crucial to test the adaptive card to ensure that it functions as expected and delivers the desired experience. Follow these steps to validate the setup and confirm that everything works smoothly:

1. Click **Save** and then **Test** in the top ribbon.
2. In the **Test Flow** panel, select **Manually**:

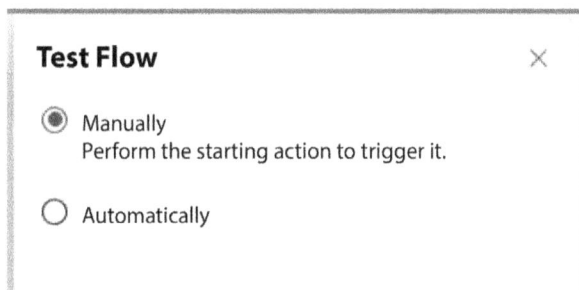

Figure 9.35 – Selecting the Manually checkbox to test the flow

3. Open the model-driven app we created in Chapter 6.
4. Create a new event by clicking **New** under the **Events** tab on the left:

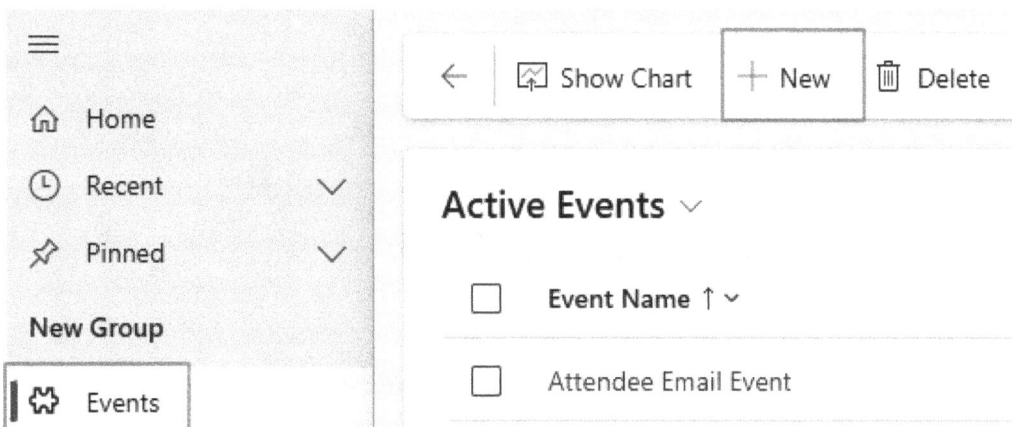

Figure 9.36 – The New button under the Events tab in the model-driven app

5. Add the following details to the **New Event** form:

    A. **Event Name**: `Teams Innovation`

    B. **Event start date and time**: **5/10/2024** at **8:00 AM**

    C. **Event end date and time**: **5/10/2024** at **5:00 PM**:

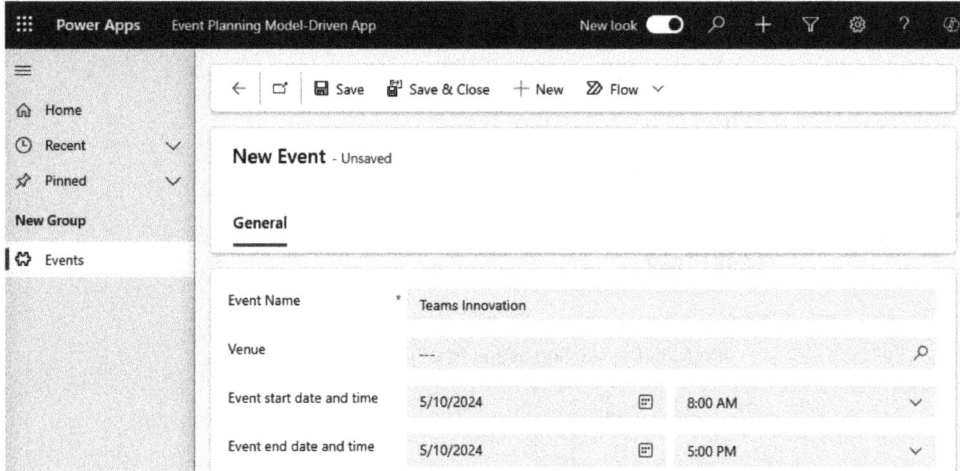

Figure 9.37 – The New Event form in the model-driven app

6. Click **Save**.

7. Click on the **Related** tab and select **Attendees**:

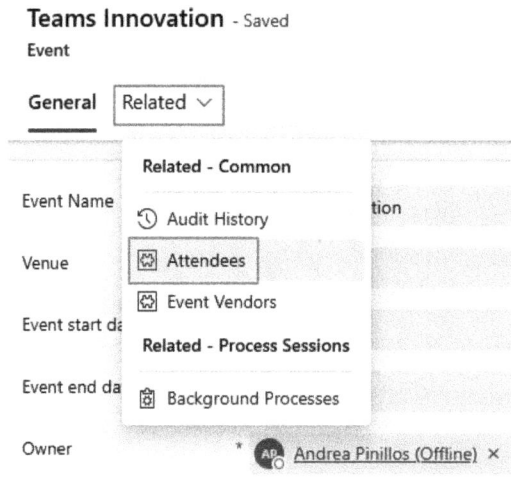

Figure 9.38 – Attendees under the Related tab

8. Click **New Attendee** and update the **New Attendee** form like so:

   A. **Attendee Name**: Andrea Pinillos

   B. **Attendee Email**: *<add your email here>*

9. Click **Save & Close**:

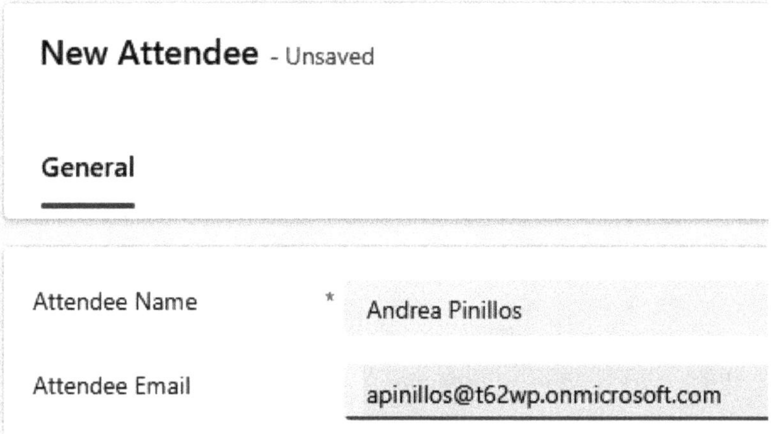

Figure 9.39 – Example New Attendee form with data

10. Navigate back to Power Automate and watch as the flow runs:

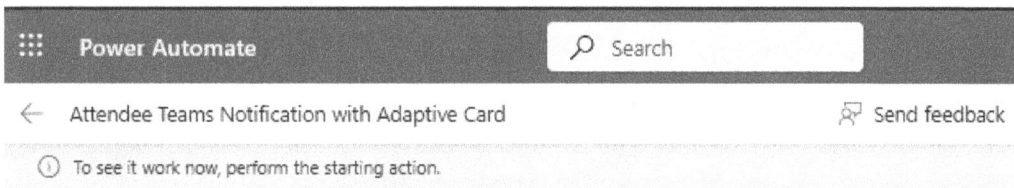

Figure 9.40 – Banner for testing our Power Automate flow

11. If the flow is successful, green checkmarks will appear on each step:

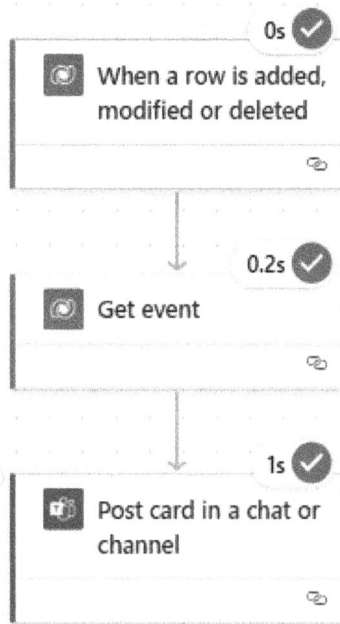

Figure 9.41 – A successful test for our Power Automate flow

12. Navigate to Teams to see if you received the notification regarding the adaptive card:

> **Callout**
> The adaptive card will be sent through **Workflows**.

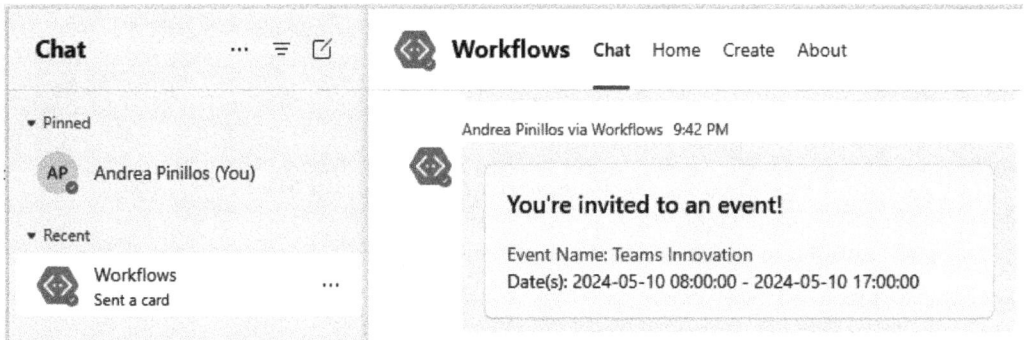

Figure 9.42 – Microsoft Teams ping example

The adaptive card should have the event name and dates, including the time. Feel free to adjust the format of the dates to your requirements and add any other dynamic information from the attendee or event table.

Summary

In this chapter, we embarked on a journey to explore the art of integrating Power Apps with Power Automate, Teams, and Outlook. We learned how to send Outlook emails and calendar invites, as well as how to send Teams notifications using Power Automate with adaptive cards. These skills are invaluable for modern workplaces, enabling tasks to be automated, streamlined communication, and optimized workflows.

Throughout this chapter, we not only grasped fundamental concepts but also immersed ourselves in practical applications. By the end of this chapter, we possessed the skills to orchestrate the flow of data and actions between Power Apps, Teams, and Outlook with finesse, revolutionizing how we work and collaborate.

In the next chapter, we'll delve deeper into advanced integration techniques, exploring how to integrate Power Apps with Power BI. By building upon the foundation laid in this chapter, we'll further expand our integration capabilities and unlock even more possibilities for automating and optimizing business processes.

Integration with Power BI

In the previous chapter, we covered integrating Power Apps with Power Automate, Microsoft Teams, and Outlook. Building on that foundation, this chapter delves into the integration capabilities with Power BI.

Users often require additional information, such as visual dashboards, detailed reports, and insightful analytics, which can enhance the functionality of their applications. While Power Apps offers some analytical and graphing capabilities, these can be limited in terms of overall reporting functionality and visual aesthetics. Integrating Microsoft Power BI with Power Apps provides a robust solution, combining the flexibility of Power Apps with the powerful reporting and visual analysis features of Power BI. Furthermore, the integration with Power BI works in both canvas and model-driven apps. Furthermore, this integration also works in the opposite direction by allowing you to embed a Power App directly in a Power BI report.

In this chapter, we're going to cover the following topics:

- An overview of Power BI and its capabilities
- Embedding Power BI visuals in a canvas app
- Embedding Power BI visuals in a model-driven app
- Embedding a Power App in a Power BI report

Technical requirements

To successfully engage with the materials in this chapter, you'll need the following:

- **Microsoft Power Apps**: Access to Power Apps for creating and managing your applications.
- **Microsoft Power BI**: Access to reports published on the Power BI service. This may require a Power BI Pro or Premium license depending on where the report is published. If you are unsure, please contact your Power BI administrator to ensure you have the appropriate license and access.

You can find the files used in this chapter on GitHub: `https://github.com/PacktPublishing/Power-Apps-Tips-Tricks-and-Best-Practices/tree/main/Chapter10`

An overview of Power BI and its capabilities

Microsoft Power BI is a business analytics tool that is part of the overall Power Platform and enables individuals to visualize and share insights from their data. It offers robust data visualization capabilities, allowing users to create interactive and dynamic reports and dashboards. Power BI can connect to a wide range of data sources, transforming raw data into meaningful information for better decision-making. These capabilities make it a fantastic addition to your Power App functionality. Let's highlight some of Power BI's capabilities:

- **Data visualization**: Power BI excels in transforming data into insightful visualizations. Users can create interactive reports and dashboards using a wide variety of visual elements, such as charts, graphs, maps, and custom visuals. These visualizations help in identifying trends, outliers, and patterns in data, facilitating better decision-making.

- **Data connectivity**: Power BI supports hundreds of data sources, enabling users to connect to on-premises, cloud-based, and streaming data. Common data sources include Excel, SQL, SharePoint, text/CSV, and many more. This flexibility ensures that users can consolidate data from multiple sources into a single, coherent view.

- **Data modeling**: Power BI provides robust data modeling capabilities, allowing users to create complex data models with relationships, hierarchies, and calculated measures. The **Data Analysis Expressions** (**DAX**) language enables the creation of advanced calculations and aggregations, enhancing the analytical capabilities of reports and dashboards.

- **Collaboration and sharing**: Power BI facilitates collaboration through its integration with other Microsoft applications, including Excel, SharePoint, Teams, PowerPoint, Power Apps, and Power Automate, to name a few. This allows users to share reports and dashboards with colleagues. The Power BI service also enables the publication of reports to the web and embedding in other applications, broadening the accessibility of insights.

- **Advanced analytics**: Power BI integrates with advanced analytics tools such as R and Python, enabling users to incorporate sophisticated statistical models and machine learning algorithms into their reports. This integration enhances the depth of analysis and provides predictive insights.

Overall, Power BI provides a set of capabilities that, when combined with other Power Platform solutions, such as Power Apps and Power Automate, enable organizations to not only visualize and analyze their data but also to build custom applications that leverage these insights. The integration of Power BI's analytical capabilities with Power Apps' customizability opens a world of possibilities for creating data-driven solutions that drive business value. Here is a good way to think about these applications: Power Apps allows you to create data (as well as change or delete data). Power BI allows you to visualize data. Think about your own day-to-day job duties and how often you need to do both activities. For example, reviewing a monthly report (visualizing the data) requires you to capture notes or commentary (creating data). Now, imagine having one application that provides both activities.

Now that we've touched on an overview of Power BI, it is time to dive into adding this to your app. We will begin with embedding Power BI visuals in a canvas app.

Embedding Power BI visuals in a canvas app

Embedding Power BI visuals in Power Apps allows you to access and interact with rich data visualizations directly within the app. This seamless integration enhances the functionality of Power Apps by providing better insights and analytics, empowering users to make informed decisions without leaving the app environment.

In this section, we will explore the steps and considerations for embedding Power BI visuals in Power Apps. In addition, there are several approaches to how this can be done, ranging from embedding in simple Power BI dashboard tiles to complete Power BI reports. Before digging into implementing this, let's first talk about the benefits of using Power BI in your Power App.

Benefits of embedding Power BI visuals

The following are the benefits of embedding Power BI visuals:

- **Enhanced user experience**: Embedding Power BI visuals in Power Apps creates a unified experience where users can view and interact with data visualizations without switching between different tools. This integration helps in maintaining focus and improves the overall efficiency of the workflow.

- **Real-time data access**: By embedding Power BI reports that use direct connections, users gain access to real-time data and insights. This capability is crucial for making timely decisions based on the most current information available.

- **Interactive reports**: Power BI reports are inherently interactive, allowing users to drill down into data, apply filters, enable dynamic interactions between graphs and other visuals, and explore different aspects of the dataset. Embedding these reports in Power Apps enables users to interact with the data within the context of the application.

Now that we have covered some of the benefits, let's get into the details of adding Power BI components to your Power App.

Embedding a Power BI dashboard tile in a canvas app

Power BI offers two options for embedding into canvas apps. The first option is integrating with dashboard tiles. The second approach is where the full Power BI report is embedded. We will cover both approaches in depth. Embedding either of these within the app is going to be very similar, but it is important to understand the difference. We will cover embedding a Power BI dashboard tile first.

> **Note**
>
> When working with Power BI reports, it is common for a report to be referred to using the term *"dashboard"*. Many reports are used to summarize data, which can be considered a *"dashboard"*. For example, a Power BI developer may be asked to build a financial reporting *dashboard*. Although the term *"dashboard"* is used, the developer might actually create a Power BI report. For this section, a Power BI dashboard is a specific item within Power BI and is a separate entity from a Power BI report.

What is a Power BI dashboard?

While most people who work with Power BI are familiar with published reports, there is additional functionality around creating Power BI *dashboards*. You might consider a **dashboard** as a summary report. However, a Power BI dashboard is a single canvas that consists of visuals that have been *pinned* from one or more reports. The pinned visuals in a dashboard are referred to as **tiles**. Let's provide an example.

Example of creating a Power BI dashboard with tiles

Assume that we create a report in Power BI to track amounts paid to vendors. This report is published to the Power BI service. A sample report is shown in *Figure 10.1*:

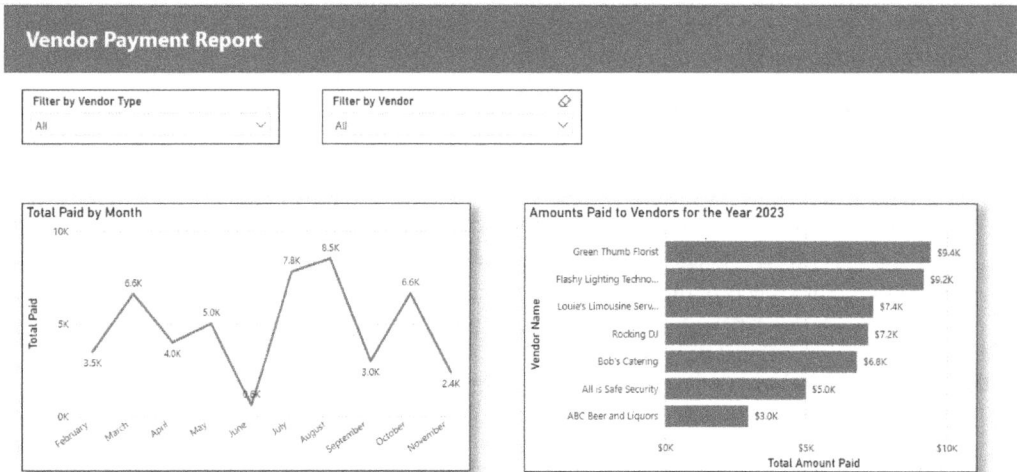

Figure 10.1 - A Power BI report on vendor payments

When viewing any visual, placing your cursor near the top-right corner will provide the option **Pin visual**. Clicking on this will then allow you to pin that visual to a dashboard. This is shown in *Figure 10.2*.

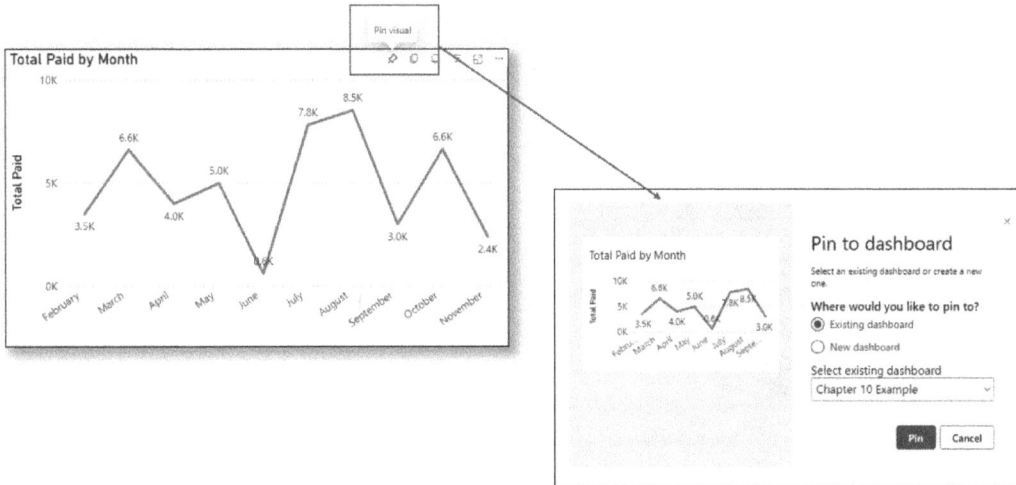

Figure 10.2 - Example of pinning a visual to a dashboard

Looking at *Figure 10.2*, a dialog box appears that allows you to pin the visual to either an existing dashboard or a new dashboard. Once you complete this, a dashboard will now appear in your Power BI service and it will include all the visuals that were pinned. *Figure 10.3* provides an example showing both the report and the newly created dashboard.

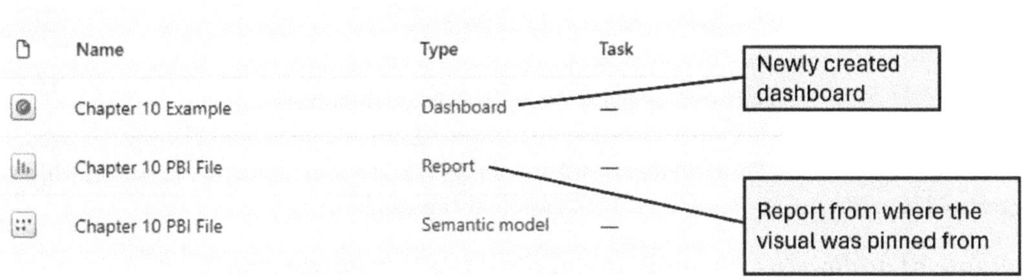

Figure 10.3 - Showing the new Dashboard file separate from Report in Power BI

One big advantage of Power BI dashboards is that additional visuals can be pinned to the same dashboard, whether they are within the same Power BI report or from separate reports. In addition, other elements, such as text, can also be added to the dashboard. *Figure 10.4* provides an updated dashboard where an additional visual was pinned along with a text box added in the bottom-left corner.

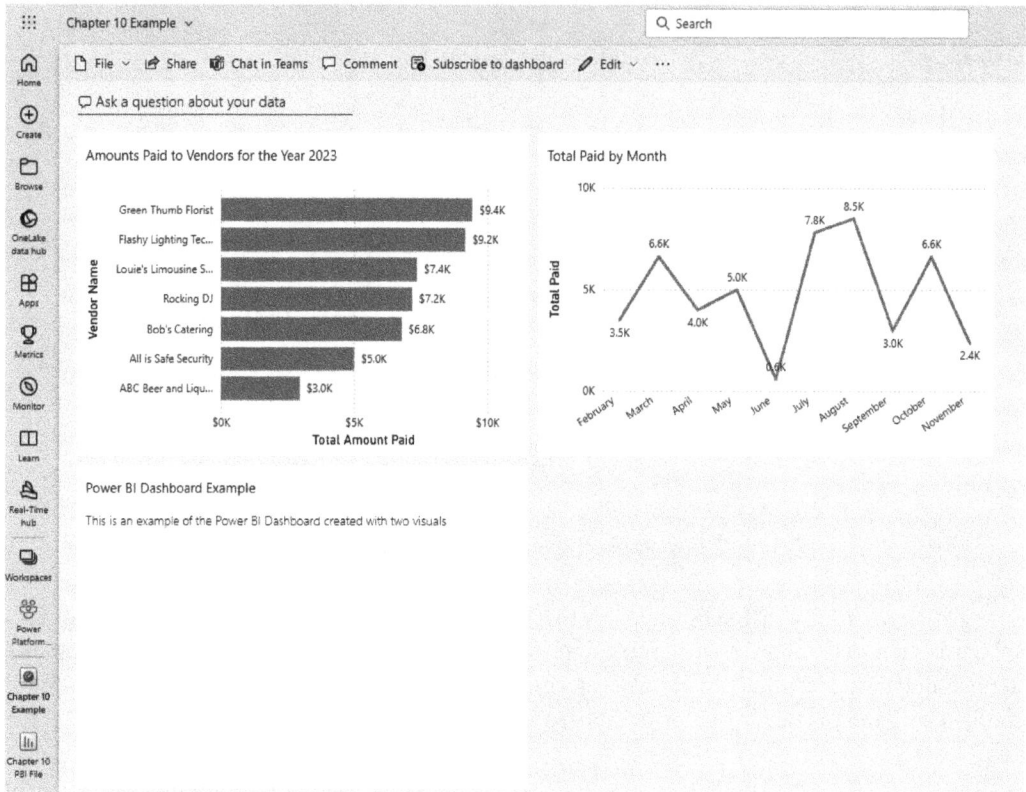

Figure 10.4 - Example of a completed Power BI dashboard

Now that we have created a Power BI dashboard, we can easily add any of these specific visuals, which are referred to as *tiles*, to our Power App. Let's show how this can be done.

Example of adding the Power BI dashboard tile in a Power App

For this example, we will first move back to our Power App. To add a Power BI visual from a Power BI dashboard, go to the **Insert** tab and select **Power BI tile** from the list of controls. This is shown in *Figure 10.5*.

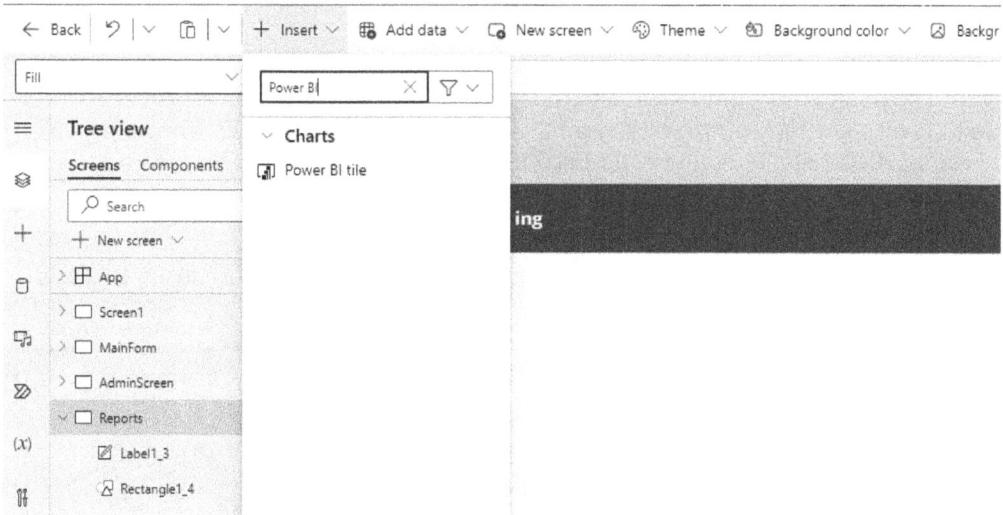

Figure 10.5 - Inserting a Power BI tile into a Power App

After selecting the **Power BI tile** option, a dialog box will appear that will require three items:

- **Workspace**: This is the name of the Power BI workspace where the dashboard is located. Please note that only the workspaces that you have access to will appear.

- **Dashboard**: This is the name of the dashboard.

- **Tile**: This is the name of the tile that you want to embed.

An example of this box is shown in *Figure 10.6*.

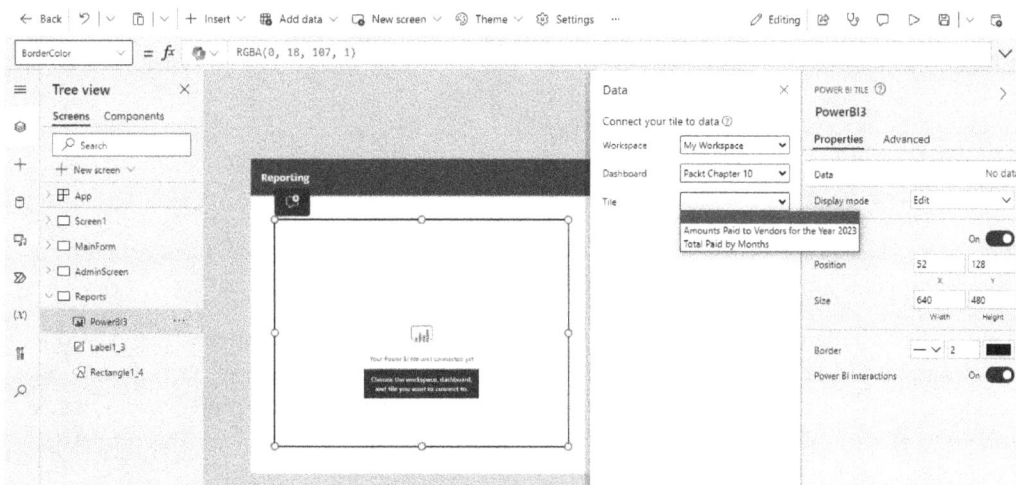

Figure 10.6 - Selecting Power BI tile from a Power BI dashboard

Note

After selecting your workspace and dashboard, you may notice that the *tile* area remains grayed out. While we are unsure why this is, to fix this we have found that simply going to the specific tile on the Power BI dashboard, clicking on the ellipses (...) in the top-right corner, and then selecting **Edit details**. Then, update the tile title, even if you simply add and remove a space. Return to Power Apps and try again.

Once the tile is selected, the Power BI visual will appear in your Power App. An example of this is shown in *Figure 10.7*.

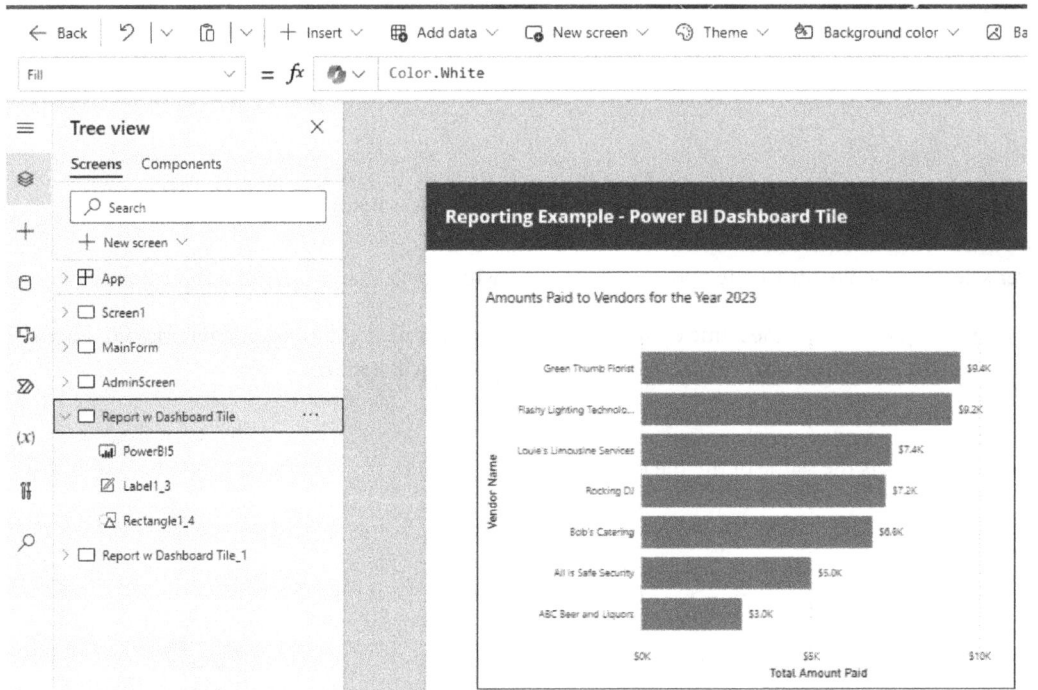

Figure 10.7 - An embedded Power BI tile

This process will add a specific visual from the dashboard. If additional tiles from the dashboard are needed, you should simply follow the process again for each one.

It is important to note now that the embedded visual(s) can be clicked on when being viewed in your app and it will take the user directly to the specific Power BI report. However, this feature can be removed. Simply select the Power BI visual, go to the **PowerBIInteractions** property, and change this to false. While the visual will still appear, clicking on it will have no effect. This property is shown in *Figure 10.8*:

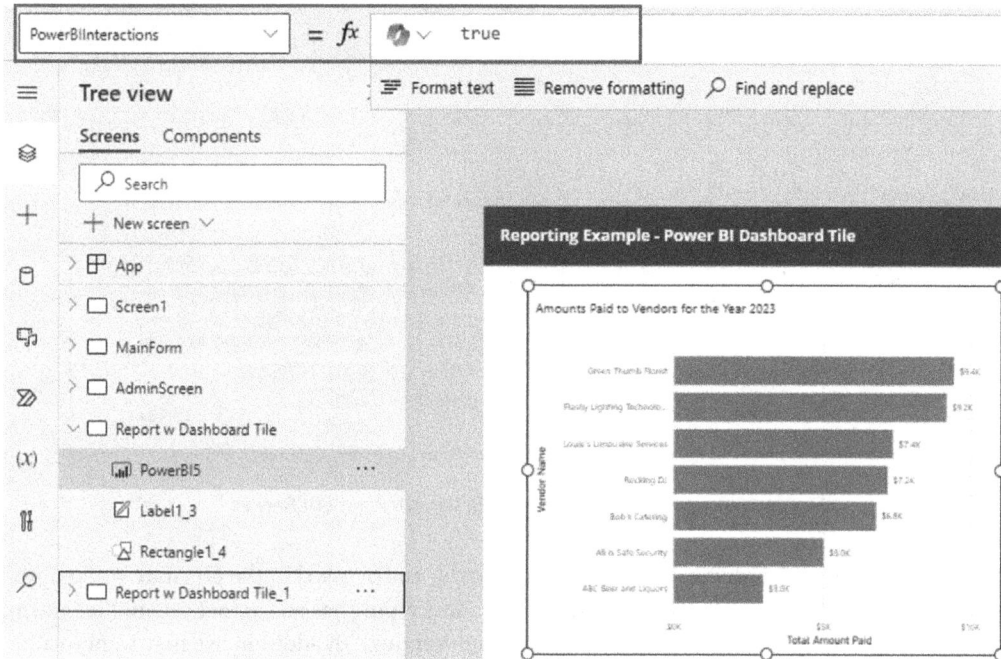

Figure 10.8 - Toggling the Power BI visual interaction

Embedding dashboard tiles is a great way to quickly add a graphical visual to your Power App. However, this approach does limit you to adding one visual at a time. In addition, Power BI dashboards can be quite limiting in terms of the overall layout. Although you can embed multiple visuals within your app, there is no interactivity between them like you have in a Power BI report. Thus, an alternative approach is embedding the full report itself. This also provides you with a greater ability to create custom solutions. We will discuss that approach next.

Embedding a Power BI report in a canvas app

Adding a Power BI report with a Power App provides significant reporting capabilities. As noted in the previous section, one advantage to doing this over embedding in individual Power BI dashboard tiles is the ability to leverage the full scope of your Power BI report. Users will be able to fully interact with the report directly from your Power App. Adding a report within the app is very similar to embedding a Power BI dashboard, with just a minor change.

We previously referenced a report that was created and then used to pin visuals to create a dashboard. Let's revisit that report and display it again here in *Figure 10.9*.

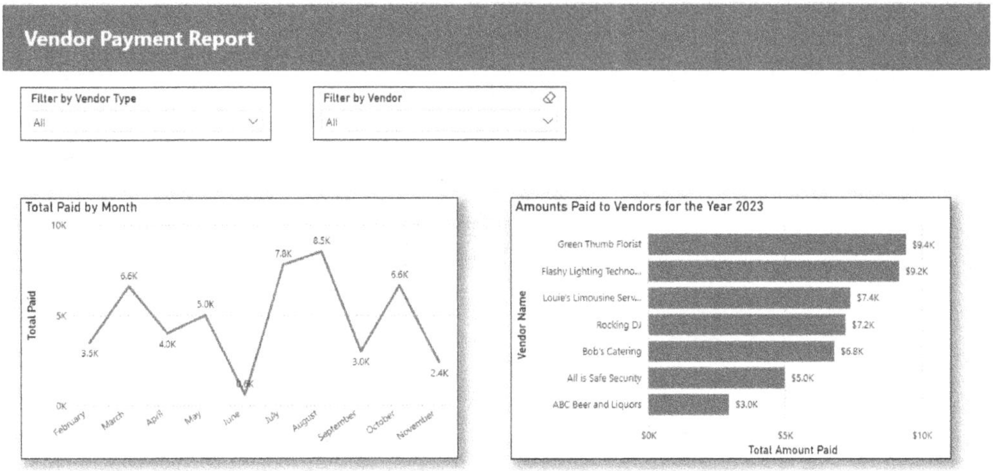

Figure 10.9 - Example of our Power BI Vendor Payment Report

Rather than pinning each individual visual to a dashboard, as discussed in the previous section, we instead want to add this full report to our Power App. This report has additional capabilities, such as the two filters within the report (by **Vendor Type** and **Vendor**). In addition, we may want to add additional visuals and/or pages later. Thus, embedding this full report will be much more advantageous.

Like embedding a Power BI dashboard tile, we want to return to our Power App and select the Power BI tile. This is shown in *Figure 10.10*.

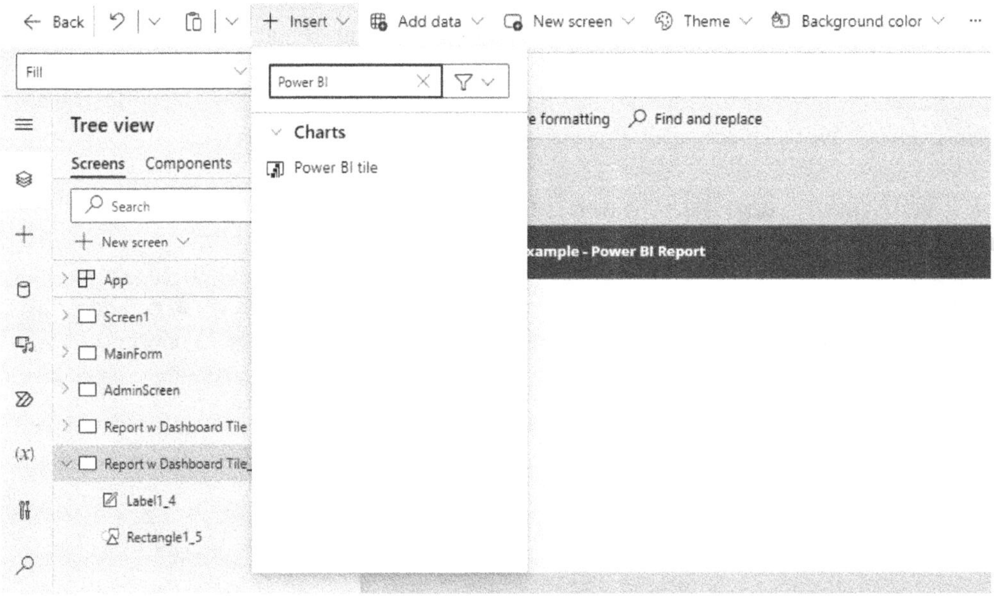

Figure 10.10 - Using the Power BI tile to embed a full Power BI report

This will once again bring up a dialog box where you can select your workspace, your dashboard, and the tile name. Simply close this, as it won't be needed. This is shown in *Figure 10.11*.

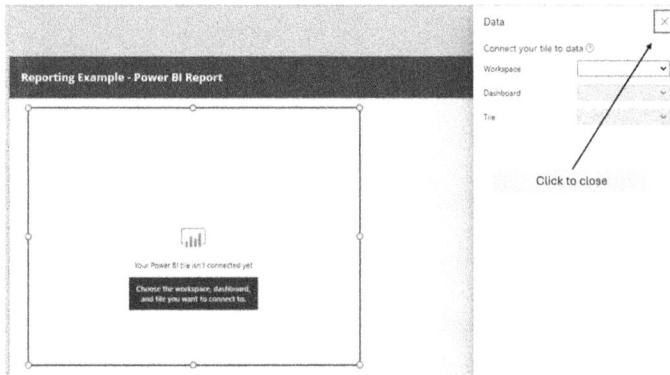

Figure 10.11 - Closing the dialog box

Open your published Power BI report, which is in the Power BI service. Select **File** and then **Embed report** and then select **SharePoint online**. This is shown in *Figure 10.12*.

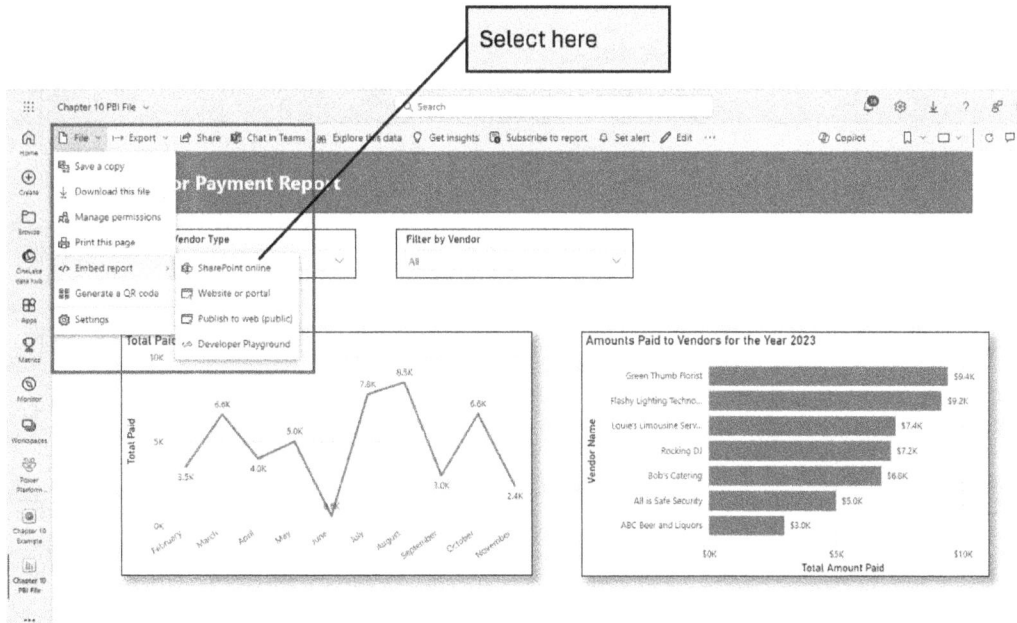

Figure 10.12 - Obtaining a URL link to embed your Power BI report

This will then allow you to copy a hyperlink to use. This is shown in *Figure 10.13*.

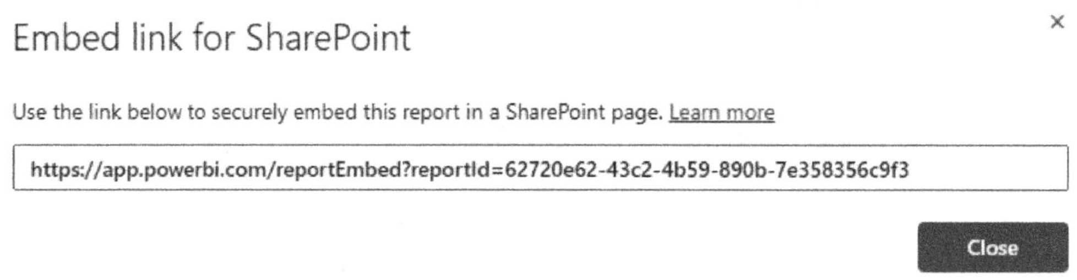

Figure 10.13 - Example of a URL link to embed a report

Return then to your Power App and select the new control that has been added. Go to the **TileUrl** property and paste in the URL that you copied. *Figure 10.14* provides an example.

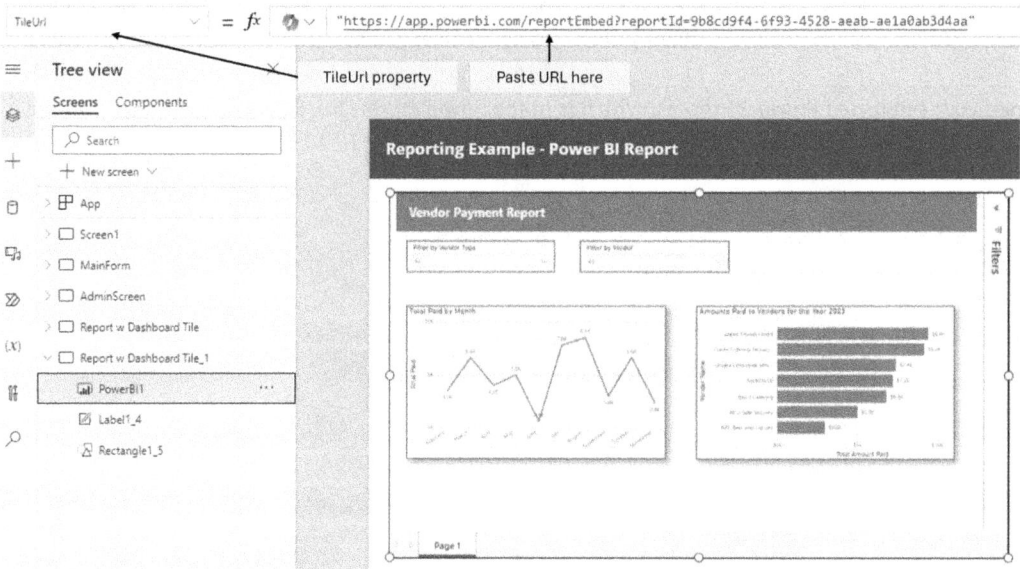

Figure 10.14 - A Power BI report embedded into the Power App

You now have a fully integrated Power BI report embedded within your Power App. If you want to disable the report interaction, simply change the **PowerBIInteractions** property to `false`.

> **Important note**
>
> When building a canvas app, there is the option to build a mobile or tablet layout. You will need to consider the size of your Power BI report to ensure that it fits properly within your app.

That is all there is to adding Power BI dashboard tiles or reports to your canvas app. However, Power BI dashboards and reports can also be integrated into model-driven apps. We will cover that in the next section.

Embedding Power BI visuals in a model-driven app

In the previous section, we showed how you can embed Power BI visuals within a canvas app. However, Power BI can also be embedded in a model-driven app. We will cover the steps in this section.

Before we begin, this section will presume that you already have a model-driven app created. If you do not, please refer to *Chapter 4, Choosing the Right Tool – Navigating Canvas Apps, Power Pages, and Model-Driven Apps*. In addition, users must have appropriate access to Power BI. This includes having access to the appropriate report and workspace, as well as having an appropriate Power BI license.

> **Important note**
>
> When embedding Power BI visuals into a model-driven app, you are required to use a solution. We covered using solutions in *Chapter 2, Working with Solutions*.

To begin, let's first cover the items that we will need to have ready, which are as follows:

- A solution that includes our model-driven app. We covered solutions in *Chapter 2, Working with Solutions*.

- A Power BI dashboard or report that has been published to the Power BI service. In our example, we will use the Power BI dashboard that was previously shown in *Figure 10.4*.

An example of our solution is shown in *Figure 10.15*. This solution contains a model-driven app.

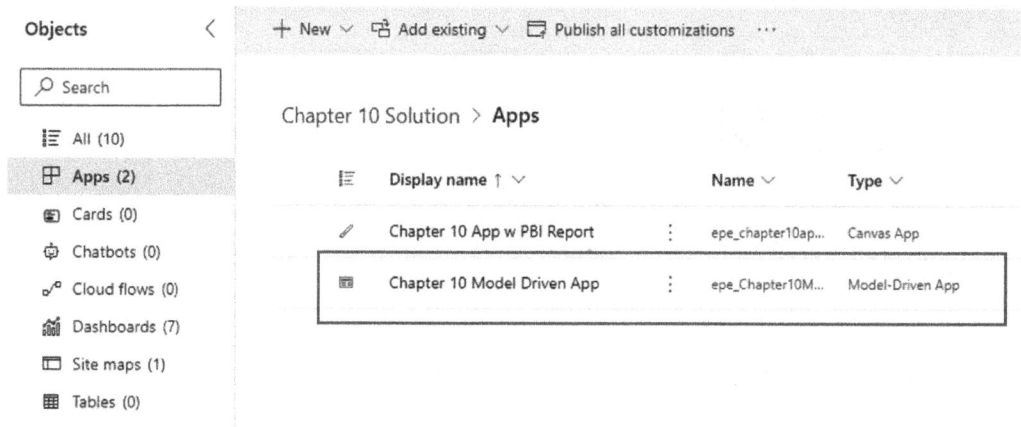

Figure 10.15 - Overview of our sample solution

Our first step is to add the Power BI dashboard to our solution. To do this, we will use the **New** option in our solution and then choose **Dashboard | Power BI embedded**. This is shown in *Figure 10.16*.

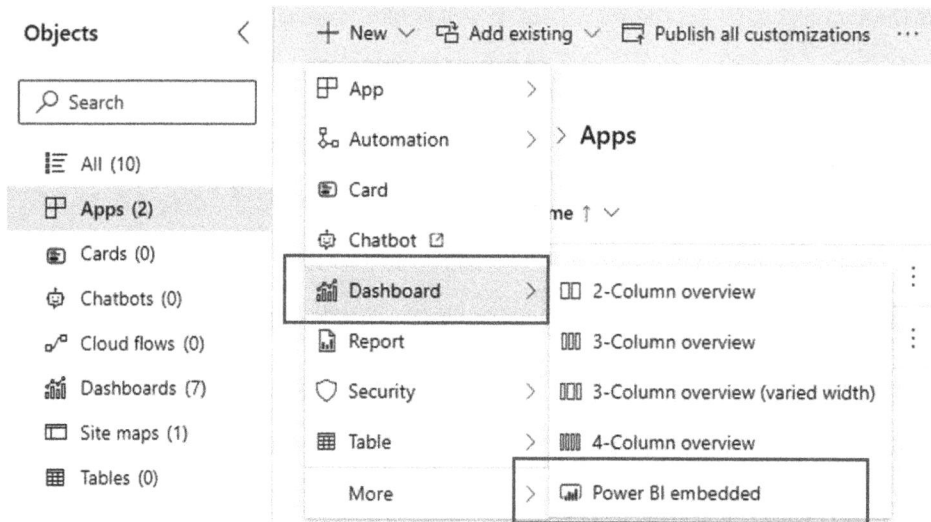

Figure 10.16 - Adding an embedded Power BI report

This brings up a dialog box where you can select various options. Let's review the key options:

- **Display name**: This is the name that will be displayed in your solution.

- **Type**: This is where you select either the Power BI report or dashboard. We covered Power BI *dashboards* in the previous section.

- **Power BI workspace**: This is where you select the applicable workspace on the Power BI service.

- **Select Power BI workspace/dashboard**: Select the applicable report or dashboard.

A complete overview of these options is displayed in *Figure 10.17*.

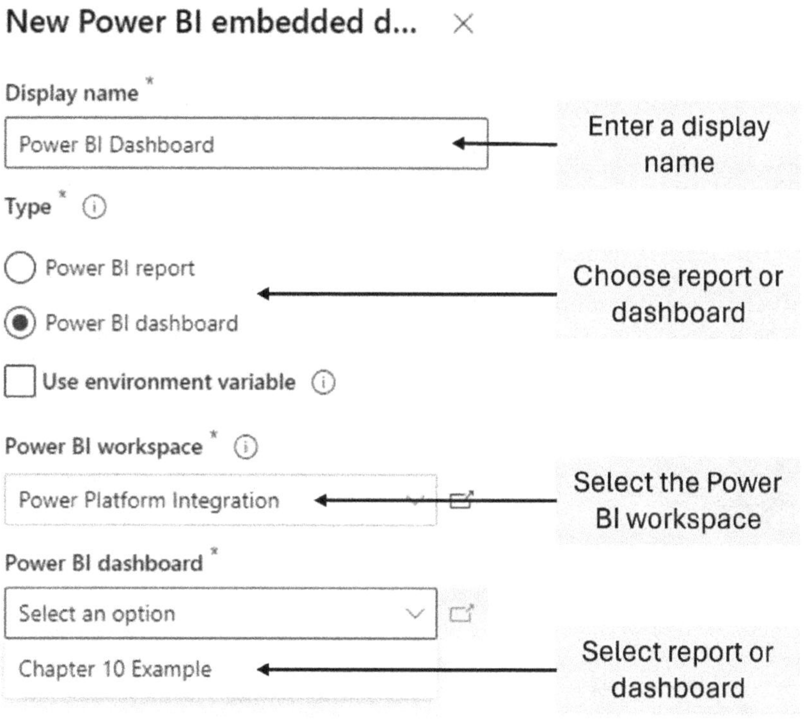

Figure 10.17 - Selecting the applicable Power BI report or dashboard

This will add the Power BI dashboard or report to your solution as a Power BI embedded type. In our example, we'll select the Power BI *dashboard* that was previously shown in *Figure 10.4* and add this to our solution. See *Figure 10.18* for an example after this is selected. Our solution now contains our model-driven app called *Chapter 10 Model Drive App Example* along with a Power BI embedded dashboard called *Chapter 10 Dashboard*.

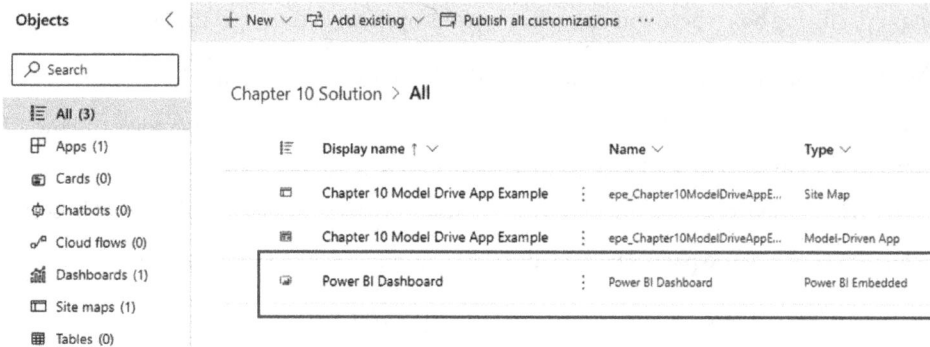

Figure 10.18 - Example of the Power BI embedded dashboard in the solution

> **Note**
>
> When adding a Power BI dashboard as a new Power BI embedded dashboard, the full dashboard is added. This is different than embedding individual dashboard tiles in a canvas app, which was discussed in the previous section.

If you want to embed a report, simply follow the same steps, but select the report option and then select the applicable report. A model-driven app can contain multiple embedded Power BI dashboards and reports. We'll now return to editing our model-driven app. Here, we will add a new page and then select **Dashboard** as the option. See *Figure 10.19*.

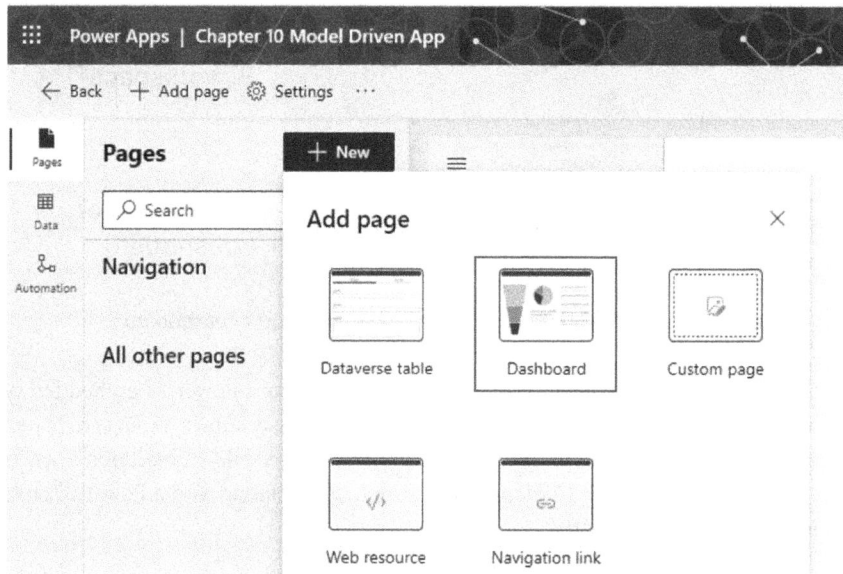

Figure 10.19 - Adding a new dashboard page to our model-driven app

A new dialog box opens, which allows us to select the dashboard from the available ones. Please note there may be system dashboards or other dashboards available. For our situation, we want to change the filter to **Power BI Dashboard**, as shown in *Figure 10.20*. There is also an option to show this in the navigation pane of your app.

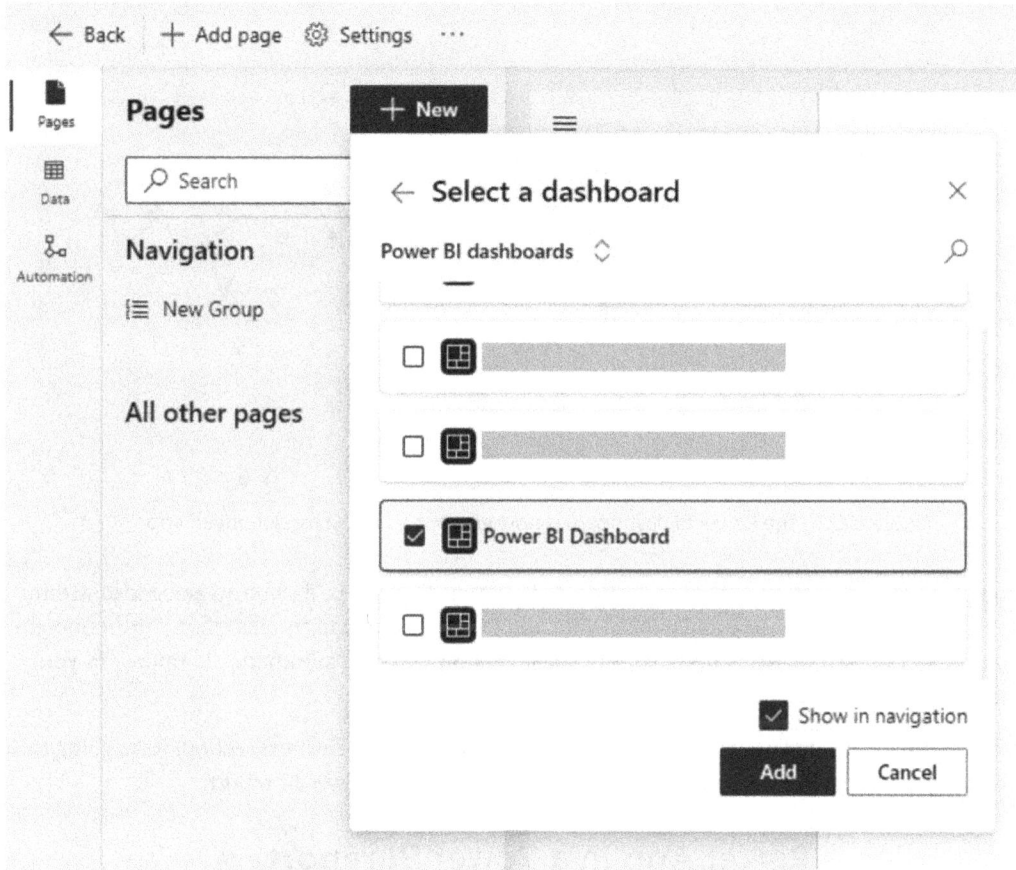

Figure 10.20 - Changing the filtering options to select our dashboard

This will add the selected report/dashboard to your app. If you select the **Show in navigation** option, the name of the embedded Power BI dashboard will display on the left side, as shown in *Figure 10.21*.

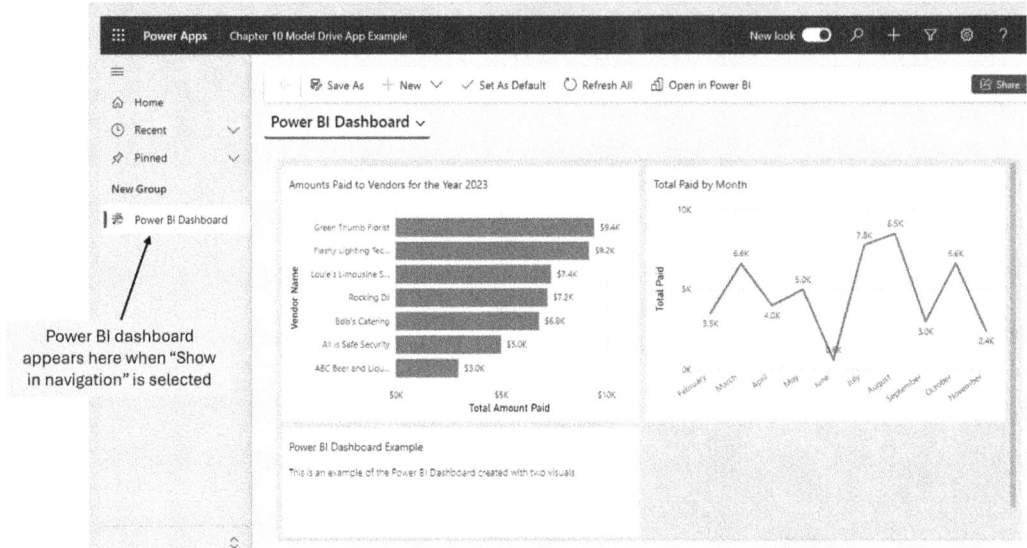

Figure 10.21 - The Power BI dashboard is now embedded in the model-driven app

By completing these steps, you should now have your Power BI report or dashboard embedded within your Power App. Be sure to save and publish your app to ensure the changes take effect. Furthermore, as your report or dashboard is updated, the latest version will also automatically update in your model-driven app.

Now that we have covered integrating Power BI into a Power App, in the next section, we're going to switch this and discuss how a Power App can be embedded into a Power BI report.

Embedding a Power App in a Power BI report

In the previous sections, we focused on embedding Power BI dashboard tiles and reports within either a canvas app or a model-driven app. However, there is also the ability to integrate in the other direction, which is embedding a Power App directly in a Power BI report. This integration also enables the ability to directly pass data from a Power BI report, including as visuals are being interacted with, immediately to the embedded Power App.

We are going to discuss two scenarios related to embedding a Power App directly in a Power BI report. To help with these two scenarios, we're going to provide a specific use case example. In *Figure 10.22*, a simple Power BI report that provides a table of our current vendors is displayed:

Vendor Table

Vendor Name	Vendor Type	Description
ABC Beer and Liquors	Catering and Food Services	They provide full bartender including beer, wine, and spirits
All is Safe Security	Security	Security Services
Bob's Catering	Catering and Food Services	Offers overall food and catering services
Flashy Lighting Technologies	Music and Entertainment	Provides lighting and other visual effects.
Green Thumb Florist	Floral and Decor	A full service floral business
Louie's Limousine Services	Transportation	Offers up limousine and other transportation services
Rocking DJ	Music and Entertainment	Offers DJ and music

Figure 10.22 - Example of a Power BI report showing a table of vendors

In this example, the data for this report is coming from a SharePoint list. Currently, all our Power BI report does is display information that is currently on the list. In order to add new vendors or make changes to existing data, users have to go to a separate location. This could be by going directly to the SharePoint list, accessing a Power App in a separate location, or some other method of data entry. The bottom line is that users are required to move away from the report and access a different application.

However, our desire is to make this report a one-stop shop and provide functionality to not only view the current data but also make additions or changes to the data. This is where embedding a Power App directly in the report makes this possible. Let's dive into the first scenario.

Embedding a Power App – scenario 1

In this first example, we'll show how you can quickly and easily integrate a Power App directly into your Power BI report. The desired end result is shown in *Figure 10.23*, where an embedded Power App is added to the report and directly accessible.

The first thing we are going to do is create a simple Power App that we can use. In *Chapter 5*, we covered how you can automatically create a Power App from a SharePoint list in the section called *Using SharePoint lists as a data source*. But, as a refresher, let's walk through the steps.

First, we go to our SharePoint list. From there, we select **Integrate | Power Apps | Create an app**. This is shown in *Figure 10.23*.

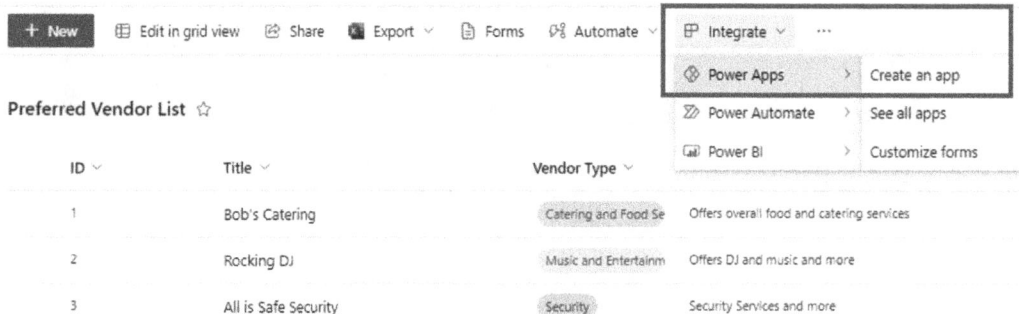

Figure 10.23 - Creating an automatic Power App from a SharePoint list

This will create a three-screen app. We will save this app to an environment. We covered Power Platform environments in *Chapter 3*.

We will now open Power BI Desktop so that we can add this newly created Power App. Within the visualization pane, there is a standard visual icon to add to the Power App, as shown in *Figure 10.24*. This will add a blank Power App visual and we will resize it as shown in the example.

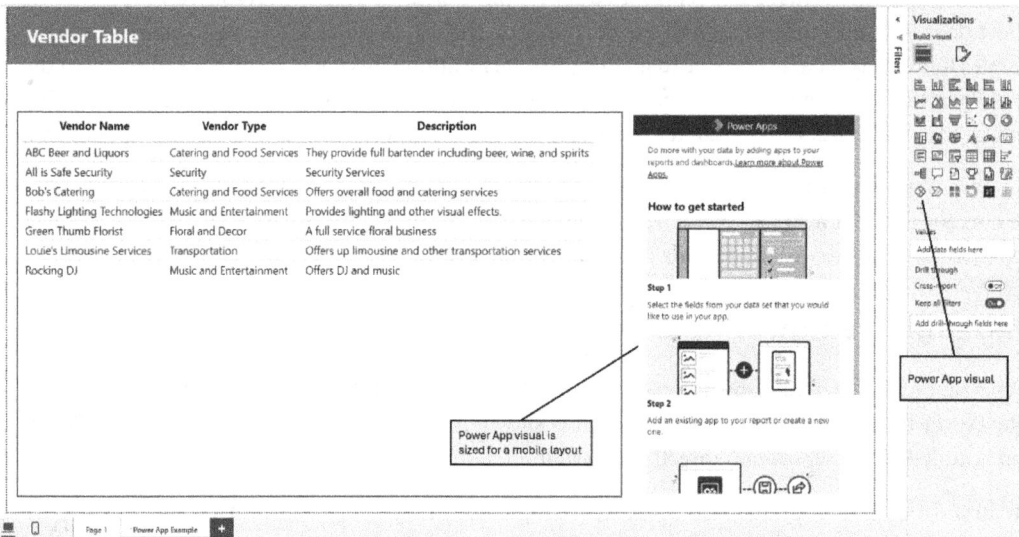

Figure 10.24 - Adding the Power App visual to the Power BI report

When the visual is added, some brief instructions are shown. *Step 1* in the instructions indicates that we need to add fields from our data to be used in the app. For the purposes of this integration, it is irrelevant what field(s) we use. You will see, however, where this becomes important when we move on to scenario 2. But this visual does require at least one field to be added in order to add in a Power App. So, we will simply pick an ID field from one of our tables. This is shown in *Figure 10.25*. You will also note that there is an area where you select the applicable Power Platform environment that is to be used. In our example, the default environment is the correct location.

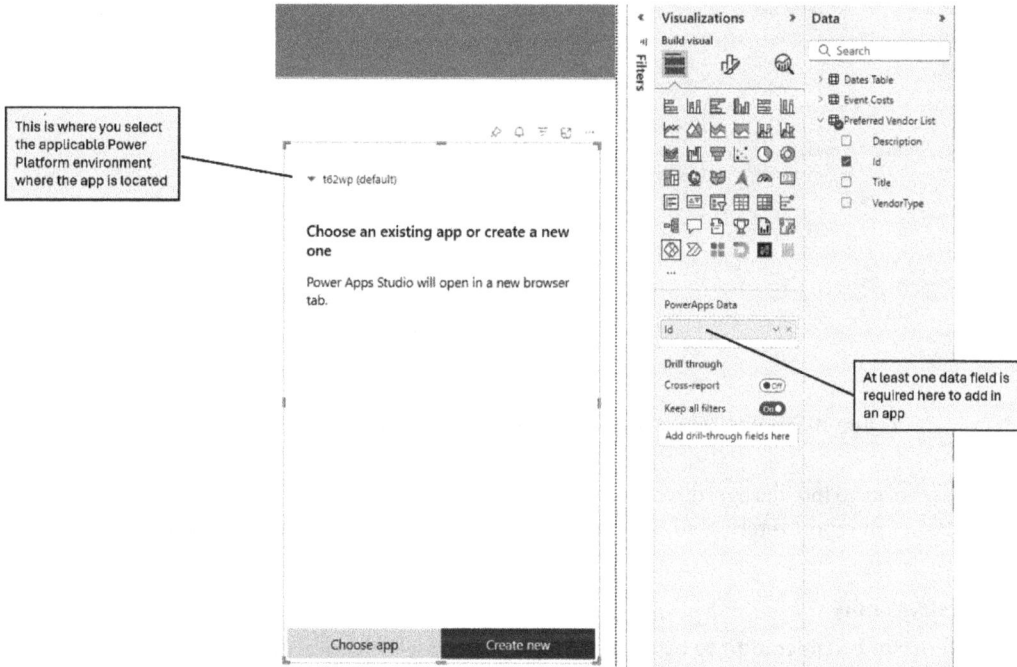

Figure 10.25 - Adding a data field in order to select our app

Because we have already created this app, we will select **Choose app** and then be presented with a list of apps available in the selected environment. We will select the app that we just created and click the **Add** button.

> **Reminder**
>
> It is important to remember that when you create an app, you have to publish the app first in order for it to be available to be selected.

You may be offered the option to go to Power Apps to customize the app. In our example, we'll just cancel this. The end result is the app is added to our report and is fully functional. This is shown in *Figure 10.26*.

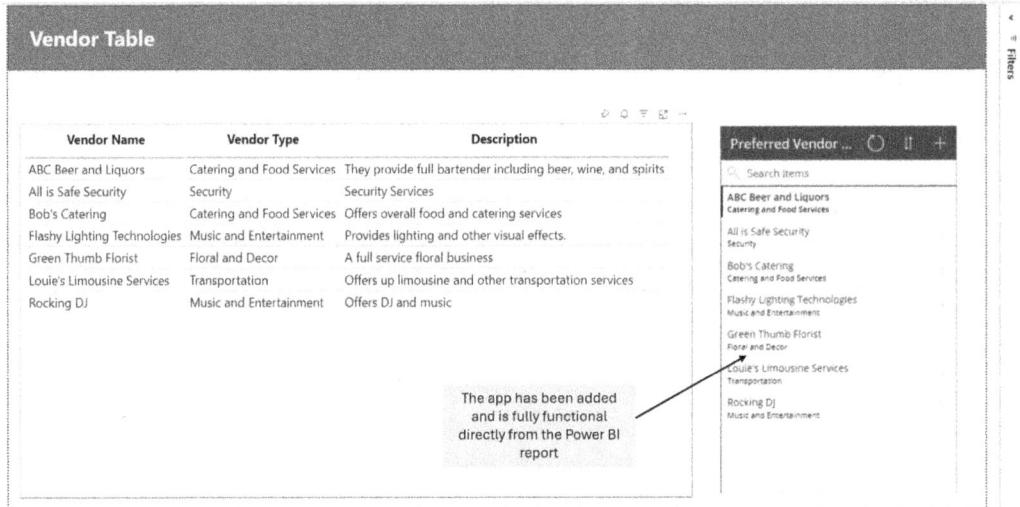

Figure 10.26 - A fully functional Power App embedded in a Power BI report

Now, our users can go into the app, directly from the Power BI report, to make additions and/or changes. Once the Power BI report is refreshed, any changes to the data will be reflected in the Power BI report.

> **Data refresh note**
>
> An important consideration to be aware of when working with embedded Power Apps in a Power BI report is data refreshes. When a Power App is embedded, as shown in *Figure 10.26*, the data connections between the Power App and its underlying data source are direct in that as data is changed in the Power App, those changes will immediately be seen in the Power App. However, the data in any Power BI visuals will not be reflected until the Power BI data is refreshed, unless the data connection method is a direct query. Discussions on Power BI data connections are beyond the scope of this book.

In this scenario, we have embedded a Power App directly within our Power BI report. But the Power BI table on the left and the Power App on the right are two separate independent entities. Wouldn't it be great if we could create an interaction between the two? For example, what if we wanted our users to be able to select a specific vendor from the table on the left and have the app automatically display that record in the app. The great news is that this interaction is available, so let's cover that in scenario 2.

Embedding a Power App improved – scenario 2

Let's build upon the first scenario and add interactivity between our Power BI report and the Power App. We are going to use the same app that was created in scenario 1, with just a few minor changes. But first, we want to discuss the two key items related to the integration. The first is providing a detailed explanation of the *PowerBIIntegration* control. Then we will provide additional information on editing a report in Power BI.

Overview of the PowerBIIntegration control

The key to fully integrating a Power BI report and a Power App, which allows for direct interactivity between the two applications, is a control, which is added to Power Apps through the Power BI visual icon.

Recall that in the previous discussion, and shown in *Figure 10.24*, when developing a Power BI report, there is an icon on the Power BI visualization pane to add a Power App. When this is done, Power Apps will add a new control to the app itself called **PowerBIIntegration**. This is shown in *Figure 10.27*.

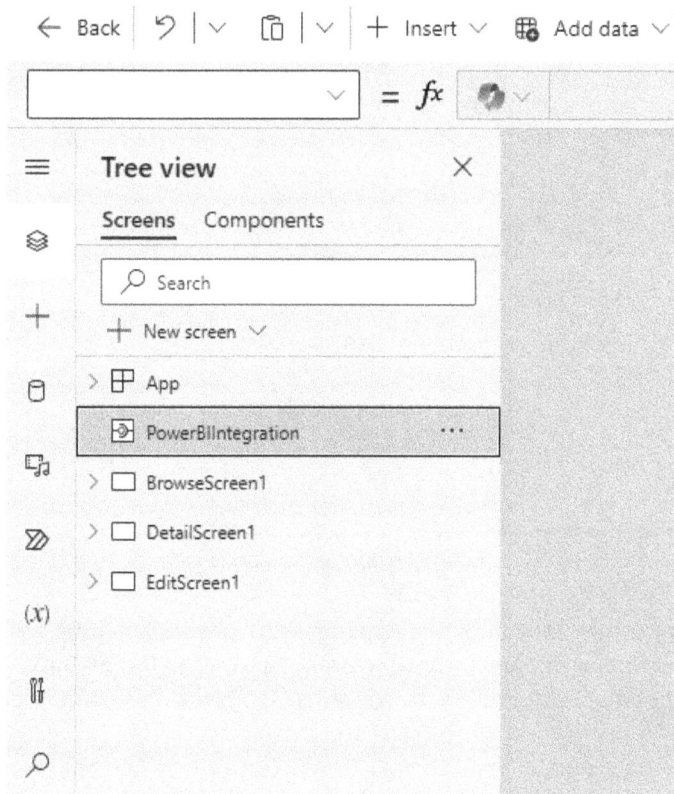

Figure 10.27 - Showing the PowerBIIntegration control in a Power App

This control has a specific property, `PowerBIIntegration.data`, which is a table that holds all the available data from Power BI at any given time. There are two important concepts to be aware of in this table:

- The table will contain all the data fields that were added in Power BI, which are attached to the Power App icon. These fields will not change unless they are updated in the Power App visual.

- The table will contain all the rows of the Power BI table that are available at that time. This information will change dynamically as the Power BI report is interacted with.

Sound confusing? Let's provide some visual examples to help clarify this. We're going to take a step back and start fresh in Power BI. We are displaying in our report a standard table visual that shows seven records. This is shown in *Figure 10.28*. Note that in this example, we've added the **Id** column, which displays a numerical ID for each record. Keep this field in mind for now!

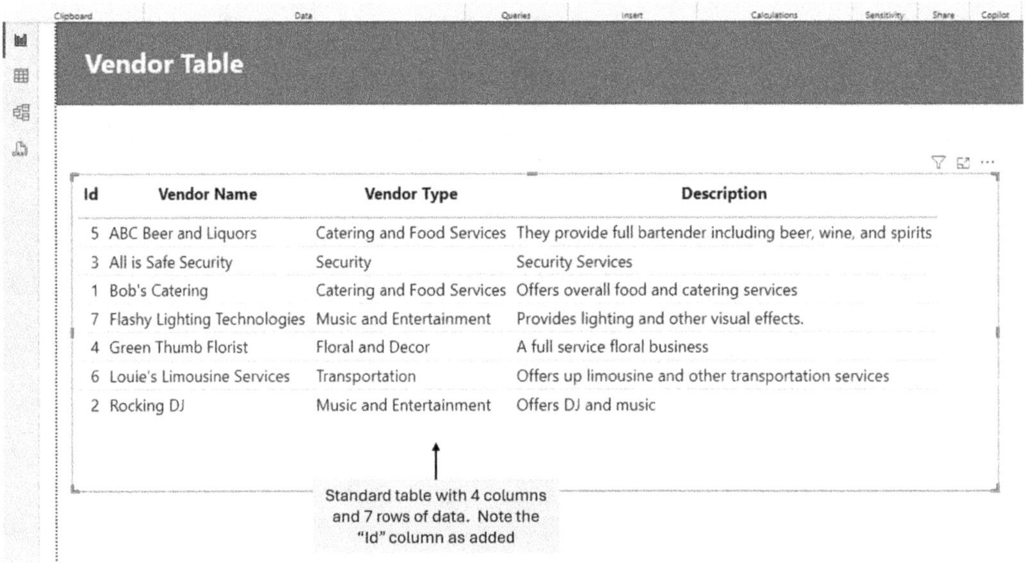

Figure 10.28 - A Power BI report with a simple table visual

We will add in our Power App using the Power App icon in the visualization pane. For the data fields, we will add the same columns that are in our table visual. However, in this instance, we're creating a new one (and using the default environment, as we won't be saving this). This is shown in *Figure 10.29*:

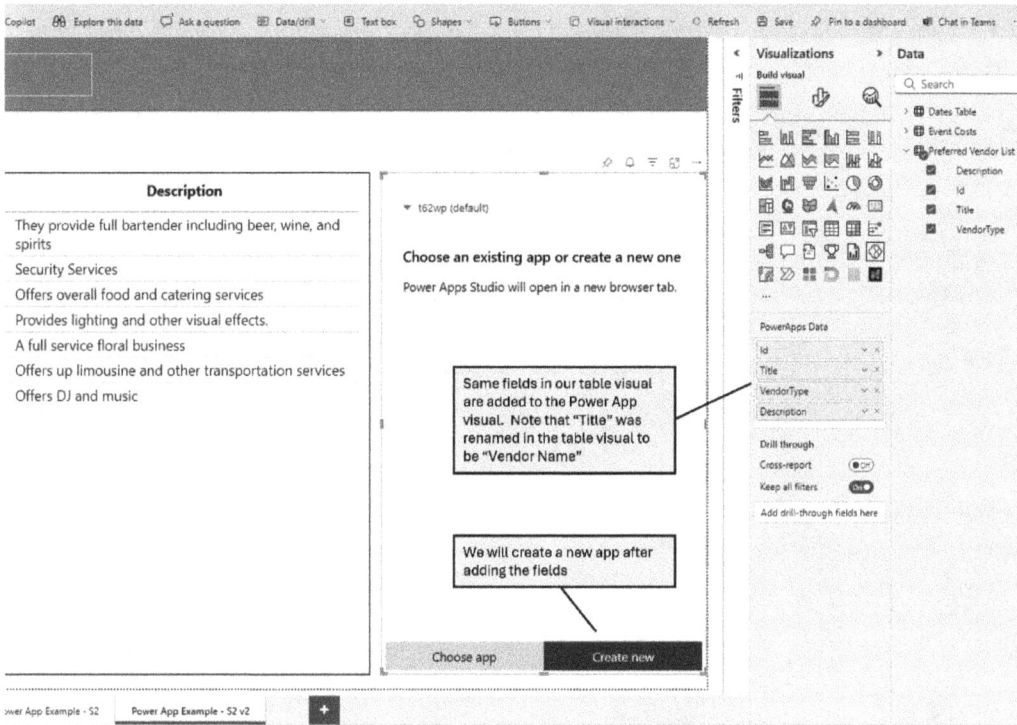

Figure 10.29 - Our Power App visual in which all necessary fields have been added

Power Apps will then create a new app with one screen. In addition, an initial gallery, which is already connected to our Power BI report through the **PowerBIIntegration** control, will be added. In our example, all seven data elements from our Power BI table appear in this gallery. By selecting **Gallery1** and reviewing the **Items** property, we can see that the source is **'PowerBIIntegration'.Data**. This is shown in *Figure 10.30*:

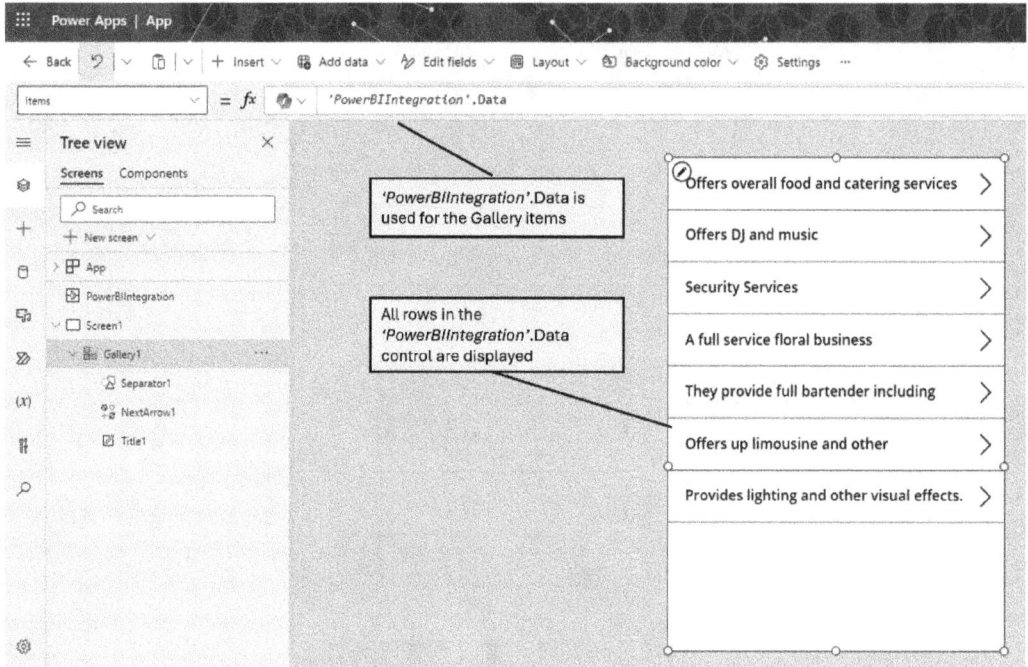

Figure 10.30 - The gallery displaying items from the 'PowerBIIntegration'.Data control

You can further see the available data anytime in the **'PowerBIIntegration'.Data** control by clicking on the **Data** portion and then expanding the table. See *Figure 10.31*:

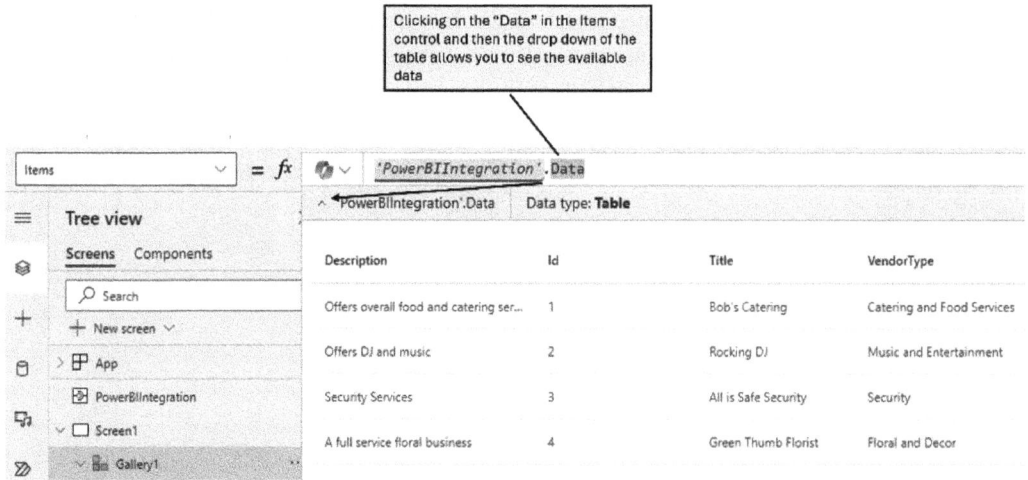

Figure 10.31 - Viewing the data elements in the 'PowerBIIntegration'.Data control

> **Note**
>
> In this section, we are discussing the **PowerBIIntegration** control. In *Chapter 11*, we will discuss integration with custom SharePoint list forms and will be introducing the **SharePointIntegration** control. It is important to note that the **PowerBIIntegration** control currently has no visible properties. However, as you will see in *Chapter 11*, there are visible properties that will be discussed.

Now that we understand the concept of the **PowerBIIntegration** control, it is time to return to improving our app.

Editing the Power BI report

When editing a Power BI report, there are two approaches that can be used. The first is using Power BI Desktop, which is what most Power BI developers use. However, once a report is published to a workspace in the Power BI service, whether that is to one's own personal *My Workspace* or another workspace, the published report can also be edited in the Power BI service. Just be aware that while you can edit a report in the Power BI service, you will have limited functionality compared to editing a report in Power BI Desktop. But see this important note regarding Power App visuals.

> **Important note on creating Power App visuals**
>
> At the time of writing this book, to embed a Power App in a Power BI report and use the **PowerBIIntegration** control, you must do the editing in the Power BI service and not Power BI Desktop. Although a **PowerBIIntegration** control will be created when adding a Power App visual to a Power BI report on Power BI Desktop, no data will be passed to the control and thus integration with your app may not work. This approach is not required for scenario 1, described previously, as the **PowerBIIntegration** control isn't used.

Adding Power BI interactivity to our app

Now that we have covered the **PowerBIIntegration** control as well as the two different approaches to editing the report, let's move on to the details of this scenario and adding interactivity between our Power BI report and our Power App. We are going to use the app that was previously created in scenario 1 and make just a few adjustments.

First, we are going to return to our published Power BI report in the Power BI service. We are going to edit our report by clicking on the **Edit** option above our report. This is shown in *Figure 10.32*:

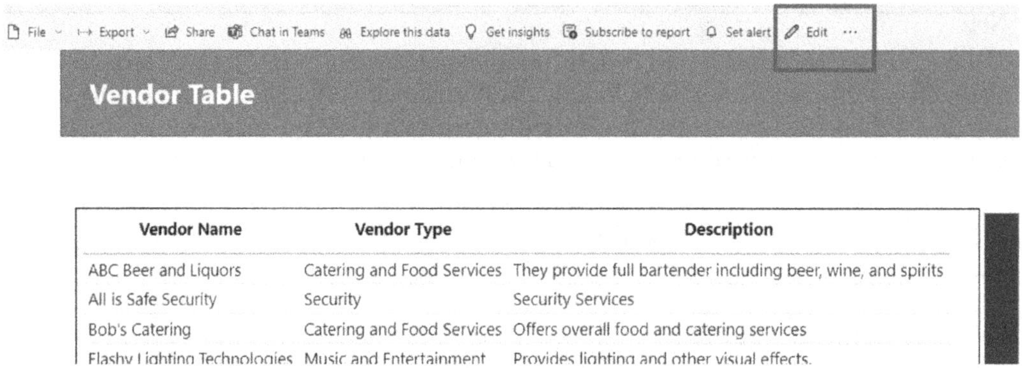

Figure 10.32 - Editing a published Power BI report

This will put our report in edit mode. We first want to ensure that all the applicable Power BI data fields that we need to pass to our **PowerBIIntegration** control, as discussed previously, are added to this visual. In our example, the only field that is needed is the **Id** field, as this field represents the ID associated with our SharePoint list item. This is shown in *Figure 10.33*:

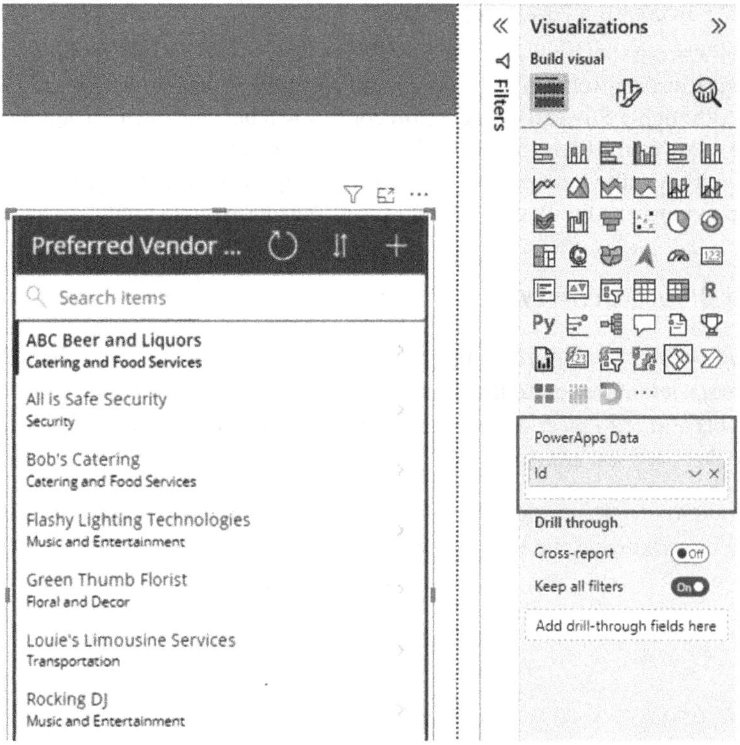

Figure 10.33 - Adding the applicable Power BI data field to our Power App

We will now go and edit the app by clicking on the ellipses (...) located in the top-right corner, just above the visual. Clicking on this will provide an **Edit** option. This will open up the current app in Power Apps.

We're going to make the following adjustments:

Step 1: Either delete the first two screens in the app (ours are named `BrowseScreen1` and `DetailScreen1`) or just move `EditScreen1` to be the first screen. This ensures that this screen is opened up first.

Step 2: Find the form that is contained in `EditScreen1`. In our example, the default form name is `EditForm1`. Find the `Item` property and replace the existing code with the following:

```
LookUp('Preferred Vendor List', ID=First(PowerBIIntegration.
Data).Id)
```

Here, we are using the `LookUp` function to find the first item in our `PowerBIIntegration.Data` table from our SharePoint list. As users select different records in our Power BI table, this will dynamically update `PowerBIIntegration.Data` and thus update which item is displayed.

Step 3: In the top-right corner of the app is a checkmark, which is used to submit the form and save any changes. We'll simply replace the default code with the following:

```
SubmitForm(EditForm1)
```

Step 4: Lastly, we'll copy the + icon from `BrowseScreen1` and paste it next to the checkmark icon. This will allow us to add new records. For the `OnSelect` property, we will use the following:

```
NewForm(EditForm1)
```

This will clear the current form and allow for new entries to be made.

Step 5: Lastly, we'll save and publish this app. We'll then return to our Power BI report. You will need to reload the app, but this can simply be done by moving to a different page on your Power BI report and then returning to the page where the app is embedded. You may need to wait a few minutes for the updated report to appear.

With these minor changes made, as different records in the Power BI table visual are selected, the Power App will update dynamically with the new information. An example of this is shown in *Figure 10.34*:

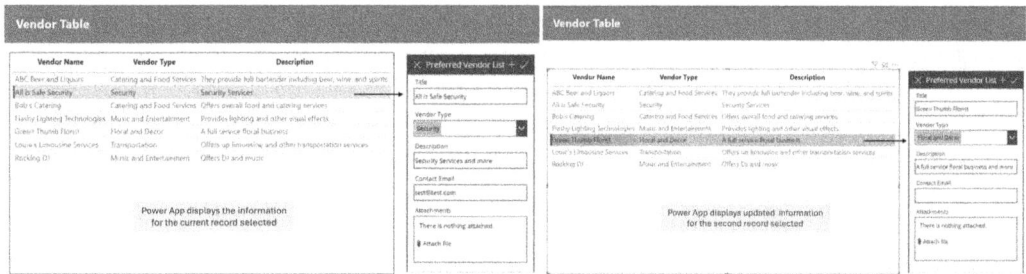

Figure 10.34 - Comparison showing the Power App displaying different information dynamically as different Power BI data is selected

> **Development tip**
>
> It is important to note that the Power App could be customized better to add additional functionality. For example, a confirmation screen could be added to the app for a better user experience. The intent of the preceding instructions was to provide a simple foundation for how the integration between Power BI and Power Apps, using the **PowerBIIntegration** control, is done.

There is one last important point for you to consider as you progress on this journey of integrating a Power App in a Power BI report. In the example just provided, we modified an existing app that was automatically generated. However, it is also possible to start with a completely new app and develop your own customized app that includes the **PowerBIIntegration** control.

Important considerations and best practices

When embedding Power BI reports within your app, there are a number of important items to consider. What follows are some key areas for you to be aware of:

- **Ensure appropriate access**: Both Power Apps and Power BI have requirements around who can access the applications. Because Power BI is separate from Power Apps, you need to ensure that any users of your Power App who need to view the Power BI visuals have the appropriate access and any required licenses.

- **Optimize report performance**: When embedding the full Power BI reports into your app, it is a best practice to ensure that your Power BI report is optimized for performance. Optimizing Power BI reports is beyond the scope of this book.

- **Maintain data security**: Power BI does allow for the use of **row-level security** (**RLS**), which ensures that users only see the data they are authorized to view. This functionality also flows down to reports that are embedded in your Power App. It is important to always test your app to ensure you have incorporated any necessary security requirements.

- **Consider cross-team collaboration**: Power Apps and Power BI are two separate applications and thus it might be common for organizations to have different development teams. As a result, close collaboration is important. It may also be possible to leverage expertise from both areas, as the Power App can be developed independently of the Power BI report.

- **Maintain good technical documentation**: Integrating Power Apps and Power BI does add complexities to the overall application. As a result, it is critical to maintain good technical documentation. It is inevitable that you will need to make improvements down the road, troubleshoot problems, or even transition the application to another developer. Good technical documentation creation in the beginning will save countless hours down the road.

Summary

Embedding Power BI reports in Power Apps enhances the capabilities of your custom applications by providing real-time, interactive data insights within the app interface. This integration not only improves user experience but also drives better decision-making through seamless access to vital information. By following the steps outlined in this chapter, you can effectively embed Power BI reports in your Power Apps, creating a powerful tool for your organization.

Furthermore, as we have shown, there is also the capability to embed a Power App directly within a Power BI report. This opens further possibilities by combining the capabilities of Power BI, for improved analytics and data visualization, with the functionality of Power Apps, to enable your users to create and/or change data.

In the next chapter, we will move on to integrating Power Apps with SharePoint.

11

Integrating Power Apps with SharePoint

In the previous chapter, we covered integrating Power BI visuals into Power Apps and Power Apps into Power BI, which allows you to add better analytics. In this chapter, we are going to discuss integrating Power Apps with SharePoint. We previously discussed connecting to SharePoint in *Chapter 5*. However, we are going to dive into additional areas.

Integrating Power Apps with SharePoint provides a great way to enhance the functionality and user experience of a SharePoint site. By embedding a Power App into SharePoint, you can create customized, responsive applications that seamlessly interact with your SharePoint data and services. Furthermore, because SharePoint sites are widely used across organizations to provide information and maintain documents, making any related apps directly accessible on the site minimizes any extra clicking that users need to do to access your app.

In this chapter, we are going to cover the following topics:

- Embedding Power Apps into a SharePoint site
- Creating a Custom SharePoint list form with Power Apps
- Understanding the `SharePointIntegration` item that integrates SharePoint and a custom Power App form

Technical requirements

To successfully engage with the materials in this chapter, you'll need the following:

- **Microsoft Power Apps**: Access to Power Apps for creating and managing your applications.
- **Microsoft SharePoint**: You will need to have access to a SharePoint site (or the ability to create one) including appropriate edit rights. If you cannot edit SharePoint pages, you may need to contact your SharePoint administrator for assistance.

Embedding Power Apps in a SharePoint site

Back in *Chapter 5*, we discussed connecting to SharePoint lists as a data source. However, SharePoint is much more than just a repository of lists. In this section, we will first provide an overview of SharePoint, including the benefits of integrating Power Apps with SharePoint. We will then cover how to embed a Power App directly in a SharePoint page. Before we get too far, let's first provide a quick overview of SharePoint.

Overview of SharePoint

SharePoint is a popular web-based collaboration platform that serves as a central area for storing, organizing, sharing, and accessing information. At its core, SharePoint offers robust document management capabilities, allowing users to create, upload, and manage documents within libraries, ensuring that critical information is easily accessible and securely stored.

Beyond document management, SharePoint provides a wide range of features designed to improve collaboration and communication within a team or organization. Users can create and manage sites, which act as customizable portals for teams, projects, or departments, fostering a collaborative environment where members can share documents, calendars, tasks, and other resources. SharePoint also integrates with Microsoft 365, which extends its functionality, allowing seamless connectivity with tools such as Microsoft Teams, OneDrive, and Outlook. It also has powerful search capabilities, workflow automation, and extensive customization options, SharePoint helps organizations enhance productivity, maintain control over their data, and drive efficient business processes. Let's next discuss the benefits of integrating a Power App into a SharePoint site.

Benefits of integrating Power Apps with SharePoint

Integrating Power Apps with SharePoint offers a multitude of benefits that can significantly enhance an organization's efficiency and collaboration capabilities. One of the primary advantages is the centralized storage and management of documents and data. By integrating Power Apps with SharePoint, organizations can create custom applications that seamlessly interact with SharePoint lists and libraries, enabling users to easily access, update, and manage information in real time. This integration reduces redundancy, ensures data consistency, and provides a single source of truth for critical business information.

Another significant benefit is the enhanced collaboration and communication facilitated by SharePoint's robust features. Integrated solutions allow teams to work together more effectively by providing shared access to resources, streamlined workflows, and automated processes. With Power Apps, users can develop tailored applications that address specific business needs, automate routine tasks, and improve overall productivity. Additionally, SharePoint's security and compliance features ensure that data is protected and access is controlled, allowing organizations to maintain compliance with industry standards and regulations. The combination of Power Apps and SharePoint empowers organizations to create dynamic, user-friendly solutions that drive innovation, improve decision-making, and foster a collaborative digital workplace.

Lastly, one additional benefit is minimizing how users access your app. One of the biggest challenges in Power Apps is providing easy access. The more mouse clicks users have to perform, or web links to remember or bookmark, the less likely your app will be used. So, easy accessibility should be a high priority. Let's provide an example of this in action.

Example of an embedded Power App on a SharePoint page

Now that we have covered the benefits, let's jump right into adding your Power App to a SharePoint site. *Figure 11.1* shows an example of a SharePoint page that has an embedded Power App. This app is fully functional directly on the site.

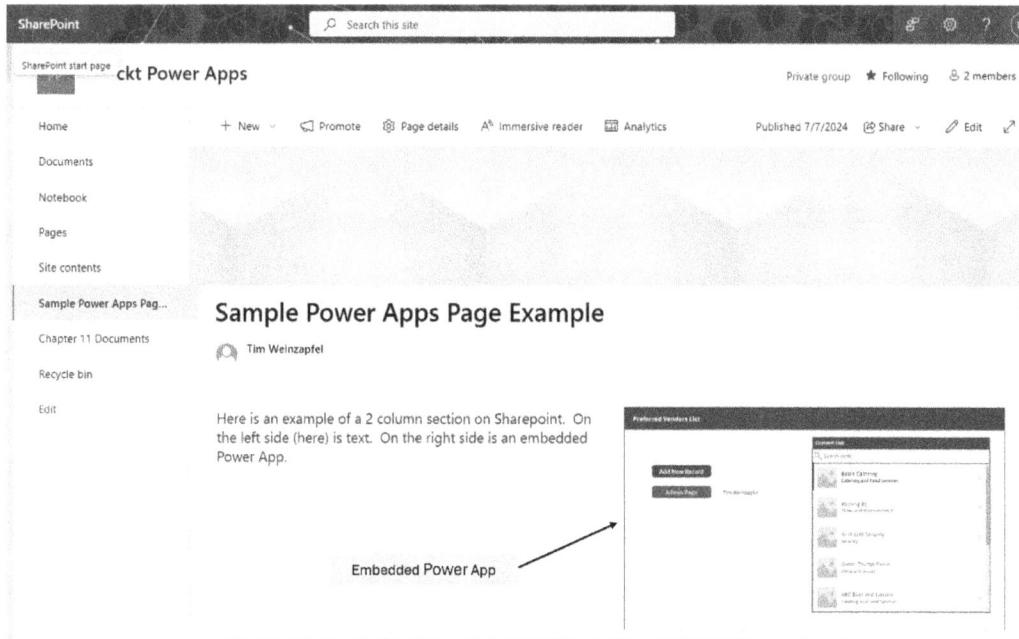

Figure 11.1 – Example of an embedded Power App

One advantage to this approach is that the Power App is immediately available to users. Let's dive into specifically setting this up. Even if you are new to SharePoint, you will see just how simple this is to do.

Adding a Power App to a SharePoint site

Adding a Power App to a SharePoint page is easy. However, to be able to do this, you will need to have the appropriate access to your respective SharePoint site to create and edit pages. If you do not have this ability, you will need to contact your SharePoint administrator for the proper access.

In our example, we are going to begin with a new SharePoint page. This can be accessed in SharePoint by creating a new page. This can be done by clicking on the **+ New** option, as shown in *Figure 11.2*.

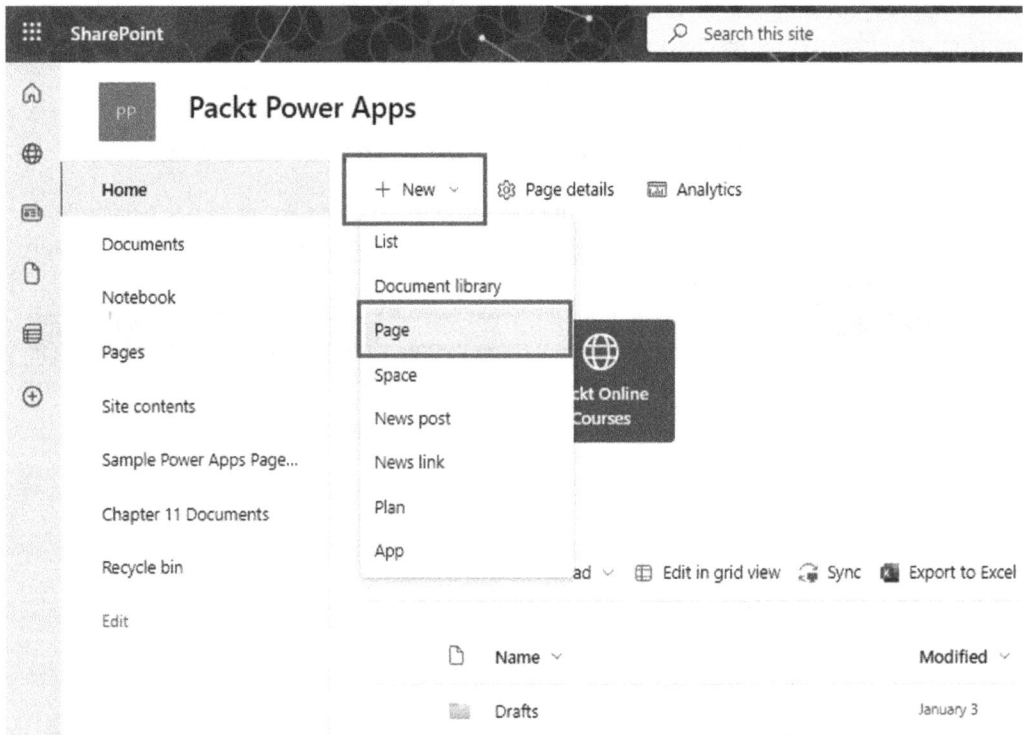

Figure 11.2 – Adding a new SharePoint page

SharePoint offers various templates. In our example, we'll pick a standard **Blank** one, as shown in *Figure 11.3*.

Page templates

From Microsoft   Saved on this site

Standard

Templates that contain the full range of web parts for general use.

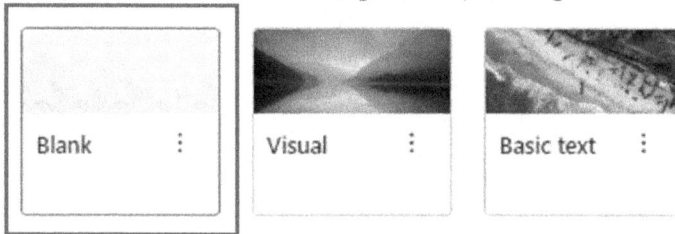

Blank ⋮

Visual ⋮

Basic text ⋮

Video

Templates for highlighting videos.

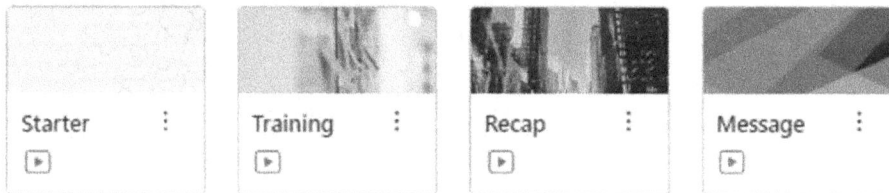

Starter ⋮
▶

Training ⋮
▶

Recap ⋮
▶

Message ⋮
▶

Figure 11.3 – Picking a SharePoint page template

We are now in the editing mode of the SharePoint page, which, in our example, includes an initial one-column section with a blank text box. We'll start by adding a simple page title.

Next, we will modify the first section and change this from a one-column section to a two-column section. To do this, click on any blank area within the section to select it. Four icons, with their actions identified, should appear on the left side, which will appear as shown in *Figure 11.4*.

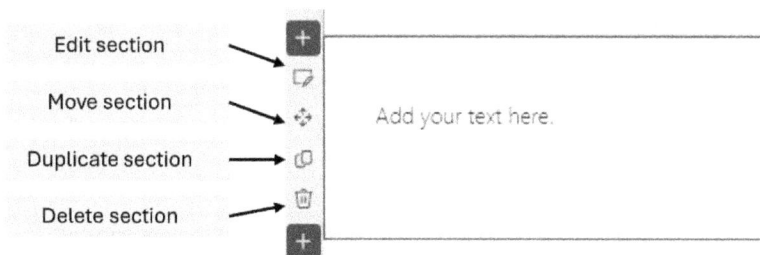

Edit section

Move section

Duplicate section

Delete section

Add your text here.

Figure 11.4 – Options specific to the SharePoint section

In this case, we'll choose the top one, **Edit section**, and change this to **Two columns**. This is shown in *Figure 11.5*.

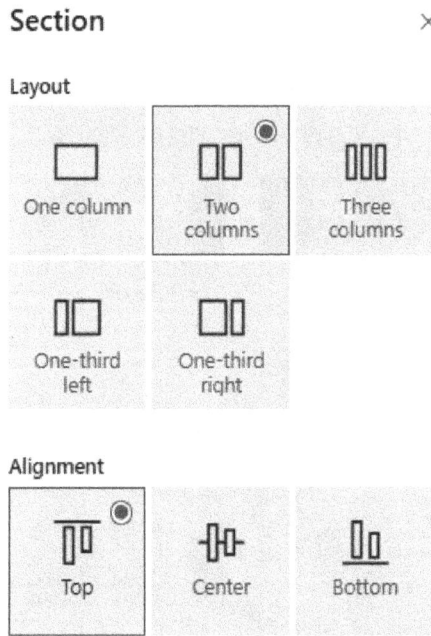

Figure 11.5 – Changing the SharePoint section to Two columns

What appears now is the default text box on the left and an open section on the right. We'll use the right section to embed our Power App. To add this, we move our cursor near the top of the section, which should show the ability to add a SharePoint web part. See *Figure 11.6* for the general location for this.

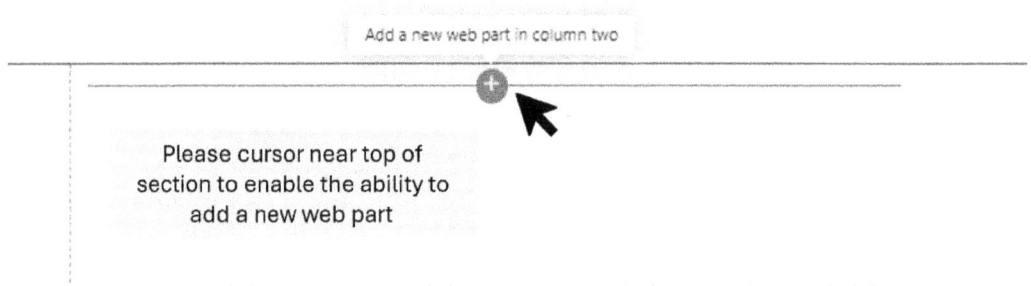

Figure 11.6 – Moving the cursor to enable the ability to add a new web part

Clicking on this will allow you to add different web parts. We want to find **Microsoft PowerApps**. There is a search bar available if it doesn't appear. See *Figure 11.7*.

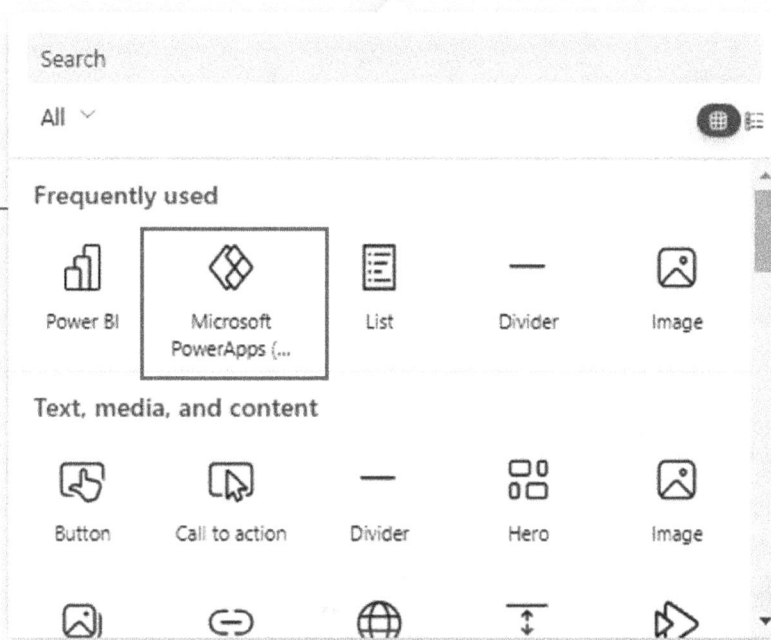

Figure 11.7 – Finding the Microsoft PowerApps web part

This will open up an option for you to enter either the web link or the ID of the app, as shown in *Figure 11.8*.

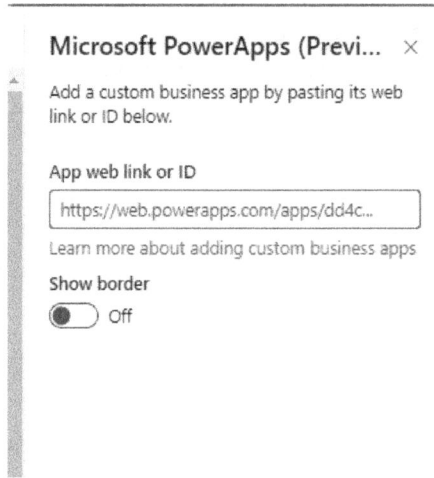

Figure 11.8 – The App web link or ID area

If you do not know where to obtain these items, they are easily found by going into the app settings and both the web link and the App ID are shown. Let's walk through how to find this.

First, when viewing your apps, click on the commands option represented by the vertical ellipses. This will allow you to access the app commands. This is shown in *Figure 11.9*.

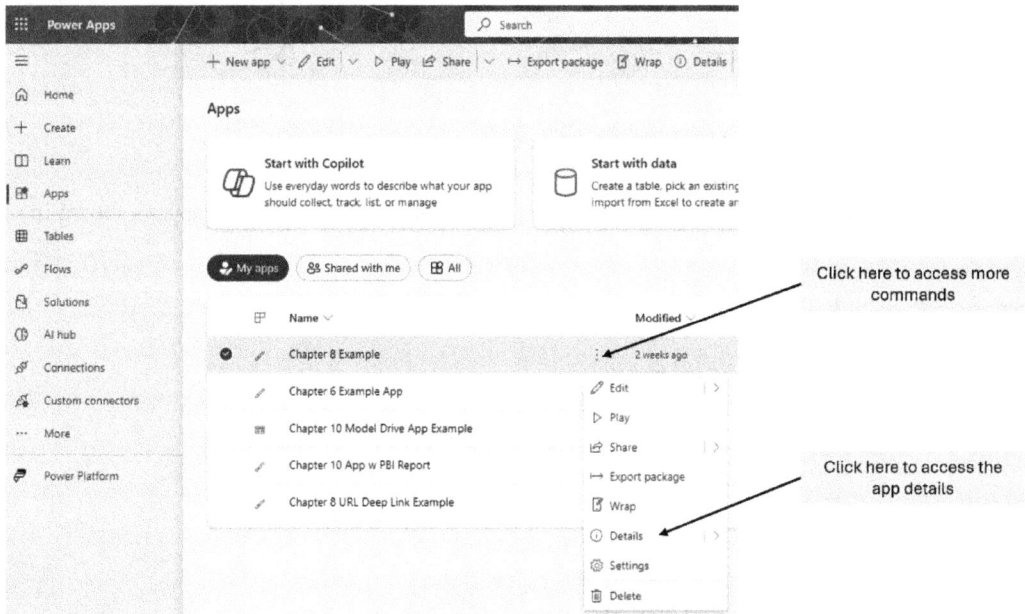

Figure 11.9 – Accessing the app commands

Next, click on **Details** within the list of available commands. This will bring up information about your app. Of importance will be either **Web link** or **App ID**. See *Figure 11.10* for an example.

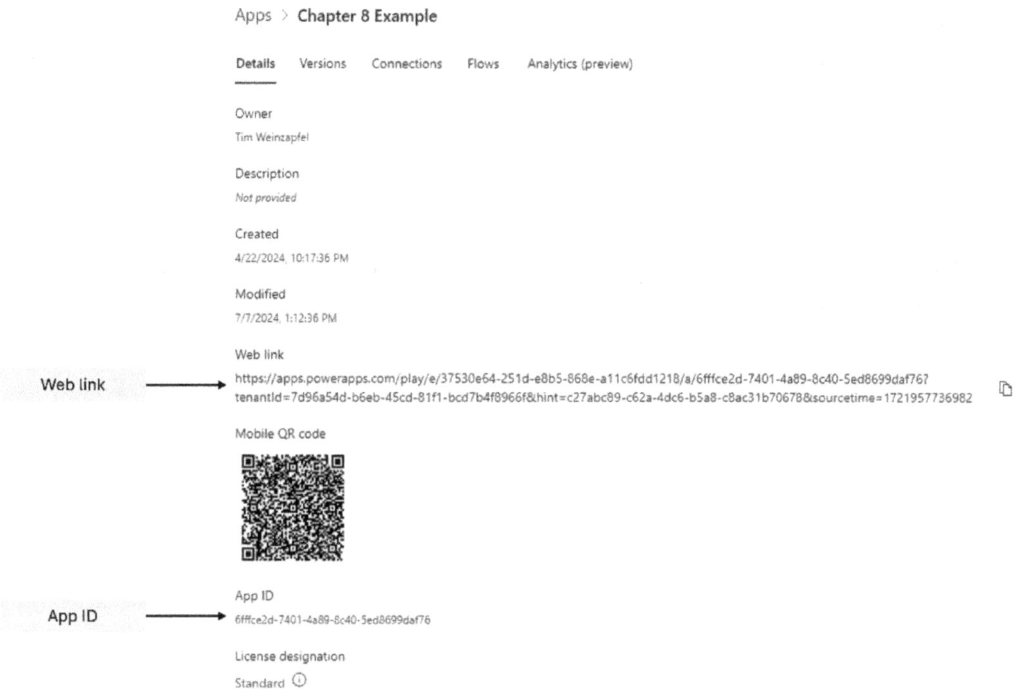

Apps > **Chapter 8 Example**

Details   Versions   Connections   Flows   Analytics (preview)

Owner
Tim Weinzapfel

Description
Not provided

Created
4/22/2024, 10:17:36 PM

Modified
7/7/2024, 1:12:36 PM

Web link

Web link ————————► https://apps.powerapps.com/play/e/37530e64-251d-e8b5-868e-a11c6fdd1218/a/6fffce2d-7401-4a89-8c40-5ed8699daf76?
tenantId=7d96a54d-b6eb-45cd-81f1-bcd7b4f8966f&hint=c27abc89-c62a-4dc6-b5a8-c8ac31b706788&sourcetime=1721957736982

Mobile QR code

App ID

App ID ————————► 6fffce2d-7401-4a89-8c40-5ed8699daf76

License designation
Standard ⓘ

Figure 11.10 – Finding the app web link and app ID from Details

Copy either the web link or the app ID and paste it into the box that was shown in *Figure 11.8*. Your app will appear with the most recently published version. Furthermore, as newer versions of the app are published, they will appear here without any further changes needed.

> **Tip**
> Be aware that when embedding a Power App into SharePoint, the most current published version of the app will appear. It is very common to make updates to the app and save them, but forget to publish the updated version.

That is all there is to embed a Power App into a SharePoint page. Let's now move to another way to integrate Power Apps with SharePoint and that is by creating custom SharePoint list forms.

Creating a custom SharePoint list form with Power Apps

In *Chapter 5*, we covered using SharePoint lists as a popular data source for Power Apps. This allows you to create a Power App where the underlying data for the app is maintained in a SharePoint list. However, there is a different way to integrate Power Apps with a list, and that involves creating a custom SharePoint list form. Let's first cover what a SharePoint list form is.

Overview of SharePoint list forms

When using a SharePoint list, there are options to add new records or edit existing records. This is then done by filling out a default SharePoint form where the new or updated data is done. Let's provide an example by displaying a simple SharePoint list of preferred vendors, as shown in *Figure 11.11*.

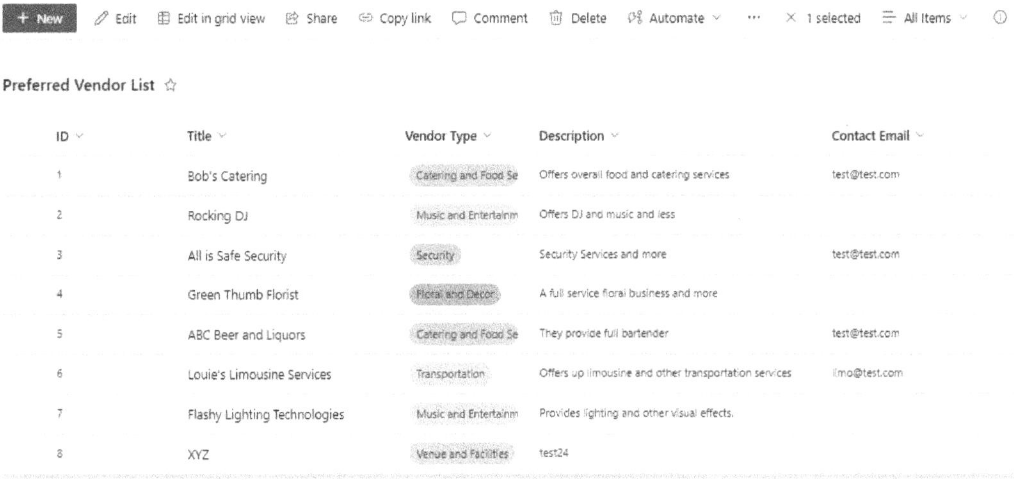
+ New
Edit
Edit in grid view
Share
Copy link
Comment
Delete
Automate
1 selected
All Items
Preferred Vendor List
ID
Title
Vendor Type
Description
Contact Email
1
Bob's Catering
Catering and Food Se
Offers overall food and catering services
test@test.com
2
Rocking DJ
Music and Entertainm
Offers DJ and music and less
3
All is Safe Security
Security
Security Services and more
test@test.com
4
Green Thumb Florist
Floral and Decor
A full service floral business and more
5
ABC Beer and Liquors
Catering and Food Se
They provide full bartender
test@test.com
6
Louie's Limousine Services
Transportation
Offers up limousine and other transportation services
limo@test.com
7
Flashy Lighting Technologies
Music and Entertainm
Provides lighting and other visual effects.
8
XYZ
Venue and Facilities
test24

Figure 11.11 – A sample SharePoint list

Clicking on the green **+ New** button opens up a default form that allows one to add a new record. Because our list has several custom columns, such as **Vendor Type**, **Description**, and **Contact Email**, these fields appear in the form. An example of this is shown in *Figure 11.12*.

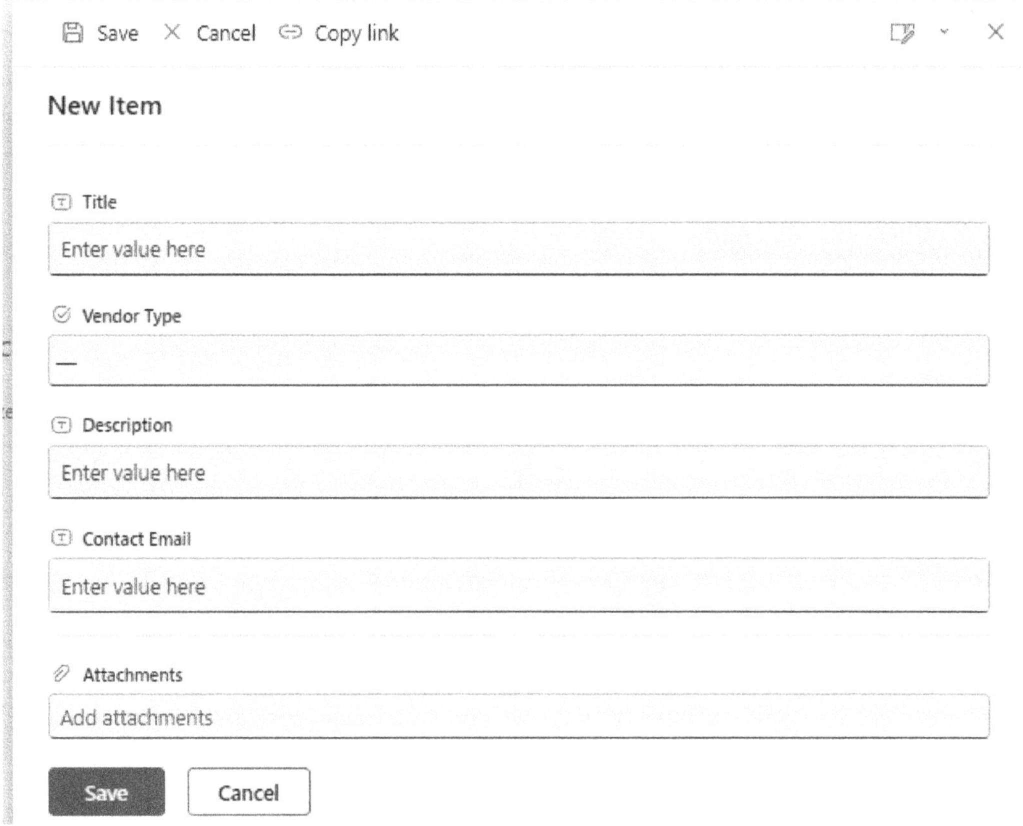

Figure 11.12 – Example of the default SharePoint form to add a new record

However, wouldn't it be great if you could customize this form, including changing the order of the fields, how they are laid out, and even changing other features such as the overall appearance and implementing data validation rules? The great news is that this is possible. *Figure 11.13* provides an example of a customized form. This is shown on the left side of the image. For comparison, we have added the standard form on the right.

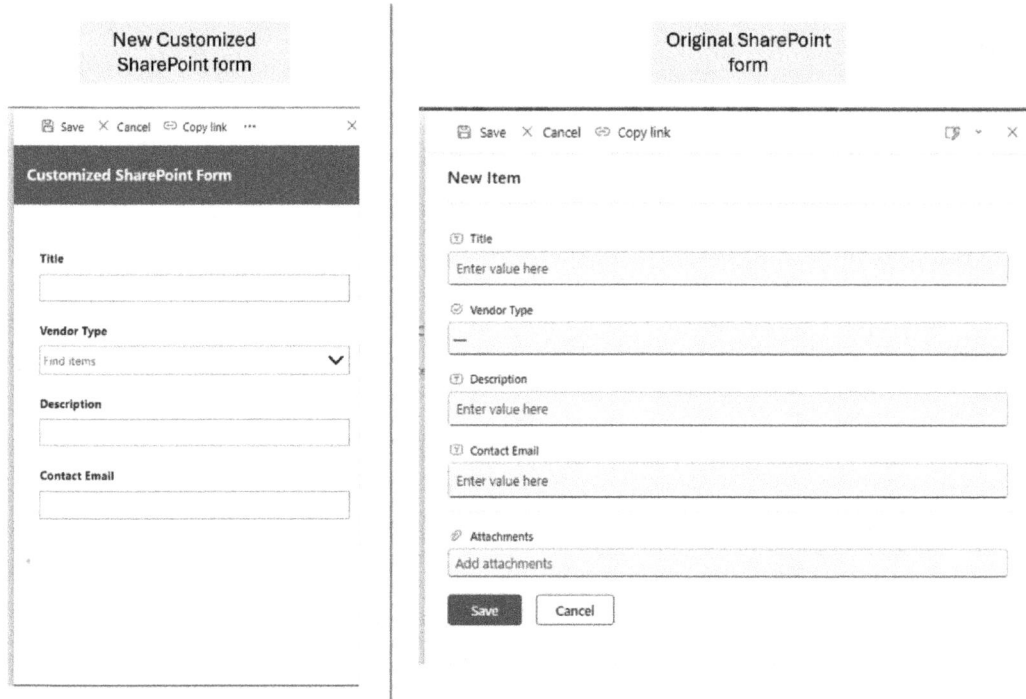

Figure 11.13 – Example of a customized SharePoint form compared to the original SharePoint form

Several notable differences include the colored banner at the top titled **Customized SharePoint Form**. In addition, not all the same fields as the standard form are showing. The **Attachments** option has been removed. Lastly, the **Save** and **Cancel** buttons are located at the top. Now, in this example, the differences aren't that significant. However, the main point is that the form on the left is completely customizable, including how it looks, what fields are displayed, and the order of the fields. Furthermore, we could have taken this further and added additional functionality, such as images and multiple screens. Overall, this opens up more possibilities. Now, let's get into how this is done.

Creating a custom SharePoint list form

Before we get too far into the details, it is important to note that your SharePoint list settings have to be updated to allow **New experience** for the list. This can be found by going to **List settings** and then **Advanced settings**. *Figure 11.14* shows how to access the SharePoint list settings.

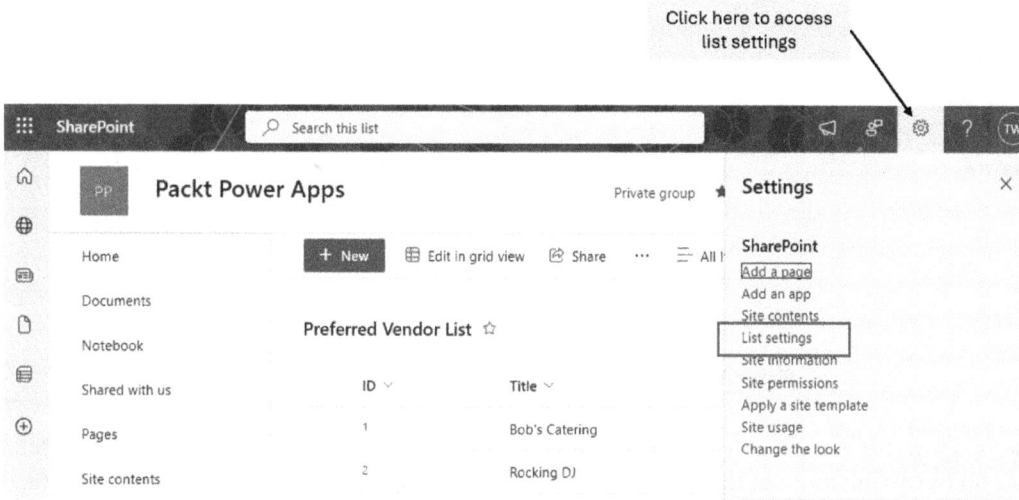

Figure 11.14 – Accessing the SharePoint list settings

Clicking on **List settings** will bring up additional options. From here, you select **Advanced settings** and then scroll down until you see **List experience**. This is shown in *Figure 11.15*.

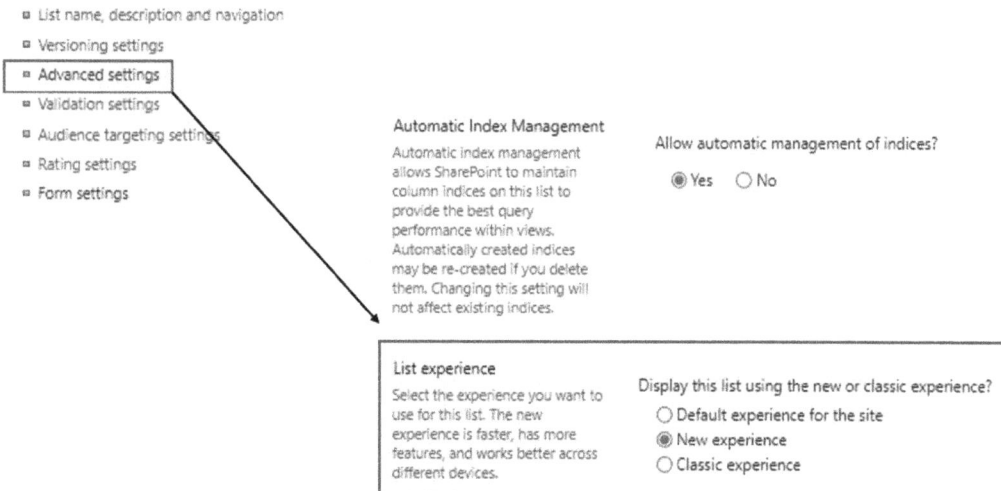

Figure 11.15 – Accessing the List experience settings

Once you have this set, let's head back to our SharePoint list so that we can customize this form. There are several different ways to access the customized form. One way to do this is when you are viewing the SharePoint list. Above the list is a toolbar with an **Integrate** option. Then, select **Power Apps | Customize forms**. This is shown in *Figure 11.16*.

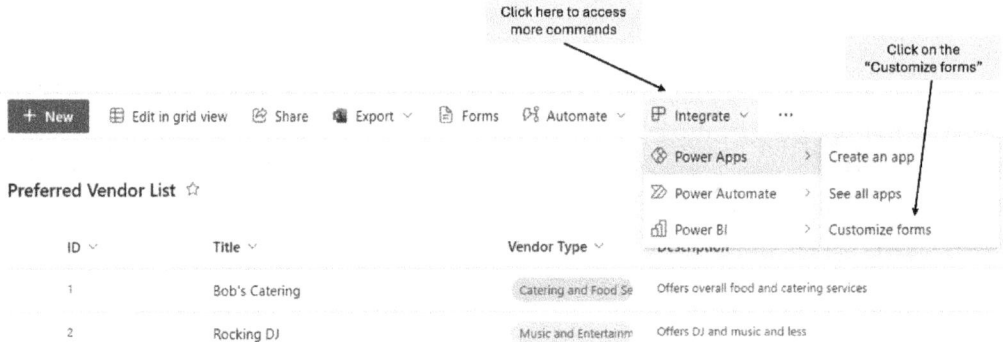

Figure 11.16 – Accessing the Customize forms screen

A second way that you can access the **Customize forms** option is by selecting an item in the list to bring up the default form. In the top-right corner of the default form is an icon that allows you to expand the actions. One of them is **Customize with Power Apps**. See *Figure 11.17* for an example.

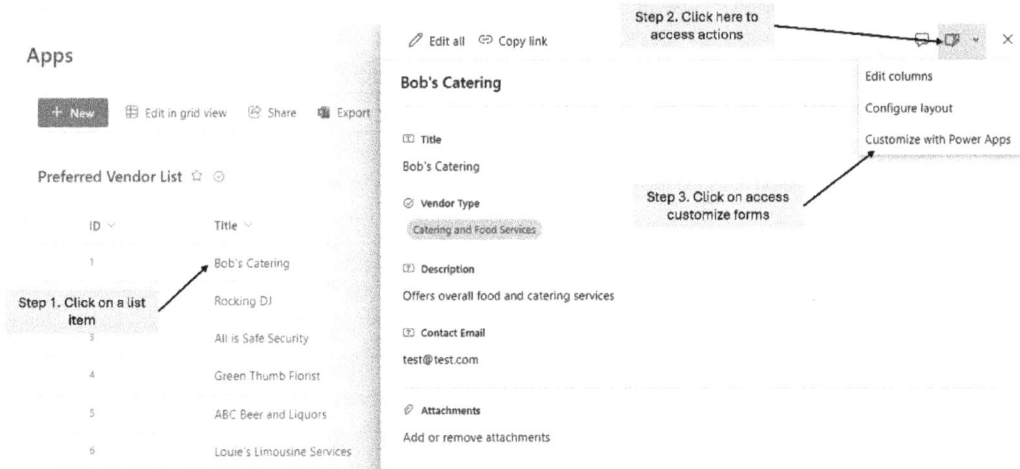

Figure 11.17 – Accessing Customize with Power Apps from the default form

Once you click on the **Customize forms** option through either of the approaches shown in *Figure 11.16* or *11.17*, Power Apps will open up and create a one-screen layout. This is shown in *Figure 11.18*.

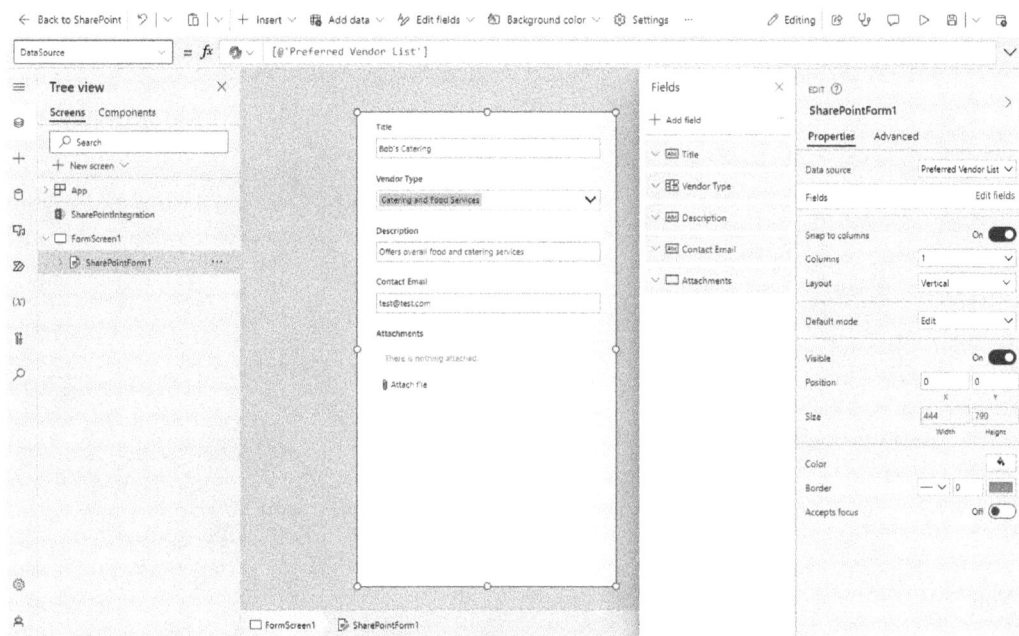

Figure 11.18 – Initial SharePoint list form created in Power Apps

Now, you are free to make updates to this form. This includes changes such as the following:

- Adding or removing fields as well as changing the order
- Changing the overall layout, the number of columns, background color, and so on
- Adding additional items including shapes, buttons, and images

In our example, we will make just a few simple changes. This will include the following:

1. Add a rectangle shape at both the top and bottom.
2. Add a label to indicate that this is a customized form.
3. Remove the **Attachments** field.

Again, because this is being done in Power Apps; you can add significant functionality to this customized form. However, in our situation, we'll keep it simple and proceed to complete the form. To do so, simply click on the **Back to SharePoint** button. You will need to save and publish your changes once prompted, as shown in *Figure 11.19*.

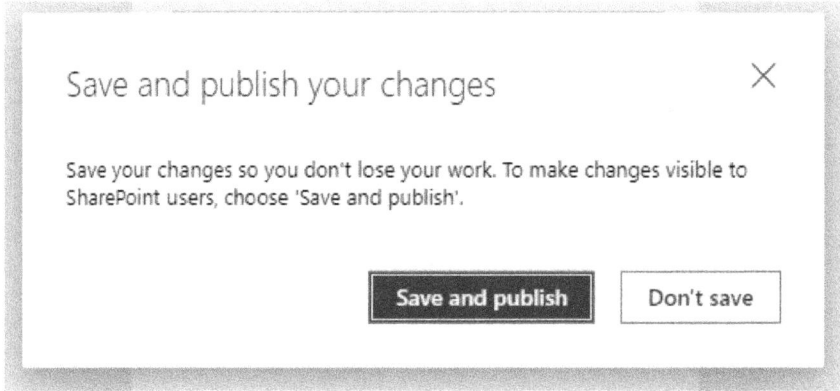

Save and publish your changes   ✕

Save your changes so you don't lose your work. To make changes visible to SharePoint users, choose 'Save and publish'.

[Save and publish]   [Don't save]

Figure 11.19 – Saving and publishing the customized form

> **Note**
>
> Two things to be aware of when going through this process. First, it may take a few minutes for the updated form to be published. In addition, you may need to refresh your web browser to ensure that you aren't viewing a cached version of the form.

When we return to our SharePoint list and either add a new record or edit an existing one, the new customized form is now available. This is shown in *Figure 11.20*.

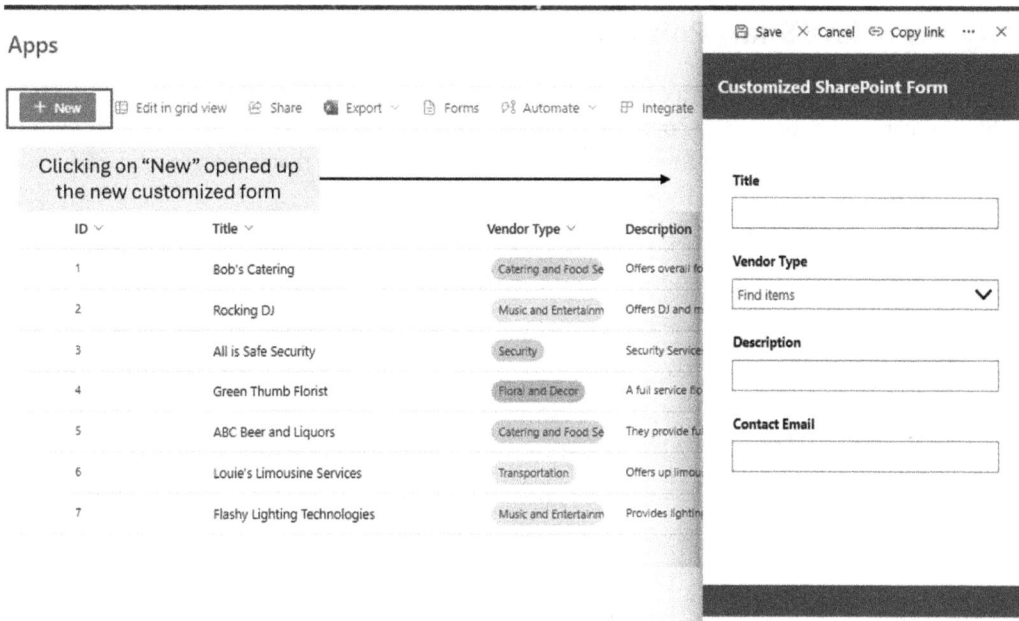

Figure 11.20 – The new customized form is available

> **Note**
> Be aware that although a Power App is created, it is saved in the default environment but will not appear or be readily accessible with other apps. The app can only be located through the SharePoint list. We cover how to edit or delete a custom SharePoint list form further in this chapter.

That is all there is to creating a customized form in your SharePoint list via Power Apps.

Now, one area that you might be thinking about is that when creating the form, there were no buttons added that submit the form (or even cancel the action). How are those actions added? In fact, if you look at *Figure 11.20*, there are options already available at the top of the customized form to save, cancel, and copy links. Those are all addressed via the `SharePointIntegration` item. We will discuss this item in the next section.

Overview of the SharePointIntegration Item

As noted previously, when creating a customized form, there isn't a need to add buttons on the form to submit or cancel the form. These options are already provided at the top of the SharePoint form and are connected to the app via a `SharePointIntegration` item that is integrated. Let's have a look at *Figure 11.21* where this is shown.

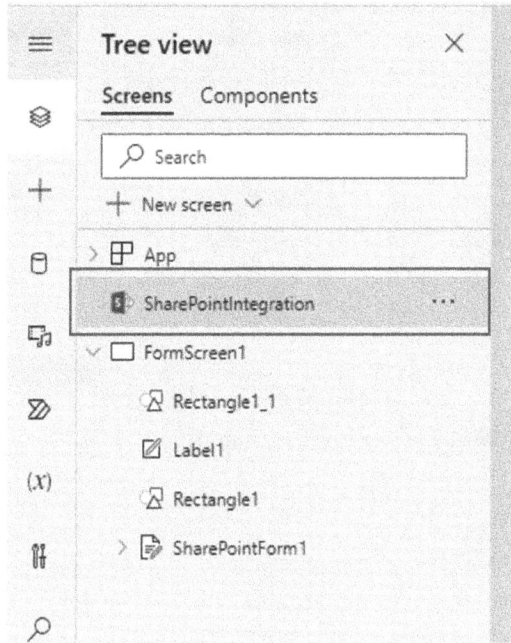

Figure 11.21 – Showing the SharePointIntegration item

Clicking on this and then expanding the properties area will reveal the various options available with this item. *Figure 11.22* provides the available list of properties in the `SharePointIntegration` control and we have displayed the default functions already embedded in several of the properties. You can see how these functions are already built in.

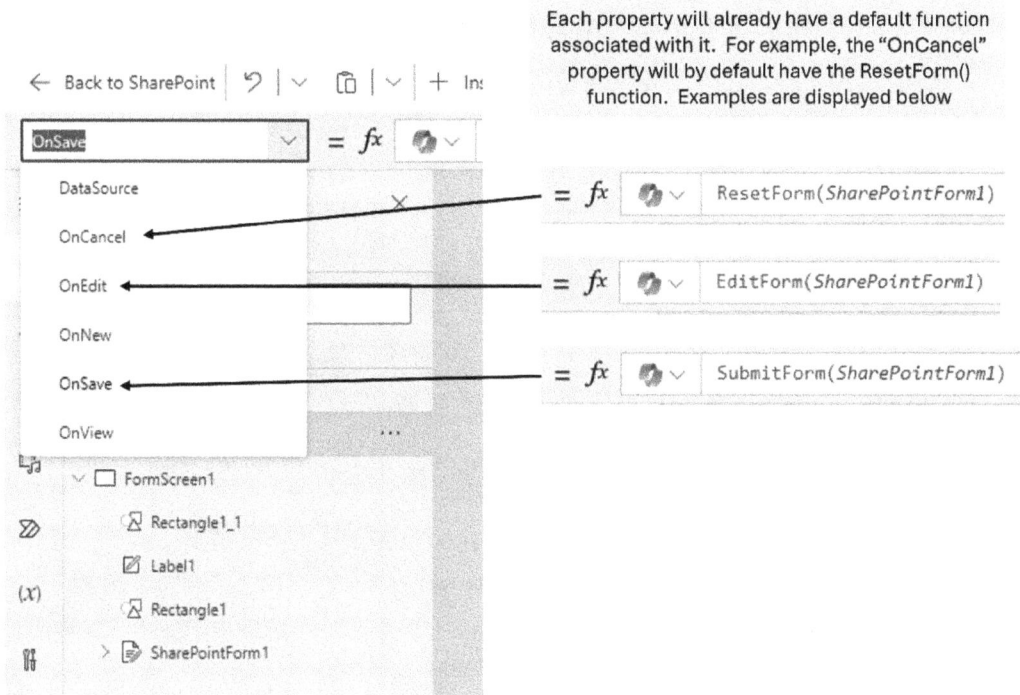

Figure 11.22 – Displaying the SharePointIntegration properties

These properties already have default functions in there, as shown in *Figure 11.22*. For example, selecting the `OnSave` property will have by default the `SubmitForm()` function, as shown in *Figure 11.22*. However, it is certainly possible for you to change this. For example, let's say that when a user wants to edit a record, you want to display a screen that provides them with additional information. This is the original:

```
EditForm(SharePointForm1)
```

You could replace the preceding with the following:

```
Navigate(NewScreen)
```

Then, have a second screen in your app that has additional information or instructions before sending them back to the edit form. While this is just one example, the main point to remember is that you have full control to customize the overall functionality of this.

Now that you have created this custom form, let's cover two final points on this. This includes how to edit a form once you have created one. The second is how to delete a custom form and revert to the default one. We will cover these in the next section.

How to edit or delete a custom SharePoint form

Once you have created a custom SharePoint form, you may either want to edit the form or even possibly delete one. Both are very simple.

To edit a form, the approach is the same as creating a new one. You can select the **Integrate | Power Apps| Customize forms** options at the top of the SharePoint list. You can also select **Customize with Power Apps** when adding or editing a record. Both approaches are shown in *Figure 11.23*.

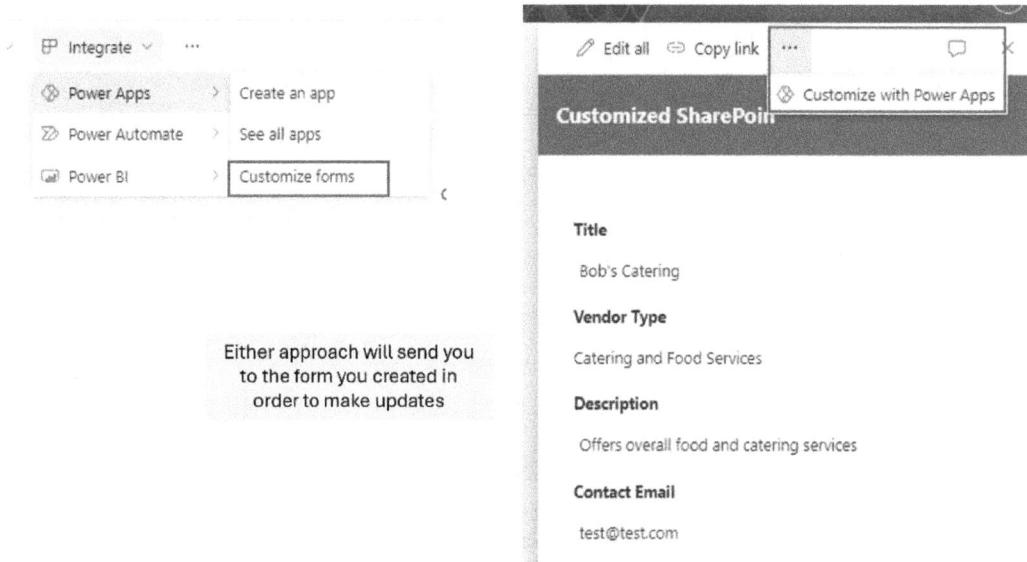

Figure 11.23 – Accessing an existing app to make changes

To go back to the default SharePoint form, this is done by accessing **List settings**, then going to **Form settings**, and then selecting the **Use the default SharePoint form** option. When selecting this, an option is presented to you to delete the custom form as well. See *Figure 11.24* for this.

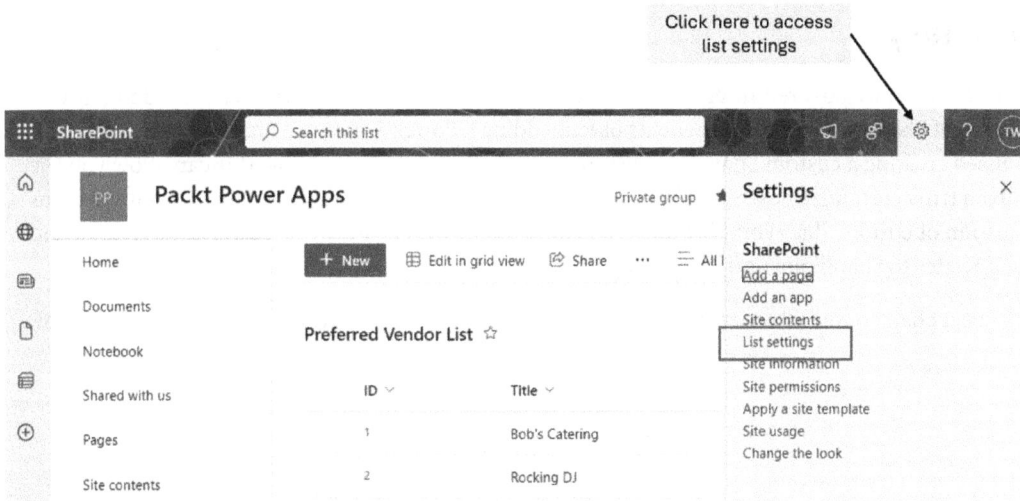

Figure 11.24 – Accessing the settings to go back to the default SharePoint form

From the list settings, select the option for **Form settings**. This will bring up the ability for you to toggle the option of **Use the default SharePoint form**. There is an additional option for you to delete the custom form. Note that you do not have to delete the custom form to switch to the default form. This allows you to switch back and forth as needed. See *Figure 11.25* for an example.

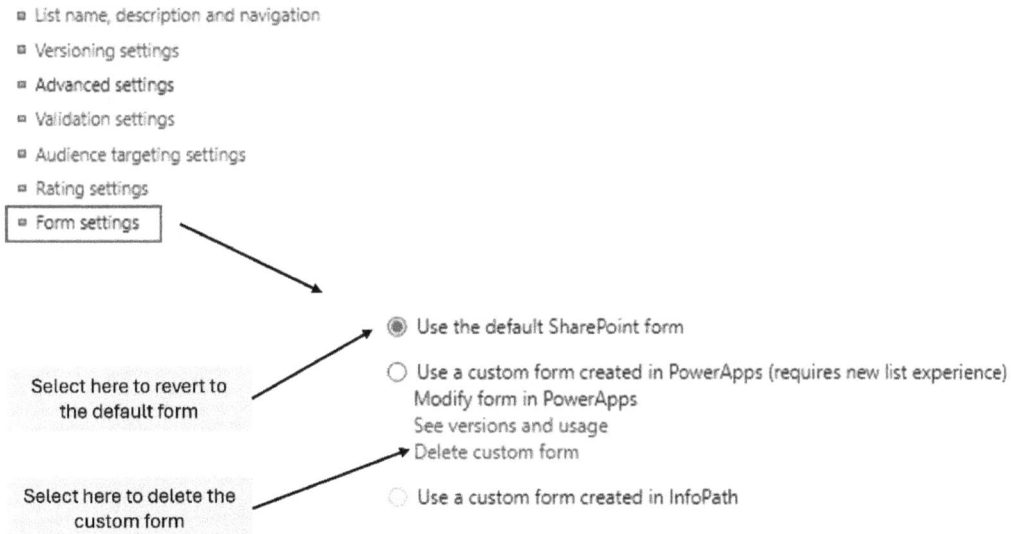

Figure 11.25 – Accessing the form options

That is all there is to updating the custom SharePoint form or going back to the default view.

Summary

In this chapter, we covered integrating Power Apps and SharePoint. First, we provided a broad overview of SharePoint and some benefits of embedding a Power App into a SharePoint site. We then discussed creating a custom SharePoint list form using Power Apps and how this can allow you the ability to truly customize how SharePoint list items can be added or edited. Finally, we continued our discussion of custom SharePoint list forms by covering the `SharePointIntegration` item and how it relates to SharePoint and Power Apps.

In the next chapter, we will cover integrating Power Apps with Virtual Agents and Microsoft Copilot.

12

Integration with Power Virtual Agents/Copilot

In this chapter, we will explore the powerful integration of Power Apps with Power Virtual Agents and AI-driven tools, such as Copilot. The goal of this chapter is to showcase the power and flexibility of these tools and demonstrate why this integration is invaluable for anyone looking to enhance their applications with intelligent, responsive user interactions.

These tools, while not similar, have complementary roles within the Microsoft ecosystem. Power Virtual Agents allows you to create sophisticated chatbots that can handle a variety of tasks, while Copilot leverages AI to assist in app creation and data manipulation, offering dynamic support throughout your development process. Understanding how to use these tools together can greatly enhance your ability to create applications that not only meet user needs but also anticipate and respond to them intelligently.

This chapter is crucial at this point in the book because it builds on the foundations we've laid in previous chapters, where we discussed creating and managing apps and flows. Now, we're taking it a step further by integrating AI and automation into these processes. This integration addresses a key challenge: how to make your applications more intelligent and user-friendly without needing to build everything from scratch. By the end of this chapter, you will understand how these tools can significantly boost user engagement and satisfaction by making your apps more responsive and adaptive.

For those who may not be familiar with Copilot or Power Virtual Agents, we'll begin by explaining what these tools are and why they are worth your time. We'll outline the practical applications they bring to your projects, making it clear how investing in understanding them can lead to tangible improvements in your work.

In this chapter, we're going to cover the following main topics:

- Exploring Copilot Studio
- Connecting to a data source
- Creating custom topics
- Building a Power App with Copilot

By the end of this chapter, you will have a comprehensive understanding of the Copilot Studio interface, how to connect it to data sources, create custom topics, and build a Power App with Copilot. These skills are invaluable for creating responsive and intelligent virtual agents and applications.

Technical requirements

- **Power Apps**: Ensure you have an active Power Apps subscription and access to the Power Apps environment: `https://makepowerapps.com/`
- **Power Virtual Agents**: Access to Power Virtual Agents within your Microsoft environment: `https://powervirtualagents.microsoft.com/`
- **Copilot**: Access to Copilot features within Power Virtual Agents: `https://copilotstudio.microsoft.com/`
- **Data source**: A sample data source (e.g., Dataverse, a SQL database, or an Excel file stored in OneDrive)
- **Microsoft Azure**: If connecting to Azure services, ensure you have the necessary permissions and subscriptions

Ensure all the preceding technologies and installations are set up and accessible before you begin working through this chapter. This preparation will enable you to follow along with the exercises and fully benefit from the hands-on learning experience.

Exploring Copilot Studio

Before we dive into the details, it's important to understand the purpose and capabilities of Copilot Studio. **Copilot Studio** is a powerful tool designed to help you create and manage virtual agents with ease. It leverages AI to assist in designing workflows, automating responses, and integrating data sources, making it an invaluable asset for anyone looking to enhance user interactions within their applications.

The goal of this section is to familiarize you with the Copilot Studio interface. Understanding the layout and features of Copilot Studio is crucial for effectively designing and managing virtual agents. By mastering the interface, you'll be able to navigate through different sections, utilize key tools, and leverage the full potential of Copilot for your projects.

Understanding the value of Copilot

Without utilizing Copilot, developers might encounter longer development cycles and more manual configuration, leading to less responsive applications. Copilot streamlines development by automating complex tasks, offering intelligent recommendations, and simplifying setup processes. Its benefits are particularly evident in scenarios that involve complex data interactions, rapid development needs, and adaptive user experiences.

Why Copilot is essential

Copilot enhances the development process by automating routine tasks, providing smart suggestions, and enabling more efficient data handling. It excels in managing intricate data queries, accelerating prototyping, and delivering intelligent responses to diverse user inputs. For developers and business analysts alike, Copilot facilitates rapid application creation and customization, making it a valuable tool for both technical and non-technical users.

Empowering self-service

Copilot's user-friendly design and intelligent automation enable self-service application development. Users with minimal technical expertise can leverage Copilot to build and customize applications independently, making it an accessible solution for those seeking to create effective, AI-driven apps without extensive development experience.

Understanding the Copilot Studio interface

Let's start by exploring the Copilot Studio interface. This involves familiarizing ourselves with the various sections, menus, and tools available within the studio. By understanding the interface, you'll gain the confidence and knowledge needed to efficiently create and manage virtual agents, ensuring that your projects are both responsive and intelligent.

Navigating the main dashboard

Here are the steps for navigating the main dashboard:

1. As a first-time user, open Copilot Studio (the link is provided in *Step 2*), and take a tour of the main dashboard.

2. Navigate to `https://www.microsoft.com/en-us/microsoft-copilot/microsoft-copilot-studio` to sign up or sign in to Copilot Studio for Power Apps.

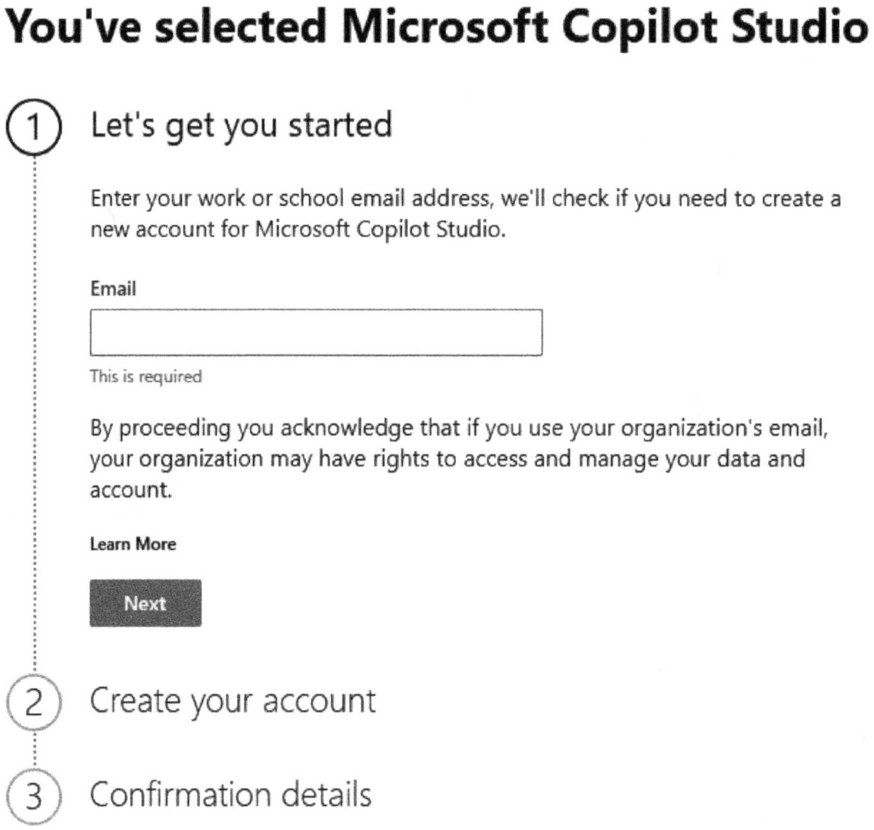

You've selected Microsoft Copilot Studio

① Let's get you started

Enter your work or school email address, we'll check if you need to create a new account for Microsoft Copilot Studio.

Email

This is required

By proceeding you acknowledge that if you use your organization's email, your organization may have rights to access and manage your data and account.

Learn More

Next

② Create your account

③ Confirmation details

Figure 12.1 – Getting started with Copilot

3. Type in your Microsoft email to sign in, and you will be redirected to `https://copilotstudio.microsoft.com/environments`.

4. Ensure you're in the **Event Planning Project Development** environment by checking the **Environment** section on the top ribbon.

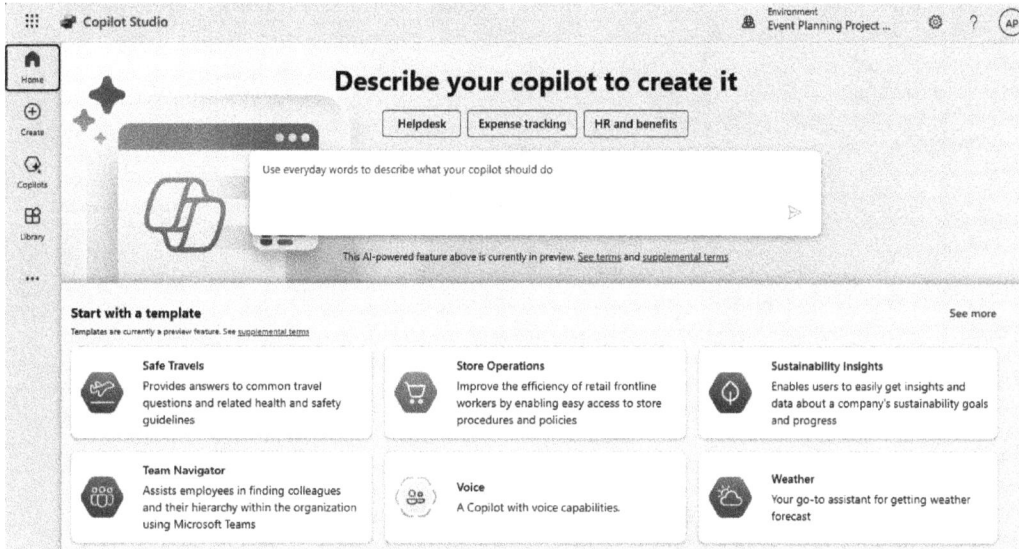

Figure 12.2 – The Environment section within Copilot Studio

Exploring key tools and features

In this section, we'll learn about the key tools available in Copilot Studio, including the designer canvas, data connectors, and prompt editors:

1.   On the left navigation, you can see the **Home**, **Create**, **Copilots**, and **Library** options.

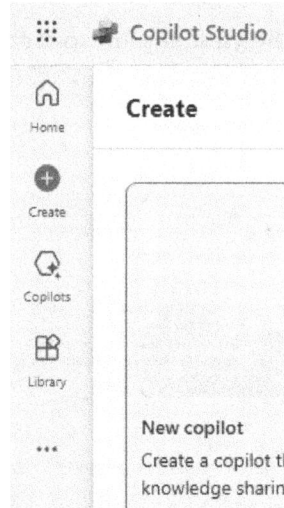

Figure 12.3 – The Create option in Copilot Studio

I.   The **Home** button will allow you to describe what your copilot should do; explore this feature by typing Event tracking in the description box.

II.   This should take you to the Copilot message interface. I started to have a conversation to help Copilot understand what I needed. See *Figure 12.4*.

III.   I included track any new events that are created in the model-driven app and then I typed I'd like to get an in app notification when there is a new event created.

> **What is an in-app notification?**
>
> An **in-app notification** is a message that appears within the model-driven app itself, rather than being sent externally via email. These notifications are designed to alert users of important updates or actions that require their attention while they are actively using the app. For instance, in this example, when a new event is created in the model-driven app, the in-app notification would immediately inform the user without needing to switch to another application. This ensures that users remain informed and can take timely actions within the app, enhancing the overall user experience.

Hi, I'm here to help you build a custom copilot. In a few sentences, how will your copilot assist your users?

3 minutes ago

Can you help me with what to include here?

3 minutes ago

Of course! You can provide a brief overview of the purpose of your copilot. This can include the tasks it will perform, the organization it is supporting, specific subjects it can help with, any further context about the role, and any additional objectives it may have as it assists.

3 minutes ago

track any new events that are created in the model-driven app

3 minutes ago

To create a copilot description, you can mention that your copilot will track new events created in the model-driven app. This can involve monitoring and providing information or assistance related to these events. Is there anything else you'd like to add to the description?

3 minutes ago

i'd like to get an in app notification when there is a new event created

3 minutes ago

Great! Your copilot will track new events created in the model-driven app and provide you with in-app notifications for any new events. Now, do you have any instructions for how your copilot should assist, for example a specific tone?

3 minutes ago

that's it!

Just now

Perfect! Your copilot will track new events and send in-app notifications. If you're all set, you're ready to move forward with your copilot setup. Let me know if you need any more adjustments or if you're ready to proceed.

Just now

I'm done

Just now

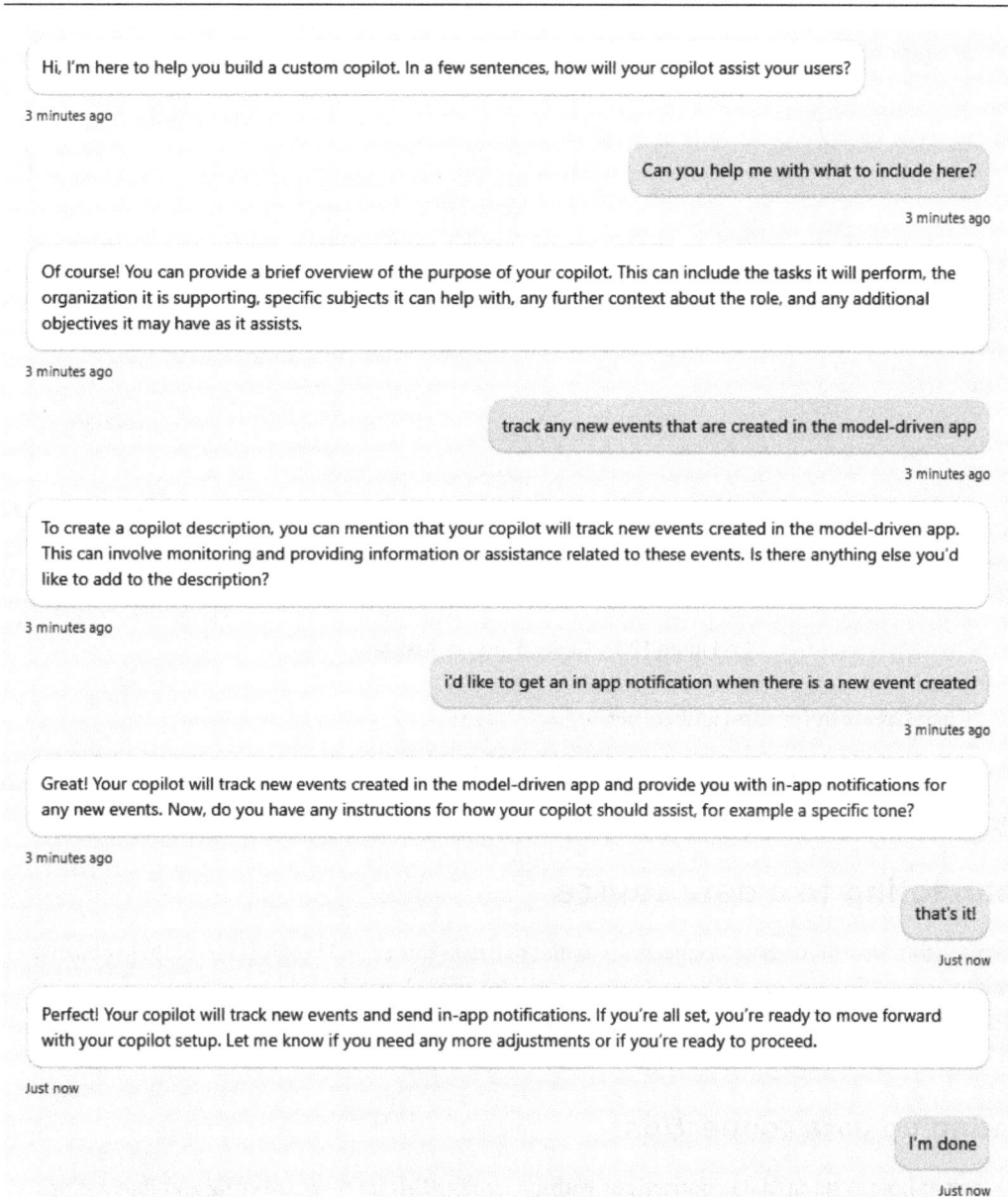

Figure 12.4 – The Create option in Copilot Studio

2. After typing I'm done in the chat window, I was prompted to go to **General Inquiry Assistant** with an option to chat on the right or an option to either **Skip to configure** or **Create**.

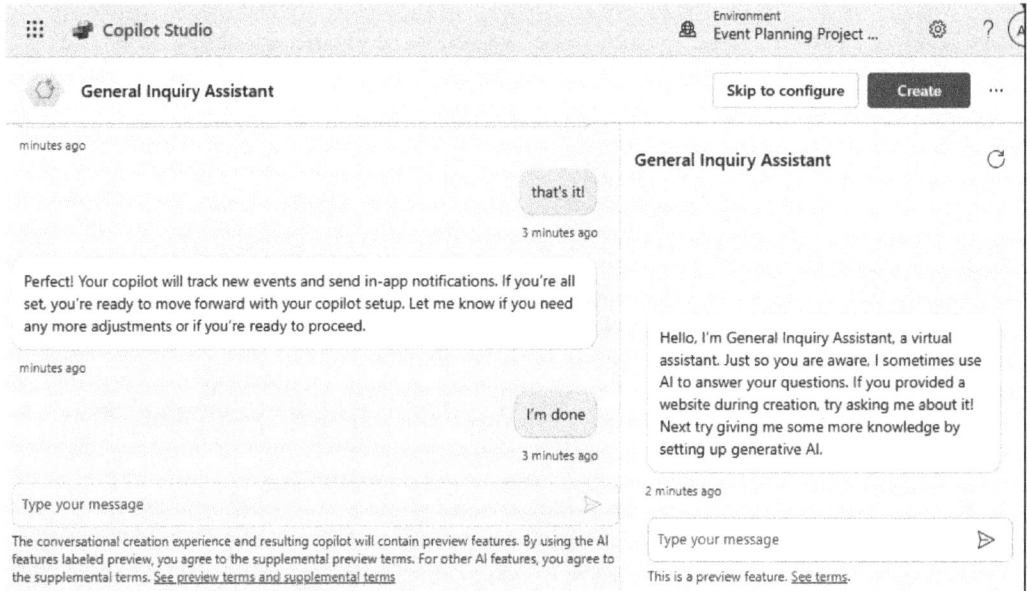

Figure 12.5 – General Inquiry Assistant

3.   Click **Create** in the top-right corner.

4.   Copilot will take you to the **General Inquiry Assistant** overview page.

Now that we have explored Copilot Studio, we'll move on to the next section.

Connecting to a data source

In this section, we will focus on connecting Copilot to a data source and building a prompt to enhance user interactions. Integrating data sources is essential for providing relevant and accurate information during conversations. You will learn how to link a data source to Copilot and create effective prompts that utilize this data to boost conversation quality.

Setting up data connections

Let's look at how to set up data connections within Copilot. This involves selecting and integrating a data source that your virtual agent will use to fetch information.

Selecting a data source

Here are the steps for selecting a data source:

1.   Choose a suitable data source, such as Dataverse, SQL database, or an Excel file stored in OneDrive.

2.   Ensure you have the necessary permissions and access credentials for the chosen data source.

3. Select the **Knowledge** tab then select + **Add knowledge**:

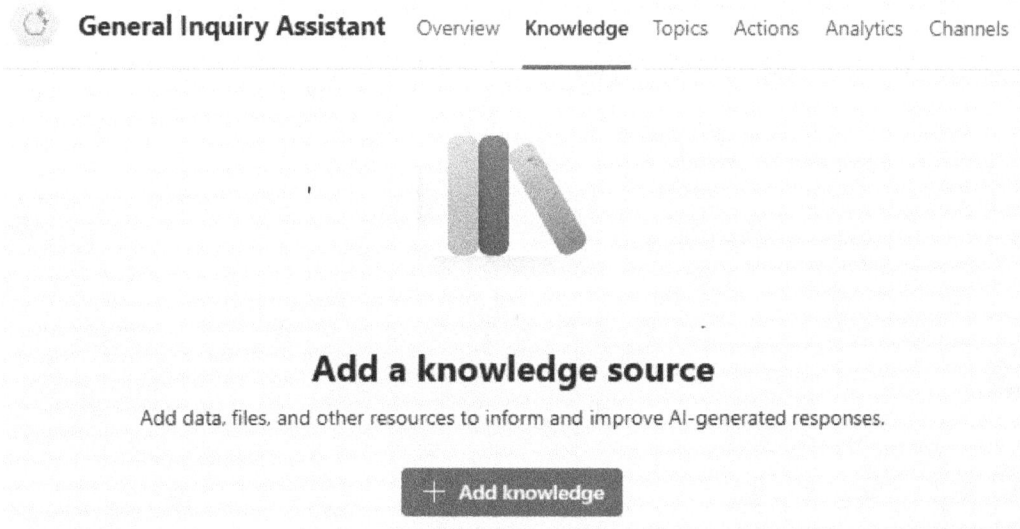

General Inquiry Assistant   Overview   **Knowledge**   Topics   Actions   Analytics   Channels

Add a knowledge source

Add data, files, and other resources to inform and improve AI-generated responses.

+ Add knowledge

Figure 12.6 – Add a knowledge source

4. In the **Add available knowledge sources** window, select **Dataverse (preview)**.

Add available knowledge sources   ✕

Users with edit permissions for this copilot can also reuse your connections for other topics within the copilot.

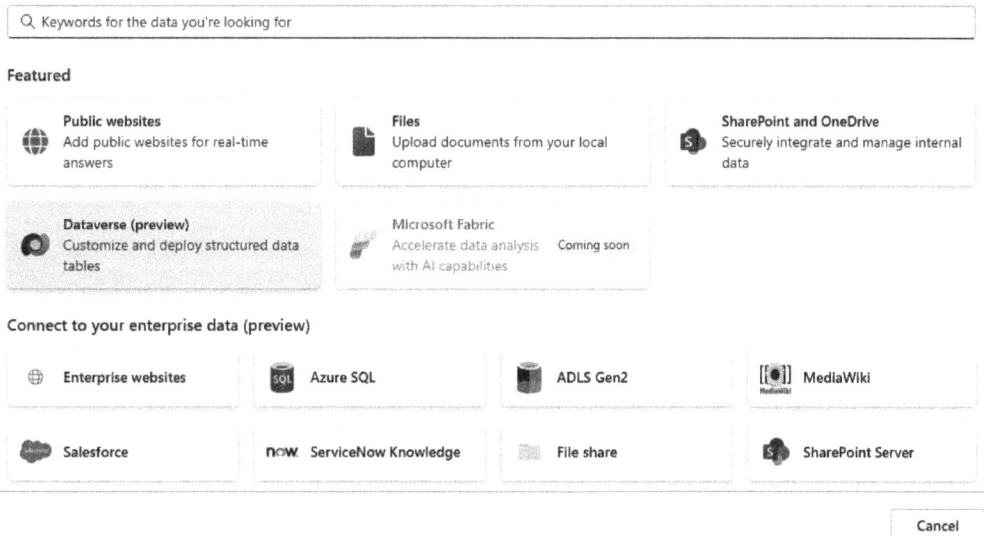

🔍 Keywords for the data you're looking for

Featured

Public websites Add public websites for real-time answers	**Files** Upload documents from your local computer	**SharePoint and OneDrive** Securely integrate and manage internal data
Dataverse (preview) Customize and deploy structured data tables	**Microsoft Fabric** Accelerate data analysis Coming soon with AI capabilities	

Connect to your enterprise data (preview)

Enterprise websites	Azure SQL	ADLS Gen2	MediaWiki
Salesforce	ServiceNow Knowledge	File share	SharePoint Server

Cancel

Figure 12.7 – Knowledge sources

Now that we've selected a data source, let's integrate it with Copilot.

Integrating the Dataverse Data Source with Copilot

Follow these steps to connect the selected data source to Copilot:

1. In **Step 1 of 3: Select Dataverse tables**, type `Event` in the search bar, select the result that includes `epe_Event`, and then click **Next**.

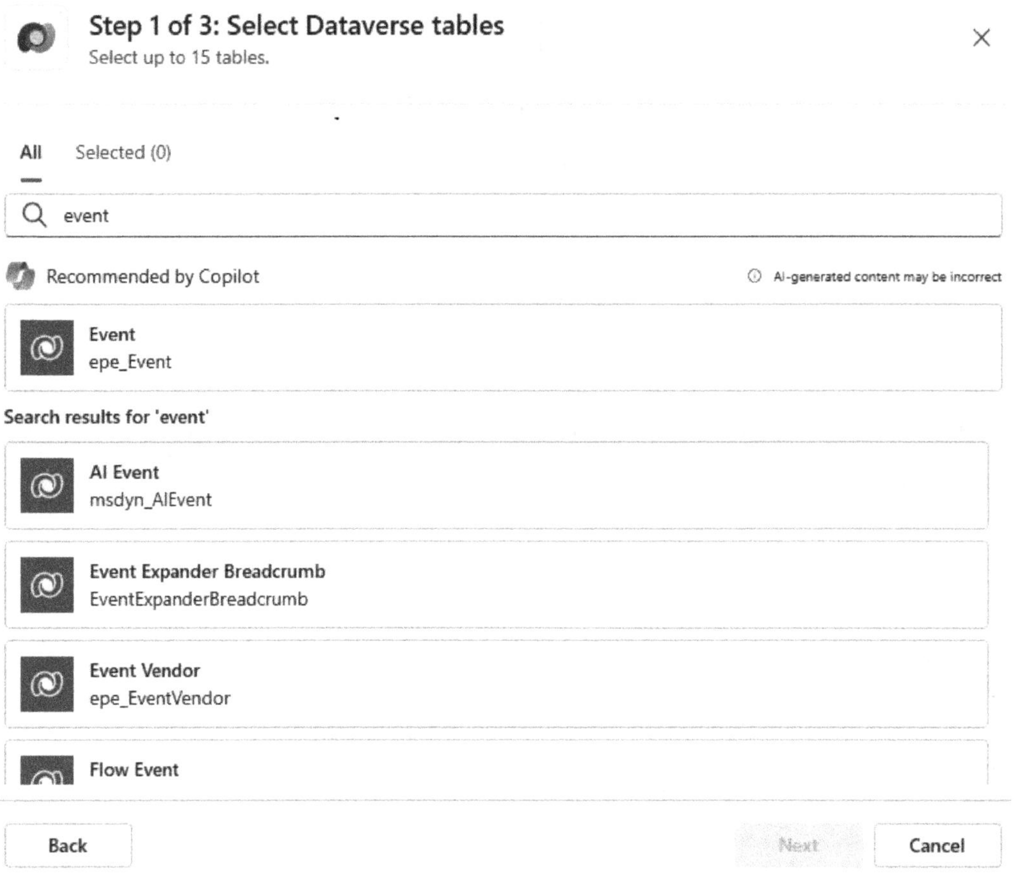

Figure 12.8 – Step 1 of 3: Select Dataverse tables

2.   In **Step 2 or 3: Preview data**, review the **Event** table and then click **Next**.

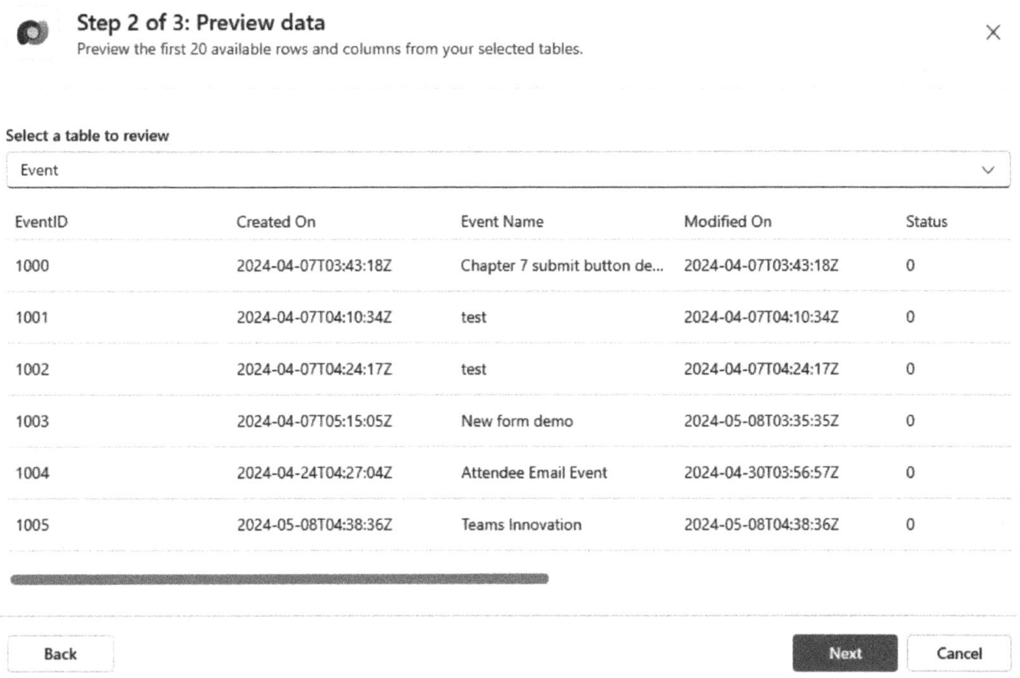

Step 2 of 3: Preview data   ✕

Preview the first 20 available rows and columns from your selected tables.

Select a table to review

Event	⌄

EventID	Created On	Event Name	Modified On	Status
1000	2024-04-07T03:43:18Z	Chapter 7 submit button de...	2024-04-07T03:43:18Z	0
1001	2024-04-07T04:10:34Z	test	2024-04-07T04:10:34Z	0
1002	2024-04-07T04:24:17Z	test	2024-04-07T04:24:17Z	0
1003	2024-04-07T05:15:05Z	New form demo	2024-05-08T03:35:35Z	0
1004	2024-04-24T04:27:04Z	Attendee Email Event	2024-04-30T03:56:57Z	0
1005	2024-05-08T04:38:36Z	Teams Innovation	2024-05-08T04:38:36Z	0

| Back | | Next | Cancel |
|------| |------|--------|

Figure 12.9 – Step 2 of 3: Preview data

3.   In **Step 3 of 3: Review and finish**, click the **Edit** button to explore both **Synonyms** and **Glossary** and then click **Add**.

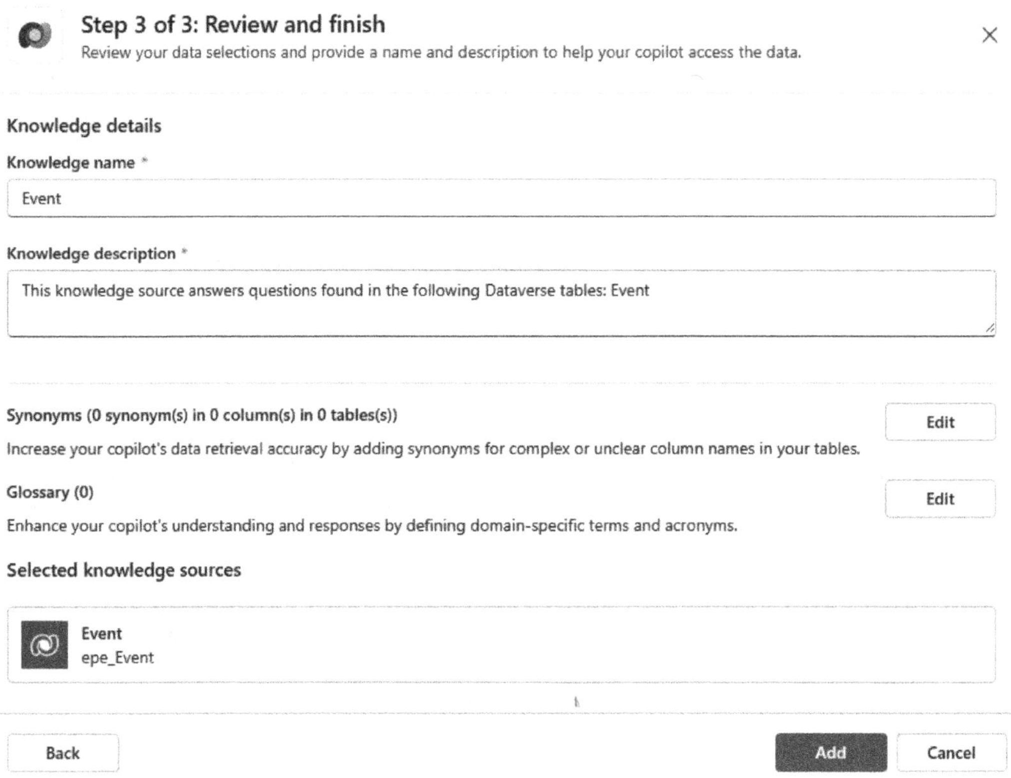

Step 3 of 3: Review and finish   ✕

Review your data selections and provide a name and description to help your copilot access the data.

Knowledge details

Knowledge name *

Event

Knowledge description *

This knowledge source answers questions found in the following Dataverse tables: Event

Synonyms (0 synonym(s) in 0 column(s) in 0 tables(s))   Edit

Increase your copilot's data retrieval accuracy by adding synonyms for complex or unclear column names in your tables.

Glossary (0)   Edit

Enhance your copilot's understanding and responses by defining domain-specific terms and acronyms.

Selected knowledge sources

Event
epe_Event

Back   Add   Cancel

Figure 12.10 – Step 3 of 3: Review and finish

4.   Test the connection to ensure data is being accurately retrieved:

I.   Type Tell me about the events on the **Test your copilot prompt** window on the right side.

II.   The test will bring back the number of events and some details of the events, including the event name and when it was created.

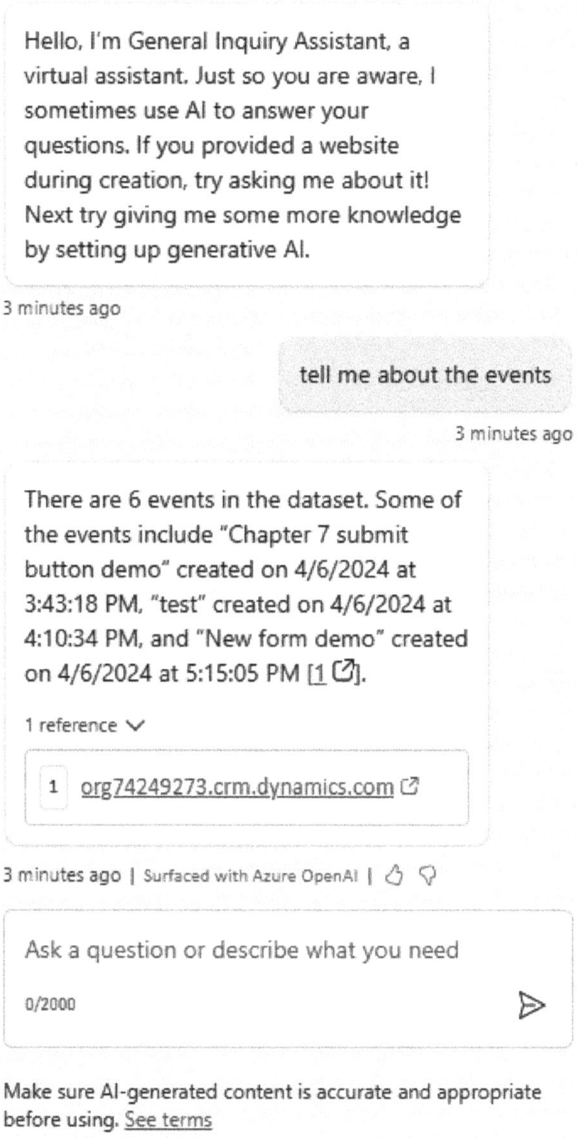

Hello, I'm General Inquiry Assistant, a virtual assistant. Just so you are aware, I sometimes use AI to answer your questions. If you provided a website during creation, try asking me about it! Next try giving me some more knowledge by setting up generative AI.

3 minutes ago

tell me about the events

3 minutes ago

There are 6 events in the dataset. Some of the events include "Chapter 7 submit button demo" created on 4/6/2024 at 3:43:18 PM, "test" created on 4/6/2024 at 4:10:34 PM, and "New form demo" created on 4/6/2024 at 5:15:05 PM [1 ↗].

1 reference ∨

1   org74249273.crm.dynamics.com ↗

3 minutes ago | Surfaced with Azure OpenAI | 👍 👎

Ask a question or describe what you need

0/2000   ▷

Make sure AI-generated content is accurate and appropriate before using. See terms

Figure 12.11 – Testing Copilot

III.   The test will also create topics based on that prompt under **Conversational boosting** in the middle pane.

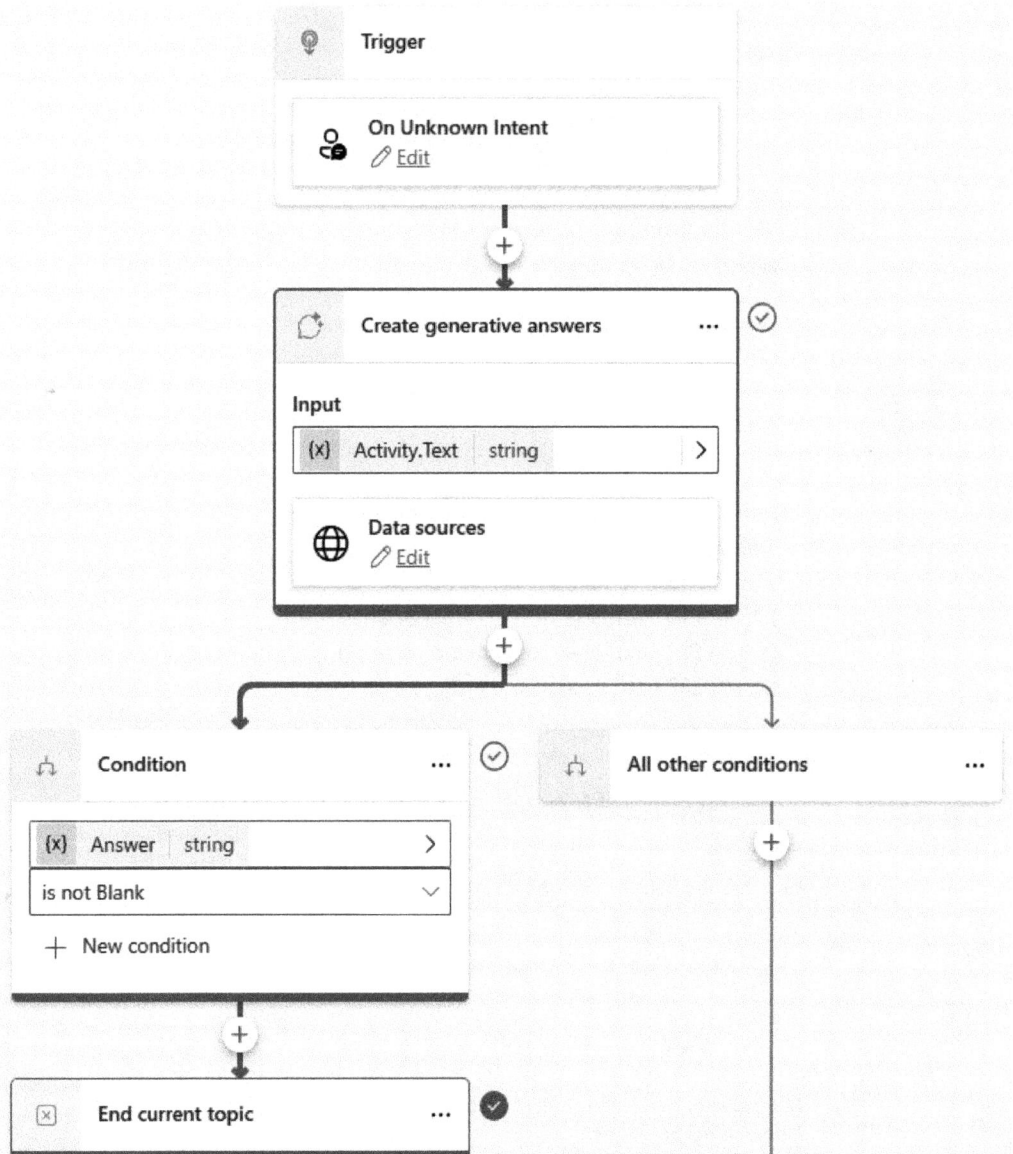

Figure 12.12 – Trigger and action in Copilot

This figure shows a workflow that handles unknown user inputs in a chatbot. When an unrecognized intent is detected, the system generates a response using the user's input. If the response is valid (not blank), the conversation continues by ending the current topic. If not, it handles other conditions. This setup ensures the chatbot can manage unexpected queries effectively.

Creating custom topics

In this section, you will learn how to create custom topics within Copilot. Custom topics allow you to tailor conversations to specific needs and scenarios, providing a more personalized and effective user experience. You will gain hands-on experience in designing, managing, and testing custom topics to ensure they meet your requirements.

Examples of custom topics

Here are some examples of custom topics:

- **Example 1: event information**:

  - **Scenario**: Users frequently ask about event schedules

  - **Custom topic**: "Event Schedule"

  - **Purpose**: This topic allows users to ask about the timing of specific events, such as "When is the keynote speech?" or "What time does the workshop on AI start?"

  - **Design**: You can set up conversation flows to provide detailed event schedules based on user queries

- **Example 2: registration status**:

  - **Scenario**: Users need to check their registration status

  - **Custom topic**: "Registration Status"

  - **Purpose**: This topic enables users to inquire about their registration status, such as "Am I registered for the conference?" or "Have I completed my registration?"

  - **Design**: Create a flow that connects to the registration database to check and confirm the user's registration status

Designing custom topics

Let's begin by designing custom topics that address specific user needs. This involves creating conversation flows and defining the structure of each topic.

Identifying user needs

Here are some points to consider while identifying user needs:

1. Analyze user requirements and identify the key topics that need to be addressed.
2. Prioritize topics based on their importance and relevance.

Creating conversation flows

Follow these steps to create conversation flows:

1. Design the conversation flow for each custom topic, ensuring it is logical and easy to follow:

    I. Ensure you are under the **Topics** tab.

    II. Select **Add a topic**.

    III. Select **From blank**.

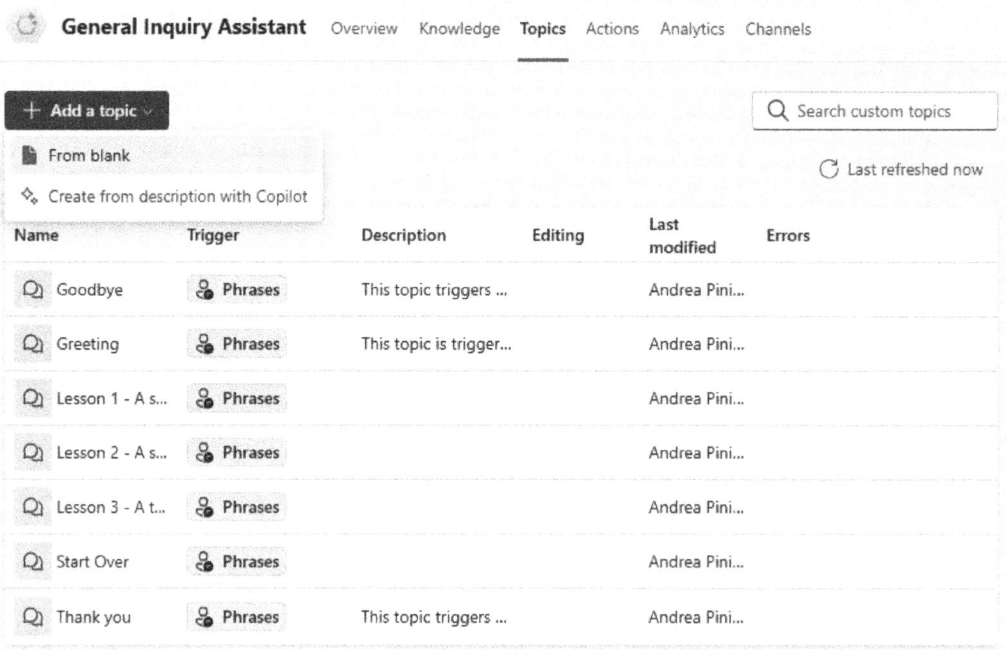

Figure 12.13 – Add a topic | From blank

2. Define the triggers, actions, and responses for each step in the conversation flow:

    I. Rename this custom topic to Event Topics.

    II. In the **Trigger** box, click **Edit** under **Phrases** and type How many total events? in the **Add Phrases** box and then click the **plus** (+) button.

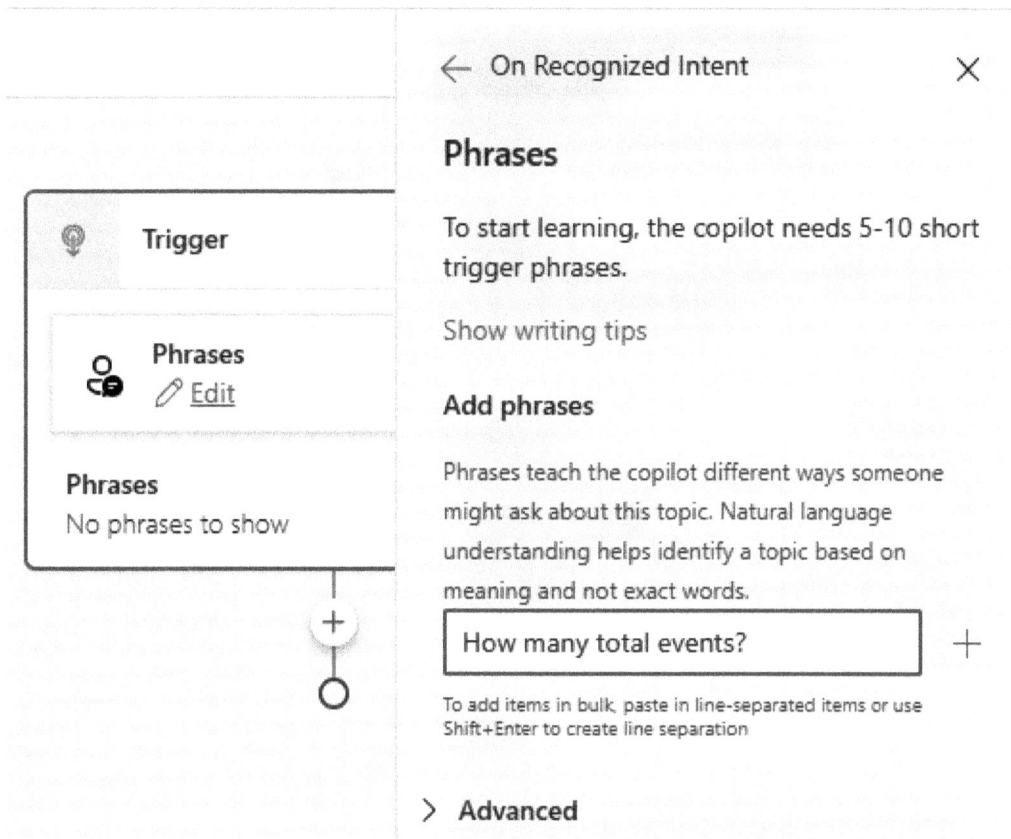

On Recognized Intent ✕

Phrases

To start learning, the copilot needs 5-10 short trigger phrases.

Show writing tips

Add phrases

Phrases teach the copilot different ways someone might ask about this topic. Natural language understanding helps identify a topic based on meaning and not exact words.

How many total events? +

To add items in bulk, paste in line-separated items or use Shift+Enter to create line separation

> **Advanced**

Trigger

Phrases
✎ Edit

Phrases
No phrases to show

Figure 12.14 – Trigger phrases

III.   That phrase should be added to the list. Type `Total events` and `Events in total` in **Add phrases** to add to this trigger so that Copilot can start learning the trigger phrases.

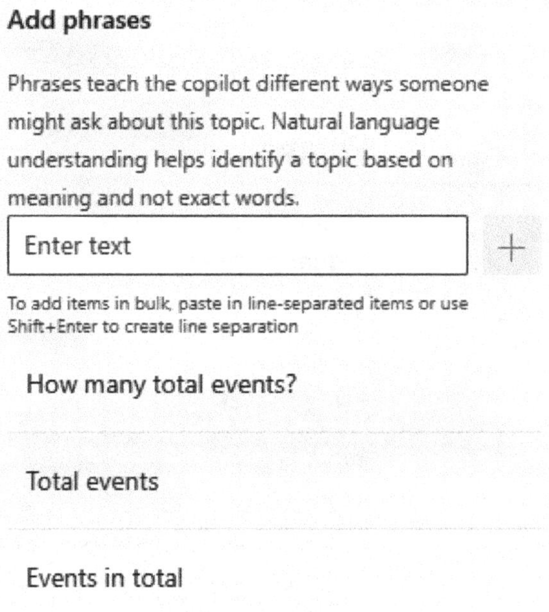

Add phrases

Phrases teach the copilot different ways someone might ask about this topic. Natural language understanding helps identify a topic based on meaning and not exact words.

| Enter text | + |

To add items in bulk, paste in line-separated items or use Shift+Enter to create line separation

How many total events?

Total events

Events in total

Figure 12.15 – Add phrases

The trigger box should look like the following:

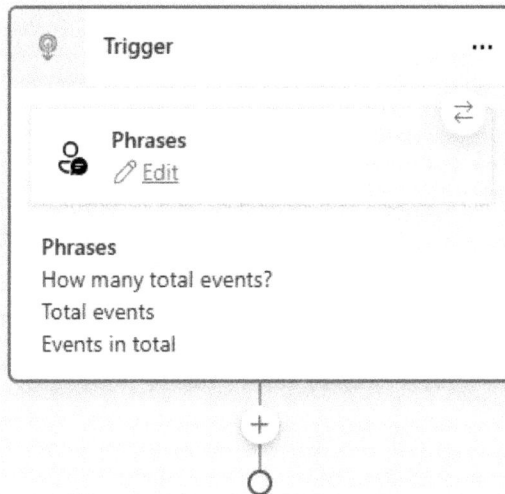

Figure 12.16 – Trigger phrases

IV.   Click on the + button under the **Trigger** box and select **Call an action**.

V.   Type `Dataverse` in the search bar and select **List rows from selected environment**.

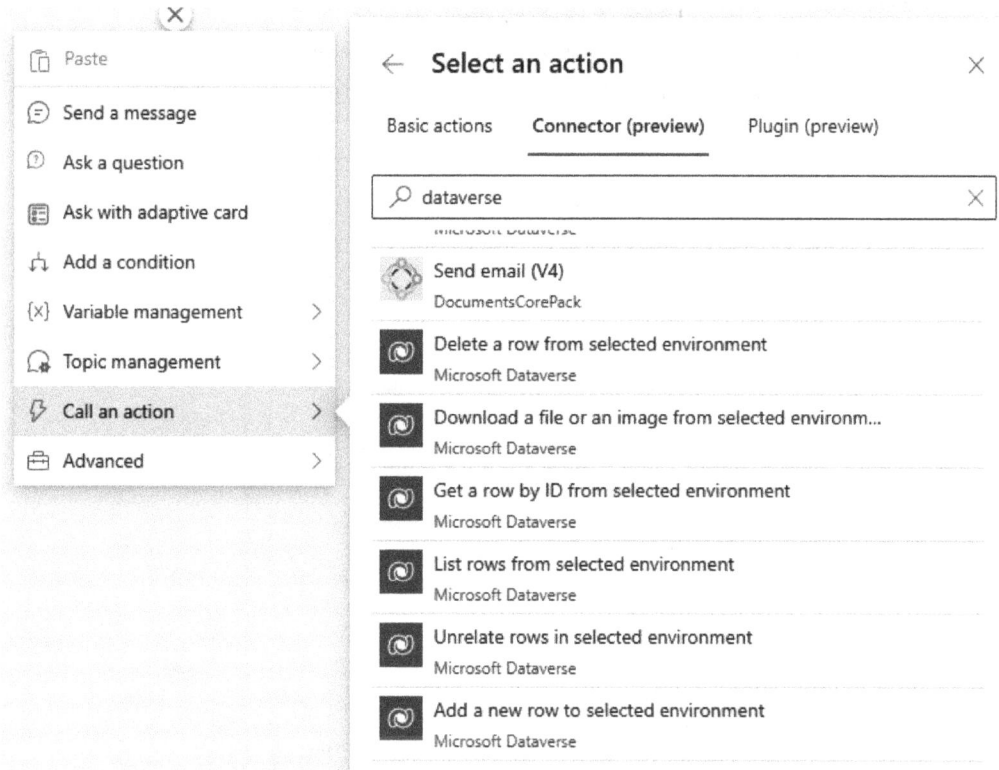

Figure 12.17 – Select an action

VI.   **Microsoft Dataverse** should be auto-selected in the **Create or pick a connection** window so click **Submit**.

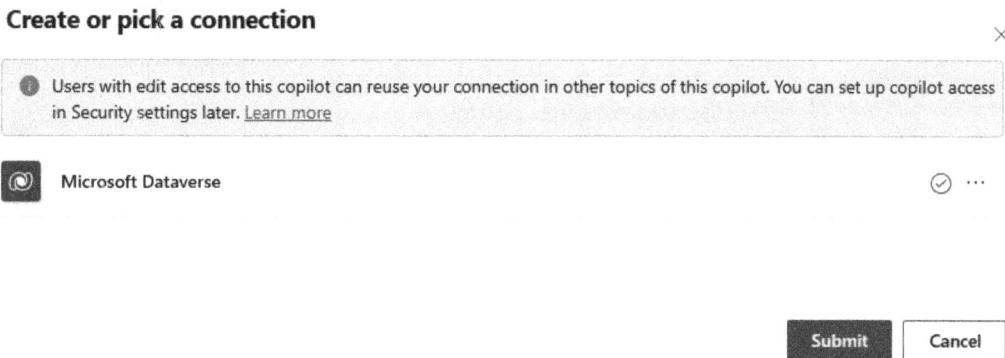

Figure 12.18 – Create or pick a connection

VII. Click in the **Inputs** box for **Environment (String)** and select the environment you've been using.

VIII. Click the **Inputs** box for **Table name (String)** and select **Events**.

IX. Click the **Outputs** box and add Var1 and Var2, similar to *Figure 12.19*.

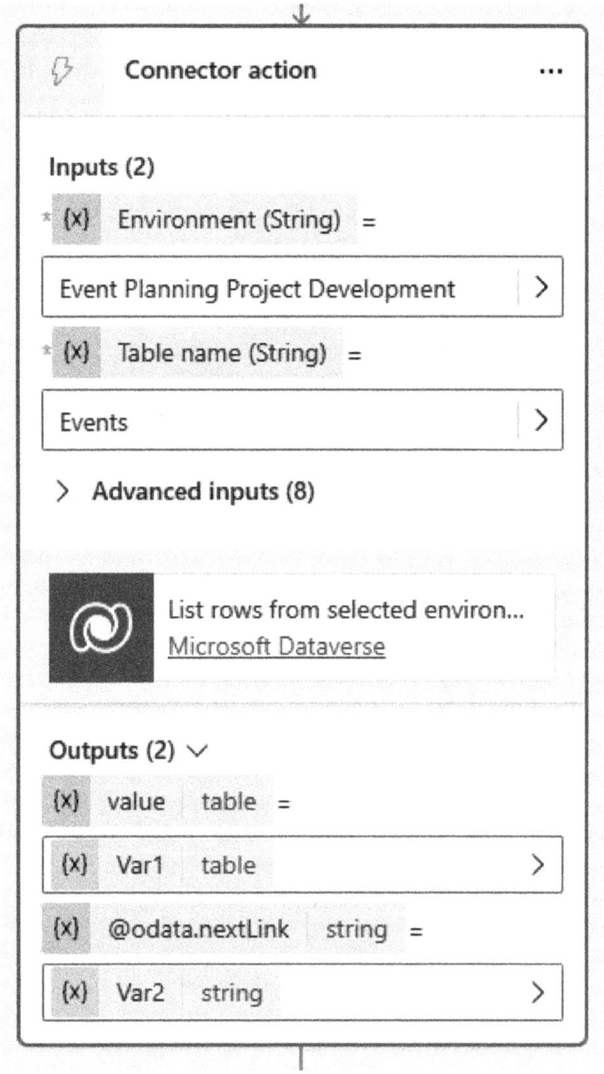

Figure 12.19 – Connection action

Managing and testing custom topics

Once the custom topics are designed, the next step is to manage and test them to ensure they function as intended. This involves implementing the topics within Copilot and conducting thorough testing.

Implementing custom topics

This is how we can implement custom topics:

1. Add the custom topics to your Copilot environment.
2. Configure the settings and parameters for each topic.

Adjust the settings to define how each custom topic should behave. This includes setting triggers, defining responses, and linking to data sources to ensure the topic functions as intended within Copilot. For more details, refer to `https://learn.microsoft.com/en-us/microsoft-copilot-studio/`.

Testing and refining custom topics

Let's conduct tests to evaluate the performance and effectiveness of each custom topic:

1. Go over to the prompt window on the right side to **Test your copilot** and type one of the prompts we created, such as How many total events?.

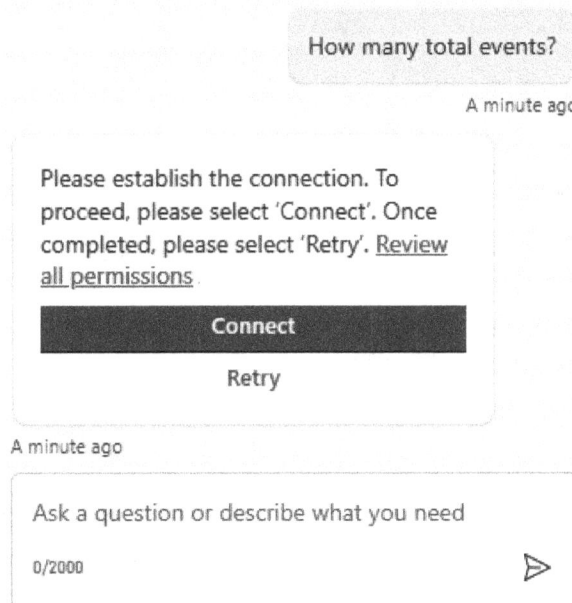

Figure 12.20 – Connect to establish the connection

2.   Make the necessary adjustments based on test results and user feedback:

    I.   The system prompted me to connect to Dataverse; so, click **Connect**.

    II.   Copilot opened a new tab and took me to the **General Inquiry Assistant** window to manage my connections.

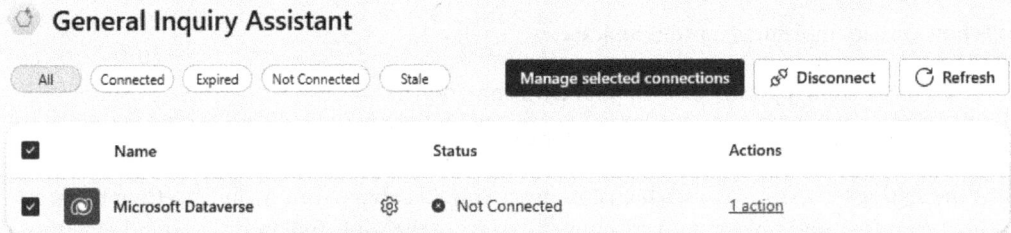

Figure 12.21 – General Inquiry Assistant Microsoft Dataverse – Not Connected

    III.   Select the **Manage selected connections** button in the top-right corner of the screen.

    IV.   Click **Submit** in **Create or pick connections**.

    V.   We should now see a green checkmark and **Connected** under the **Status** column of the **Microsoft Dataverse** row.

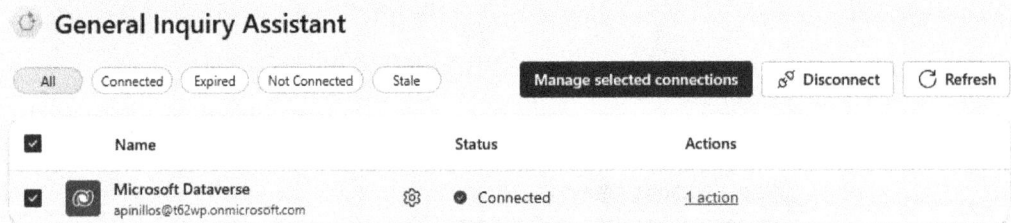

Figure 12.22 – Connected Microsoft Dataverse

    VI.   Navigate back to the **Copilot Studio** tab and click the **Retry** button in the **Test your Copilot** window on the right pane.

    VII.   You might need to use the **Refresh** button in the **Test your Copilot** window to refresh the pane.

In this section, you successfully created and managed custom topics to address specific user needs. You designed conversation flows, configured triggers and actions, and tested topics to ensure they functioned as intended. This customization enhances the user experience by providing tailored interactions and responses.

Now that you have a solid understanding of creating and managing custom topics, we will transition to the next section: *Building a Power App with Copilot*. This next section will show you how to integrate the capabilities of Copilot into a Power App, demonstrating how your custom topics can enhance the functionality of your applications and further improve user interactions.

Building a Power App with Copilot

In this section, you will learn how to build a Power App using Copilot. Integrating Copilot with Power Apps enhances your application by adding intelligent, AI-driven capabilities. This combination improves user interactions and simplifies the development process. Using Copilot with Power Apps makes your app smarter and more responsive to user inputs, while also streamlining development by automating tasks and reducing manual configuration. This integration allows you to create applications that adapt to complex queries and processes, ensuring they are both effective and engaging.

Developing the Power App with Copilot

Let's start by developing a Power App using Copilot. This involves building the app's components, integrating data sources, and configuring Copilot functionalities.

Building an app with generative AI

Here are the steps for building an app with generative AI:

1. Navigate to the home screen of the Power Apps maker portal using `https://make.powerapps.com/`.

2. Use the prompt box under **Let's build an app. What should it do?** to type `Create an app using the events table to expose events for users and let them submit new ones`. Then, click the paper plane icon arrow.

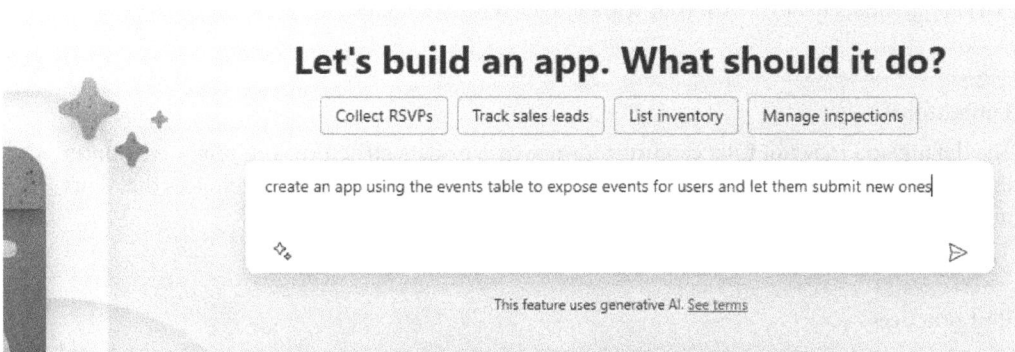

Figure 12.23 – Let's build an app (Copilot) in Power Apps

3. This will automatically create a canvas app with sample data; you can change the data type here or proceed with the table properties. Click **Create app**.

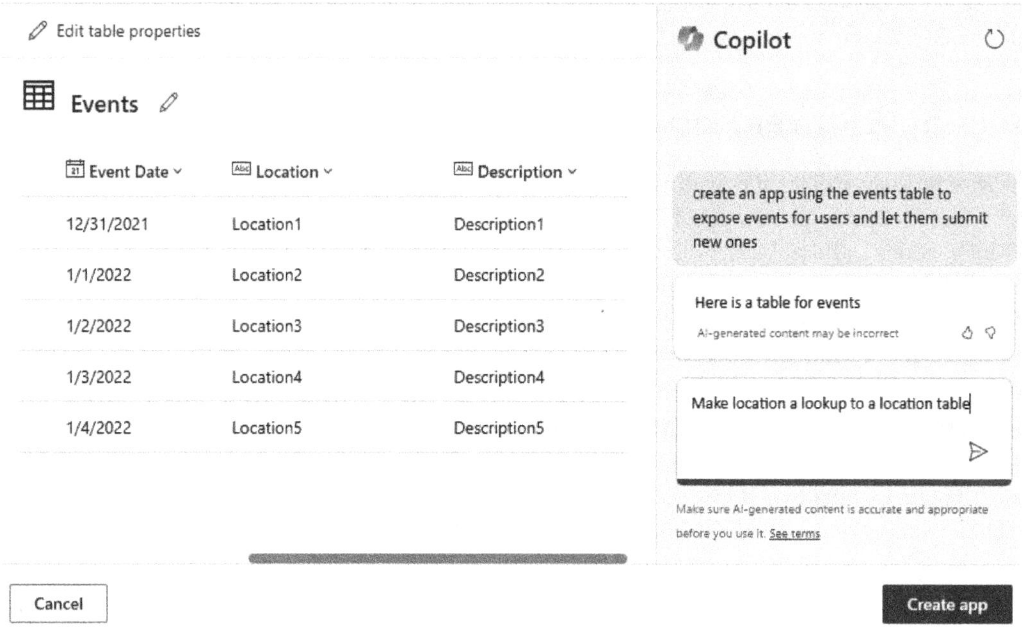

Figure 12.24 – Events for a canvas app

Showcasing screens

The initial screen generated by Copilot typically includes a default layout with placeholder data. While this provides a quick start, you'll need to customize it to better fit your specific requirements. For instance, you might need to adjust data fields or rearrange **user interface (UI)** components to align with your app's intended functionality.

Limitations

The default app may not fully capture your use case or data structure. You might encounter limitations with sample data that don't reflect real-world scenarios or layout choices that don't meet your needs.

Best practices

Take the opportunity to review and modify the generated screens to better match your requirements. Customize the layout, data bindings, and UI elements to ensure that the app delivers the user experience you envision.

Configuring and reviewing data sources

Effective app development with Copilot requires properly setting up and integrating data sources. Confirm that the necessary data sources are in place, specifically the Events table for managing event-related data.

Here are the steps:

1. **Navigate to the data source icon**: In the left panel of Power Apps, locate and click on the data source icon.

2. **Verify the Events table**: Ensure that the Events table has been created and is visible. This table will be pivotal for storing and managing your event-related data within the app.

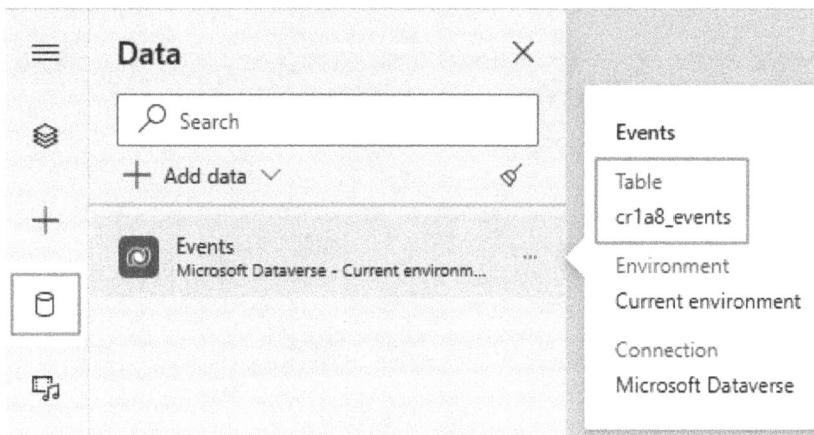

Figure 12.25 – Events data setup

Showcasing screens

Review the data integration screen where you connect your app to the Events table. Verify that all fields are correctly mapped and that the data source is linked properly.

Limitations

Common issues include incorrect data mappings or missing fields that can affect how data is displayed or interacted with in the app.

Best practices

Ensure data accuracy and completeness by validating the data connections and ensuring all required fields are correctly integrated. Regularly check data source configurations to avoid discrepancies.

Testing and deploying the Power App

The final step is to test and deploy your Power App. This ensures that the app functions correctly and meets user expectations.

Testing the Power App

Let's conduct thorough testing of all app components and Copilot functionalities by clicking the play button in the top-right corner of the screen.

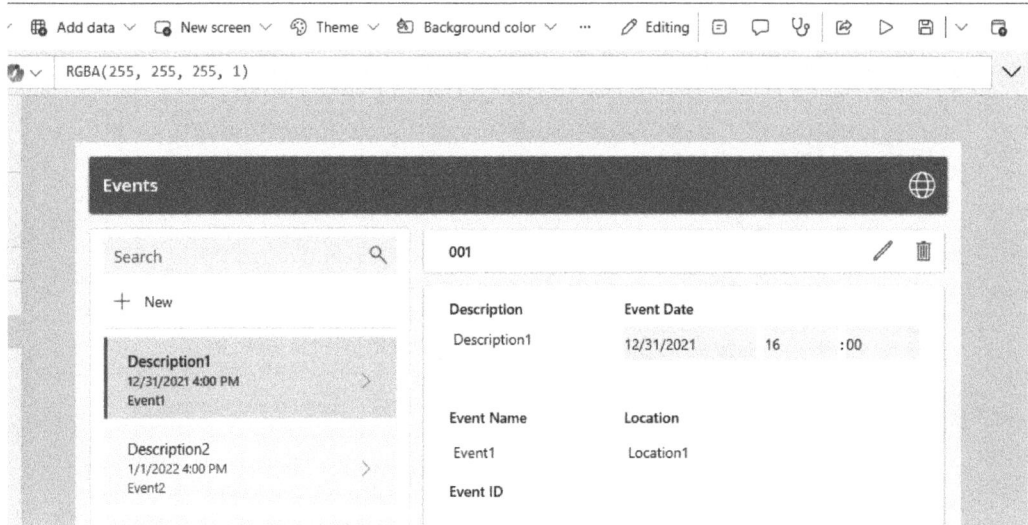

Figure 12.26 – The play button for Events

> **Showcasing screens**
>
> During testing, pay attention to how the app behaves with different inputs and user interactions. Observe any issues or areas for improvement, such as slow response times or incorrect data display.

> **Limitations**
>
> Testing might reveal bugs or design flaws that were not apparent during initial development. Address these issues promptly to enhance the user experience.

> **Best practices**
>
> Thoroughly test the app in various scenarios to ensure robustness. Gather feedback from real users to identify potential improvements and address any usability concerns.

Deploying the Power App

Here are the steps for deploying the Power App:

1. Publish your Power App to the appropriate environment.
2. Ensure that all configurations and settings are correctly applied.

Showcasing screens

Review the deployment settings screen to confirm that all configurations are accurate and complete before finalizing the deployment.

Limitations

Deployment issues can arise from incorrect settings or missing configurations, affecting app performance or accessibility.

Best practices

Double-check all deployment settings and configurations to ensure a smooth launch. Verify that all aspects of the app function correctly in the production environment.

Integrating Copilot with Power Apps enhances the app's intelligence and streamlines development. This section has guided you through building an app with generative AI, configuring data sources, and testing and deploying the app. By understanding the screens, their limitations, and best practices, you can create a responsive and effective Power App that meets your users' needs.

Summary

In this chapter, we explored how integrating Power Virtual Agents and Copilot with Power Apps can significantly enhance application development and user interactions. We began by familiarizing ourselves with the Copilot Studio interface, including the designer canvas and data connectors, which are crucial for effectively designing and managing virtual agents. By connecting Copilot to data sources and creating custom topics, we personalized interactions to better engage users. We concluded by developing and deploying a Power App that leverages Copilot's advanced capabilities.

In the next chapter, we will focus on governance and security considerations for deploying and managing Power Apps. We will explore best practices for compliance, data security, and operational integrity, providing essential guidance for safeguarding your deployments and effectively implementing these powerful tools within your organization.

Part 4:
Governance, Security, and Deployment

In this part, you'll explore crucial aspects of managing and securing your Power Apps, ensuring they operate smoothly and securely. You'll start by exploring how to establish a robust Power Platform environment strategy, implement service accounts, and set up data policies to support Data Loss Prevention (DLP). Additionally, you'll be introduced to the principles of Application Lifecycle Management (ALM) to effectively manage the lifecycle of your apps. Next, you'll discover strategies for managing errors within your apps. Finally, the book concludes with a guide on registering an app in Azure. These topics equip you with the knowledge to govern, secure, and maintain your Power Apps throughout their lifecycle.

This part has the following chapters:

- *Chapter 13, Governance, Security, and Application Life Cycle Management*
- *Chapter 14, Error Handling*
- *Chapter 15, Registering a Power App in Azure*

13

Governance, Security, and Application Life Cycle Management

In the last chapter, we covered the integration of Power Apps with Microsoft Copilot Studio and Copilot. Copilot Studio enables the creation of individual copilots, which are AI-powered conversational interfaces, while Copilot provides an AI-based tool for rapidly building a Power App. This concluded our discussion on integrating Power Apps with other applications. In these remaining chapters, we are going to shift our focus to other areas of importance. This includes talking about governance, security, application maintenance, and deployment.

Application governance and security are important topics. And, while many topics could be covered in this space, we're going to focus on three areas. The first area involves establishing a strategy for creating and using Power Platform environments. We covered environments in *Chapter 3* so be sure to revisit that as needed. A well-defined environment strategy is the foundation for effective Power Platform governance. Then, we will discuss using service accounts versus individual accounts. Utilizing service accounts versus individual user accounts can have significant implications for security, accountability, and operational efficiency. We will cover the concept of **data loss prevention** (**DLP**) in Power Platform environments and how data policies can be used to support this. Finally, we will wrap up how these areas can be applied to support an overall **application life cycle management** (**ALM**) approach.

In this chapter, we're going to cover the following main topics:

- Establishing a Power Platform environment strategy
- Understanding service accounts versus individual accounts
- DLP in Power Platform environments
- Understanding ALM concepts

Technical requirements

To successfully engage with the materials in this chapter, you'll need the following:

- **Microsoft Power Apps**: Access to Power Apps for creating and managing your applications.
- **Power Platform Admin Center**: You will need to have access to the Power Platform Admin Center. If you do not have access to this, you may need to contact your Power Platform administrator for assistance.

Establishing a Power Platform environment strategy

In *Chapter 3*, we covered the concept of Power Platform environments. Environments are containers that are used to manage the Power Platform assets, including applications, automation flows, connections, environment variables, solutions, and more. They are also important to the overall development of applications, especially if ALM principles are to be followed. We will cover ALM concepts later in this chapter.

Effective environment management is the cornerstone of a successful Power Apps deployment. Utilizing multiple environments, such as one for development, one for testing, and one for production, provides a structured approach to application development and deployment. Each environment serves a distinct purpose, ensuring that applications are rigorously tested before reaching end users, thereby minimizing the risk of introducing errors or untested features into the production environment. Let's cover some key areas regarding an environment strategy:

1. Avoid using the default environment
2. Create multiple environments for specific purposes
3. Establish a naming convention strategy

Avoid using the default environment

By default, each tenant will contain a default environment. This environment is available and shared by all users within the tenant. This can result in many users developing applications, ranging from simple test applications to organizational mission-critical applications. However, using the default environment beyond small-scale or personal applications is generally not recommended. This is especially true for mission-critical applications or those that contain sensitive company data. While there are many reasons for this, some key ones are as follows:

- **Lack of control and oversight**: The default environment is open to many users, which can result in the unregulated proliferation of applications and data. In addition, it will become hard to differentiate between Bob's personal application and an organization's mission-critical application as the list of applications within the environment will inevitably grow over time.

- **Security concerns**: The default environment may not have adequate security measures in place. This is especially important when you're working with sensitive data or critical applications. Having separate environments allows you to set up better security measures, as well as limit who has access.

- **It is a poor development practice**: When developing applications, it's a best practice to have distinct areas for development, testing, and production. This allows you to develop and upgrade applications separately from official or live applications that are currently being used. Separate environments are a much better approach that allows for this.

These are just a few of the reasons, but the bottom line is to create and use separate environments. With that said, let's provide some recommendations.

Create multiple environments for specific purposes

Before we get into developing a strategy for creating multiple environments, it is important to know the different environment types within Power Platform and their intended purpose. There are six different environment types:

- Default

- Sandbox

- Production

- Developer

- Trial

- Microsoft Dataverse for Teams

Table 13.1 provides a brief description of each type:

Environment Type	Description
Default	This is the default environment that's automatically created for every tenant. All individuals within the tenant can create applications. This is best used for smaller-scale projects or personal use; however, it is best to apply DLP measures, something we'll cover later in this chapter.
Sandbox	This environment is best used for development and testing purposes. It is similar to a production environment but can be copied or even reset.
Production	This environment is intended for permanent work. It is best used for the final deployment of completed applications.

Environment Type	Description
Developer	This is a special environment that can be assigned to an individual user to build and modify applications. In other words, individual developers will have a unique environment. This is great for developers to use for training or developing test applications. However, applications cannot be shared with other users.
Trial	This is a free, time-limited environment that's used for evaluating Power Platform capabilities. They expire after 30 days and are limited to one user.
Microsoft Dataverse for Teams	This environment is automatically created when you create an application in Microsoft Teams.

Table 13.1 – Power Platform environment types

> **Tip**
> Because the Default environment is open to all individuals in a tenant and not recommended for mission-critical or sensitive applications, one recommendation is to rename the default environment to something like *Personal Productivity Environment*. This may better describe its intended purpose.

Now that we've covered the different environment types, let's discuss developing an overall environment strategy. Remember, although every tenant comes with a default environment that people can use, having just one environment is not recommended.

This raises the obvious question – how many environments should you have?

The easy answer is – *it depends*. At a minimum, and following ALM principles, you'll want to have separate environments for development and production. However, depending on the complexity of the applications that will be developed, you may want to have a third environment for testing It's also possible to have more environments for items such as **user acceptance testing** (**UAT**), training, or other uses. The bottom line is understanding what you are aiming to build, and then right-sizing it so that it fits your needs. Let's explore various scenarios and examine different environment setups:

- **Scenario 1 – small-scale projects**

  Department A wants to create small scale projects with some Power Apps and Power Automate flows. There are a few developers within the department who will be developing applications for others to use.

  In this situation, it may be best to just have two environments. A "sandbox" environment, called "DEV", allows the developers to both create and test out their projects. Then, a "production" environment, called "PROD", where completed applications will be available to end users. One question that you may be thinking is why use a "sandbox" environment for the DEV environment. As covered previously in *Table 13.1*, a "sandbox" environment type is very similar to a "production" environment type. It is certainly possible to have both DEV and PROD environments be a "production" environment type. Using a "sandbox" environment allows you the ability to either copy or reset the environment.

One alternative to consider when working with small-scale projects is to utilize just one environment for both development and production purposes. Then utilize solutions, which we covered in *Chapter 2*, to maintain and organize your projects. This can help reduce overall costs and maintenance efforts by limiting the number of environments. Be cautious in using this approach and ensure that it doesn't grow beyond a reasonable effort to where separate environments become better.

- **Scenario 2 – mid-scale projects**

  The IT organization is creating applications that will be used across various departments and end users. There may be many applications being maintained for different users.

  In this example, it might be best to have three environments. Two separate "sandbox" environments where one can be used by the developers to create applications, called "DEV", and a second one for users to perform testing., called "TEST". Then a "production" environment, called "PROD" for maintaining completed applications that will be used by the end users. We'll consider this as **Scenario 2(a)**.

  As an alternative, you may want to consider having separate development environments for specific groups or teams (especially if they are dedicated to specific areas within an organization). However, these groups use a shared production environment. We'll consider this as **Scenario 2(b)**.

- **Scenario 3 – large-scale projects**

  An organization is creating complex and high-value applications that will be used across the organization.

  In this example, given the added importance of the project, there should be an additional "sandbox" environment for UAT in addition to the other "sandbox" environments for development and initial testing.

Environment access best practice

When having multiple environments, it's best practice to set the applicable access to the environment users. For example, developers should have the Environment Maker access in a development environment, but not testing or production environments. This prevents changes from being made to apps in those settings. Furthermore, testers should only have access to testing environments to ensure they aren't performing testing on apps in development. Similarly, end users should only have access to the production environment as well.

As you can see, how you set up your environment strategy is going to depend on your situation. We intend to provide an in-depth overview of the environment types and different strategies to help you determine what is best for your situation. *Figure 13.1* provides a summary diagram for the sample scenarios. The suggested environment type is noted in parentheses (see *Table 13.1*):

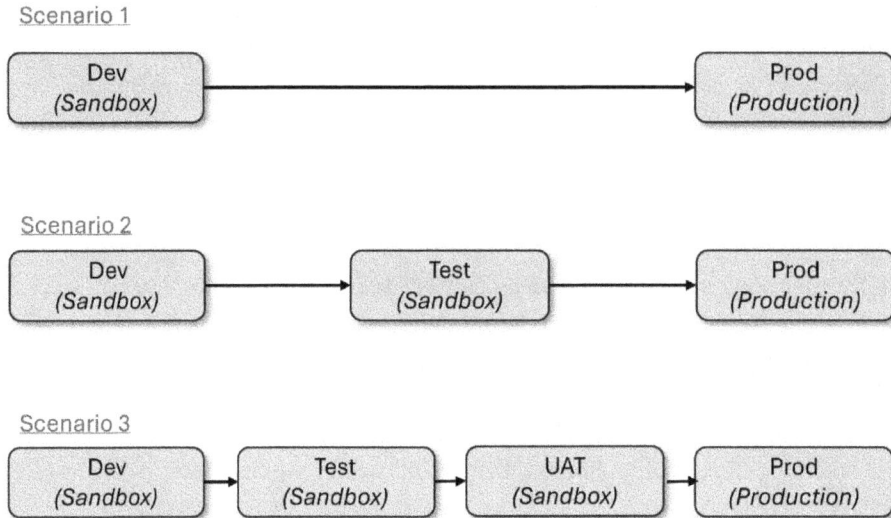

Scenario 1

| Dev
(Sandbox) | → | Prod
(Production) |

Scenario 2

| Dev
(Sandbox) | → | Test
(Sandbox) | → | Prod
(Production) |

Scenario 3

| Dev
(Sandbox) | → | Test
(Sandbox) | → | UAT
(Sandbox) | → | Prod
(Production) |

Figure 13.1 – Examples of different environment strategies

Overall, when setting up your environment strategy, there are several questions you must ask yourself:

- What are the different stages of development (for example, development, testing, and production) that you anticipate needing?

- How many different business units or departments will be using Power Platform? Will they need to have their own environments to keep data separate?

- What are the data residency and compliance requirements? Do certain environments need to be located in specific regions to comply with data residency laws?

- What is the security and access control strategy? How will environments be used to manage access and permissions for different user groups or roles?

- Are there any specific regulatory or audit requirements? Do certain applications need to be isolated due to regulatory or audit reasons?

- What is the intended use of Power Platform in different environments? Will some environments be dedicated to specific functions, such as experimentation, training, or production?

- How will you manage and govern the environments? What governance processes need to be in place to manage multiple environments effectively?

- What is the budget for managing environments? How many environments can you afford to maintain, considering licensing and operational costs?

- What are the performance and scalability considerations? Will separating environments help with performance optimization or scaling the solutions?

- How will you handle disaster recovery and backup? Should environments be separated for better disaster recovery planning?

- What are the interdependencies between applications? Do certain applications need to interact with each other, and how does that impact environment setup?

Let's expand on this by adding some best practices for establishing an environment strategy.

Environment best practices

Here are the best practices for establishing an environment strategy:

- **Share resources with Azure AD Entra ID security groups**: Security groups can be used to manage access to Power Apps, flows, Dataverse security roles, and other Office 365 services such as SharePoint Online. This removes the administrator's burden to update access to individual end users for each component. The app owners can modify that at the security group level without IT involvement (unless IT restricts access to security group management).

- **Automate environment creation**: Administrator connectors (Microsoft Power Platform for Admins) make it possible to create an approval flow where users request environments when IT has restricted environment creation to admins. Central IT can review a request and approve or reject the creation of the environment, without being responsible for manually going to the admin center and creating the environment for the user, just for validating the request details, business justification, data loss prevention (DLP) requirements, and whether enough capacity is available.

- **Create temporary development environments**: As mentioned, it's recommended to have separate development environments as much as possible, and specifically avoid simultaneous app development for critical solutions in the default environment. If environments are created for development purposes, put a deadline on how long the environment should be available to the developers and have a process in place to back up and remove those not being used anymore.

- **Less is better**: Although it's important to make sure resources are reasonably partitioned between projects and business units using environments, it's still important to find a good balance between security and feasibility. Managing shared test and production environments is a good way to facilitate more important solutions while preserving capacity and following best practices. This maintains restricted permissions because test and production have restricted environment permissions, and therefore the end users can't modify the applications.

- **Capacity**: Each environment (besides trial and developer environments) will consume 1 GB to initially provision. This might be a constraint for provisioning environments if your organization doesn't pay for premium Power Apps or Dynamics 365 licenses, and it's also a shared capacity across the tenant that needs to be allocated to those who need it. You can conserve capacity by:

  - Managing shared test and production environments. Unlike shared development environments, permissions in test and production environments should be limited to end-user access for testing.

  - Automate cleanup of temporary development environments and encourage the use of trial environments for testing or proof-of-concept work.

- **Administrator involvement**: It's not always possible to have central IT involved in every development project happening throughout the tenant, especially if the IT team is smaller or there's a larger enterprise to manage. You can reduce the burden on the administrator by:

  - Automating environment creation so the tenant admin only needs to approve the request.

  - Automating development environment cleanup with temporary environments.

- **Clearly communicate your organization's environment strategy to makers**: Set up a SharePoint or a similar site that clearly communicates:

  - The purpose of your default environment.

  - The purpose of shared team and user productivity environments, in addition to other shared environments makers might have access to (for example, training environments) and the process of how to request access to those environments.

  - The purpose of trial environments and how to request them.

  - The purpose of developer environments and how to create them

  - The process of requesting custom environments for specific business units or project purposes.

- **Understand your responsibilities as a maker**: As a maker, there are steps you can take to ensure an efficient and organized area. These include:

  - **Keep the tenant clean**: Be sure to delete your environments, apps, and flows if they are no longer needed. Use test environments if you are experimenting.

  - **Share wisely**: Watch out for oversharing of your environments, apps, flows, and shared connections.

  - **Protect organization data**: Avoid moving data from highly confidential or confidential data sources to non-protected or external storage. We will cover setting up data policies later in this chapter.

The next item we must consider concerning establishing environments is having a good naming convention. Let's touch on that next.

Establish a naming convention strategy

Establishing separate environments is important, but how you name them is also crucial. Having good naming conventions helps drive consistency, provides clarity and purpose, and greatly helps with overall management. Let's dig into some key benefits of a good naming convention.

Benefits of a naming convention strategy

Here are some benefits of having an effective naming convention for your environments:

- **Consistency**: A naming strategy ensures that all environments, resources, and components follow a consistent pattern. This uniformity makes it easier for teams across the organization to understand and navigate the system, reducing confusion and errors. It also helps simplify integration with other systems and fosters better collaboration among teams as everyone adheres to the same rules.

- **Clarity and communication**: A well-thought-out naming strategy helps in quickly identifying the purpose, scope, and ownership of an environment or resource. For example, a name can indicate whether an environment is for production, development, or testing, or which department or project it belongs to. Clear and descriptive names make it easier to communicate about specific environments or resources in documentation, meetings, and discussions.

- **Simplified management**: With a standardized naming convention, IT administrators and other stakeholders can more easily manage, monitor, and maintain environments and resources. It also aids in automating management tasks, such as deployments, monitoring, and reporting. In addition, when names follow a consistent pattern, the risk of selecting the wrong environment or resource during operations is minimized, reducing the chances of costly mistakes.

- **Security and compliance**: Naming conventions can include indicators of security levels, compliance requirements, or data sensitivity, helping to manage and enforce security policies more effectively. Furthermore, in regulated industries, naming conventions can help demonstrate compliance with industry standards or legal requirements, making audits and reviews easier to manage.

- **Troubleshooting and support**: Clear, consistent names make it easier to identify and locate environments and resources during troubleshooting, reducing downtime and speeding up resolution times. Support teams can more easily assist when names indicate the purpose and scope of each environment, reducing the need for additional clarification and speeding up issue resolution.

- **Cost management**: A naming strategy can include elements that help track resource usage and costs, such as project codes or department identifiers. This enables better financial oversight and cost management. By clearly identifying environments and their purposes, a naming strategy helps avoid the creation of redundant or duplicate environments, which can lead to unnecessary costs.

- **Governance and control**: A naming strategy supports governance by enforcing a structured approach to creating and managing environments. This helps ensure that resources are used efficiently and that organizational policies are followed. Consistent naming conventions make it easier to audit environments and resources, track changes, and ensure compliance with internal and external standards.

In summary, a naming strategy is vital for maintaining order, enhancing communication, improving efficiency, and ensuring the security and scalability of your Power Platform environments and resources. It's a foundational element of effective governance and resource management.

With these benefits in mind, let's move to some recommendations to help you establish a naming convention strategy.

Developing a naming convention strategy

When developing naming conventions for Power Platform environments, several key topics should be considered to ensure consistency, clarity, and ease of management. Here are some suggested areas to consider:

- **Environment types and purpose**: Clearly differentiate between environments based on their purpose. For example, you might use prefixes such as *PROD-*, *DEV-*, *TEST-*, or *UAT-* to indicate the environment type. For specialized environments, consider using naming conventions for environments dedicated to specific purposes, such as *SEC-* for security testing or *TRN-* for training environments.

- **Business unit or department identification**: Consider incorporating department or business unit identifiers, such as *FIN-* for finance, *HR-* for human resources, or *MKTG-* for marketing, to indicate which part of the organization the environment serves. For environments supporting multiple departments, consider a neutral or combined identifier, such as *XFN-* for cross-functional.

- **Region or geography**: Consider using regional or geographical identifiers if environments are region-specific, such as *NA-* for North America, *EMEA-* for Europe, Middle East, and Africa, or *APAC-* for Asia-Pacific.

- **Project or solution names**: If the environment is tied to a specific project or solution, incorporate a project code or short name, such as *CRM-* for a CRM project or *PAY-* for a payroll system.

- **Compliance and security**: If the environment must comply with specific regulations (for example, GDPR and HIPAA), consider incorporating this into the name – for example, *HIPAA-* or *GDPR-*. Reflect security levels in the name if there are different security requirements by using identifiers such as *SEC1-*, *SEC2-* or *LOW-*, *MED-*, *HIGH-*.

> **Important note**
>
> Although we have highlighted several possible prefixes, it is also important to ensure that the naming convention is concise and avoids being unnecessarily long or complex. To provide an example, for this book, we established a sample environment named **Event Planning Project Development**. However, as *Figure 13.2* shows, only the first part of the name appears. Had we added *DEV* or *PROD* at the end, it wouldn't be readily visible:
>
>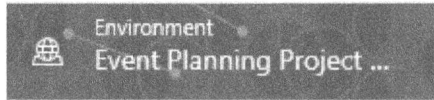
>
> Figure 13.2 – An example of a long environment name that is not fully displayed

In summary, having a robust naming convention strategy can help enhance organization, clarity, and management of Power Platform environments.

Now that we've covered key areas around developing environments let's discuss another aspect of environments: understanding service accounts and individual accounts.

Understanding service accounts versus individual accounts

In the previous section, we discussed the importance of establishing a good strategy when setting up environments for your applications. However, an equally important topic is around service accounts and individual accounts. Why is this? Here are two very simple examples.

- **Example 1**: *Tom builds an automated flow that sends automated emails to users. Because the flow is tied to Tom's Microsoft account, the email appears to be sent from Tom. Even though the email provides clear instructions for people to contact another email address for questions, because the email is sent from Tom's account, he will inevitably end up receiving questions from users who neglect to read the email fully.*

- **Example 2**: *Betty is your team's Power Platform expert and has built some amazing applications, including Power Apps and Power Automate flows. One day, Betty happens to buy the winning lottery ticket, wins millions of dollars in prize money, and shows up the next day to turn her computer and badge in for good. While it might be good for Betty, you will quickly realize that the applications that Betty built were done under her individual Microsoft account. Unless she shared her work with other users, there is a significant risk that the work could be lost completely.*

How can you avoid this? One way is to ensure people are sharing their entities (applications, flows, and so on) with other individuals. Another way is to use a service account.

What is a service account?

In the context of Power Platform governance, a **service account** is a user account that is designed to perform specific automated tasks within an environment. Unlike individual user accounts, which are tied to specific employees or users, service accounts are dedicated to executing background processes, handling integrations, and automating routine operations. These accounts are often configured to run unattended, providing the necessary credentials for applications or services to interact with Power Platform without requiring manual intervention.

Key differences from individual accounts

While both service accounts and individual user accounts can access Power Platform resources, their purposes and management differ significantly. Individual accounts are typically tied to a single user, who is responsible for performing tasks within the platform based on their role and permissions. These accounts are subject to changes when an employee joins, leaves, or changes roles within an organization, which can impact access and control.

In contrast, service accounts are not linked to any specific individual. They are created and managed specifically to perform consistent, ongoing tasks, such as data integrations, running automated workflows, or accessing APIs. This distinction ensures that the functionality these accounts provide remains uninterrupted by personnel changes, making them ideal for scenarios where continuous operation is critical.

Use cases for service accounts

Service accounts are particularly valuable in several scenarios within the Power Platform:

- **Automation**: Many processes within Power Platform, such as scheduled flows in Power Automate or background data processing tasks in Power BI, require consistent and reliable access to resources. Service accounts ensure that these tasks can run smoothly without the need for manual login or supervision. Going back to *Example 1* in this section, Tom could use the service account instead of his own account for the automated emails. The result would be emails being sent from the service account.

- **Integration**: When integrating Power Platform with other systems, service accounts provide a secure and stable means of authenticating the connection. This is essential for maintaining the integrity of data and operations as these accounts can be configured with the necessary licenses and/or permissions to access specific resources without exposing individual user credentials.

- **Maintenance and monitoring**: Service accounts are often used for monitoring and maintaining environments, such as running health checks, logging activities, or performing routine updates. By using service accounts, organizations can ensure that these critical tasks are performed consistently, without dependency on any individual user's availability.

When using service accounts, there are some areas to consider as well. These include:

- **Licensing**: Service accounts are like individual accounts and therefore require their own licensing.

- **Security risks**: Service accounts are generally used by multiple users and it can therefore be difficult to attribute actions to any specific user. Furthermore, when individual users should no longer have access to the service account, this requires changing the login information.

- **Deactivation and cleanup**: Service accounts can become "orphaned" when the original user base, application, or service they were associated with is no longer in use. These orphaned accounts may linger in the system without any clear ownership or responsibility,

In summary, service accounts can play a vital role in maintaining the stability, security, and efficiency of operations within the Power Platform. Their ability to perform tasks independently of human users makes them an essential component of any governance strategy, particularly in environments where automation and integration are key priorities. However, it is important to consider if the risks of having a service account are important to you. Some organizations may not be comfortable with using service accounts because, for example, a shared resource with admin privileges cannot be tracked to a single person. This is valid but can be mitigated with steps such as enforcing location-based conditional access, tracking the audit logs to an IP, or more extensive methods like maintaining a secure access workstation that requires user identification during use and restricting the service account access to that device.

> **Development tip**
>
> After reading about the importance of using service accounts, you might be thinking about whether it's best to do all development using a service account only and not individual accounts It is certainly possible for Power Platform entities (applications, flows, and so on) to still be developed using individual accounts, and there may be specific needs for this. However, it is a best practice to either share these entities with other users or the service account to mitigate the risk of loss when individuals leave the organization.

Now that we have covered service accounts, let's discuss the concept of DLP and how Power Platform data policies can help.

DLP – implementing data policies in Power Platform environments

DLP is a critical topic to be aware of concerning Power Apps and Power Platform. First, let's discuss what DLP is. While there are varying definitions of DLP, it is essentially a concept of identifying and preventing sensitive data from being shared unsafely or inappropriately. Furthermore, certain regulations such as HIPAA and GDPR have strict requirements around the use and safeguarding of sensitive data.

Within Power Platform, DLP is done by establishing **data policies**. Data policies act as guardrails to prevent users from exposing sensitive data. They define which connectors can be used either within Power Apps, Power Automate, or other components within Power Platform and within all (or specified) environments. A simple example of this is preventing a user from accessing sensitive business data in SharePoint (via the SharePoint connector) and accidentally publishing it on social media (such as through the LinkedIn connector).

Let's consider some key concepts around data policies.

Key concepts of Power Platform data policies

When working with data policies, there are three key areas you should be familiar with:

- **Connectors**: Connectors are the Power Platform API tools that allow information to be passed back and forth between an application or service. A few popular connectors include Office 365 Outlook, Dynamics, SharePoint, Teams, OneDrive, and many other Microsoft applications. However, connectors are also available to many other third-party applications, such as Salesforce, Dropbox, X (formerly Twitter), and Google services. There are over 1,000 connectors available.

- **Data classification categories**: When setting up data policies, three classifications are available:

  - **Business**: Connectors that handle business-related data, such as SharePoint or Dynamics

  - **Non-Business**: Connectors for personal or social media services

  - **Blocked**: Connectors that are entirely restricted to prevent data exposure

> Key note
>
> Be aware that applications, such as Power Apps or Power Automate, cannot utilize connectors that are in different data classification categories. For example, data policies prevent you from implementing a Power Automate flow that uses a connector that is classified in the *Business* category alongside a connector that is in the *Non-Business* category. We will cover this topic in more detail shortly.

- **Data policy scope**: Data policies can be implemented to cover all environments within a tenant or specific to certain environments.

We will cover each of these as we walk through setting up data policies.

Establishing Power Platform data policies

Let's walk through the steps of creating a data policy. Please note that you will have to have access to the Power Platform Admin Center to do this.

Step 1 – access the Power Platform Admin Center

This can be accessed either from Power Apps or Power Automate under the **Power Platform** option via the left navigation area. See *Figure 13.3*:

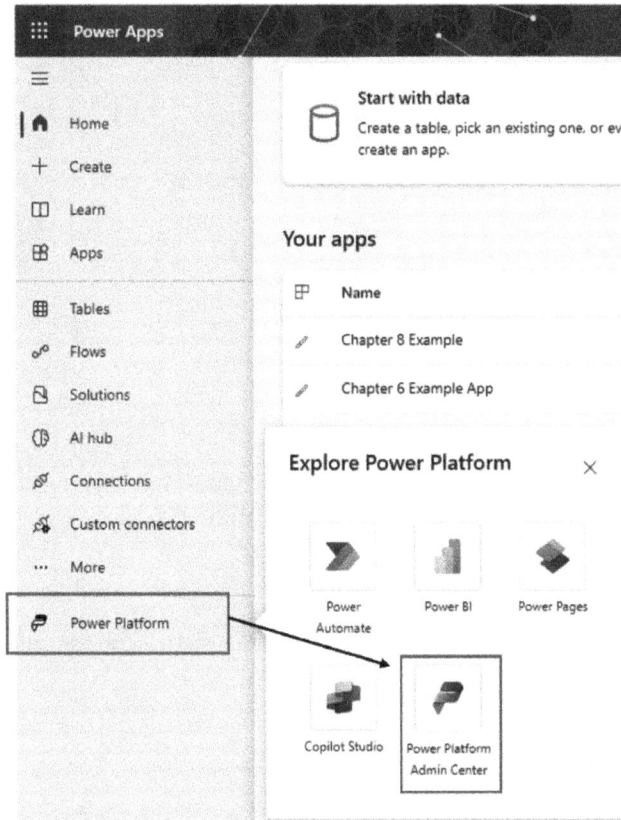

Figure 13.3 – Accessing the Power Platform Admin Center

Step 2 – create a new data policy

From the left navigation area, select **Policies** and then **Data policies**. This screen will display all existing policies. Here, you'll have the option to create a new policy. See *Figure 13.4* for an example:

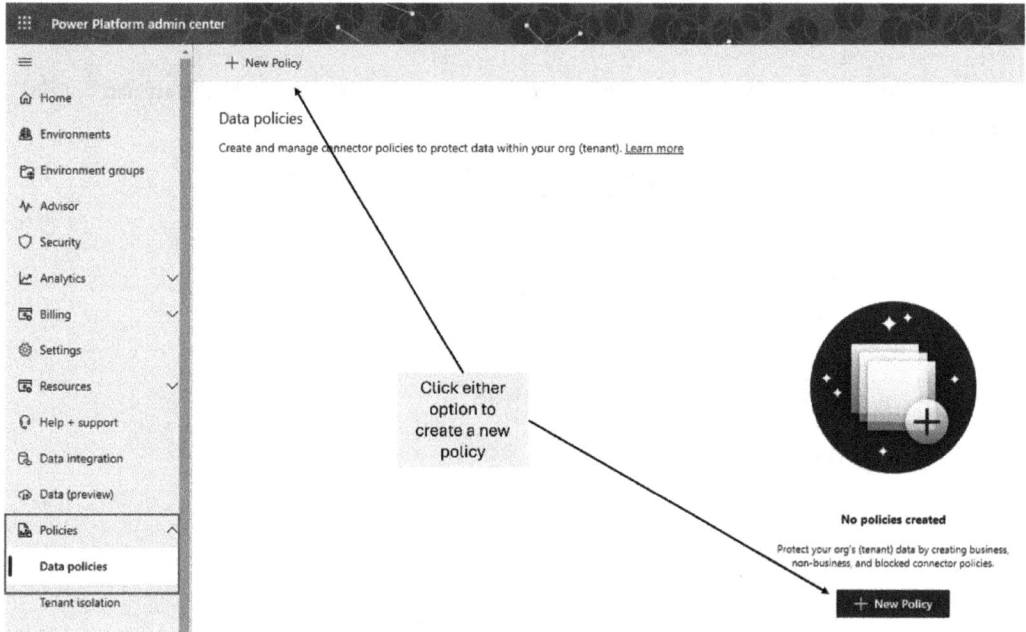

Figure 13.4 – Adding a new data policy

Step 3 – give the data policy a name

Choose a name for the data policy. As an example, *Figure 13.5* shows a policy called **Sample Policy**. Click on the **Next** button to continue:

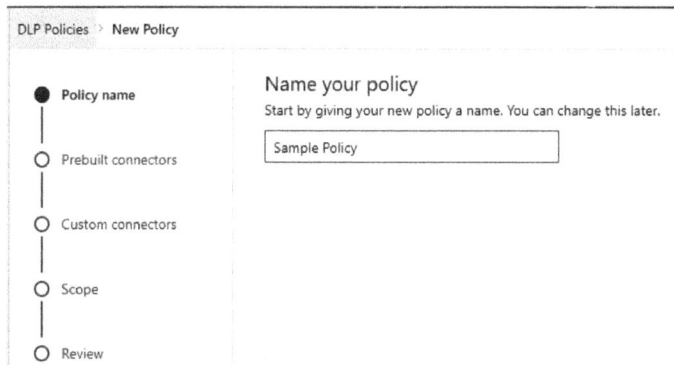

Figure 13.5 – Creating a name for the data policy

Step 4 – assign prebuilt connectors

This is the step where connectors are assigned to one of three categories:

- **Business**: These are connectors for sensitive data
- **Non-business**: These are connectors for non-sensitive data
- **Blocked**: These connectors cannot be used in Power Apps or Power Automate

Figure 13.6 provides an example. All connectors are currently in the **Non-business** category:

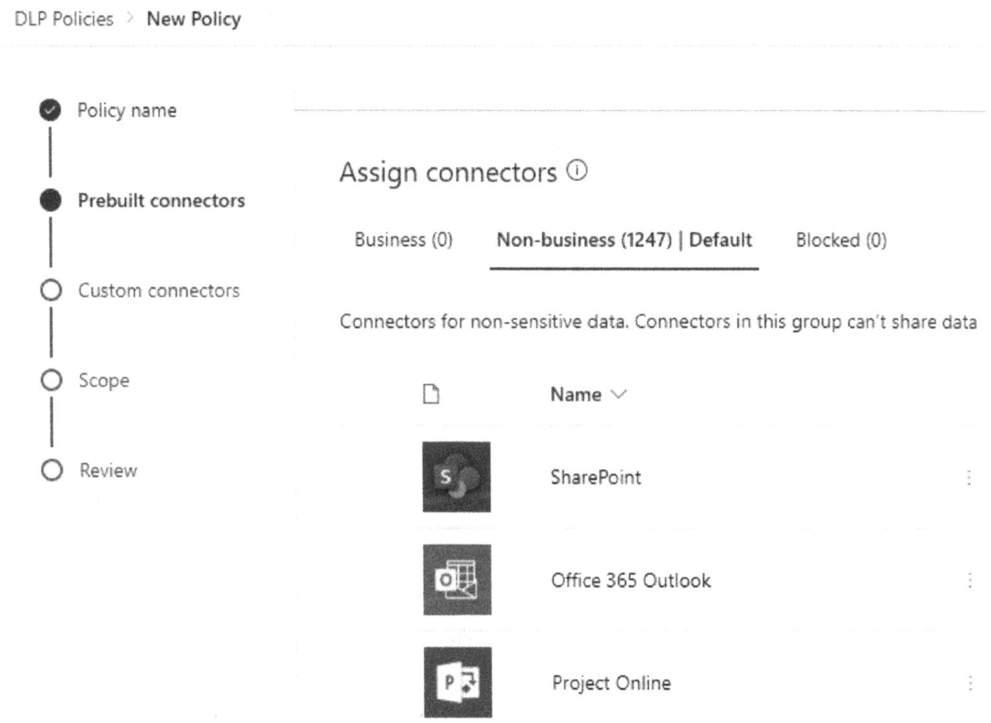

Figure 13.6 – Example of the three types of connectors

> **Important note**
> The key to data policies is that connectors in one category cannot be used with connectors in another category.

In our example, we want to establish several data sources that will be used for business purposes. For this, we will select the connectors for **SharePoint**, **Office 365 Outlook**, **Power BI**, and **Salesforce**. Once we've selected these, we can click on the **Move to Business** button. This is shown in *Figure 13.7*:

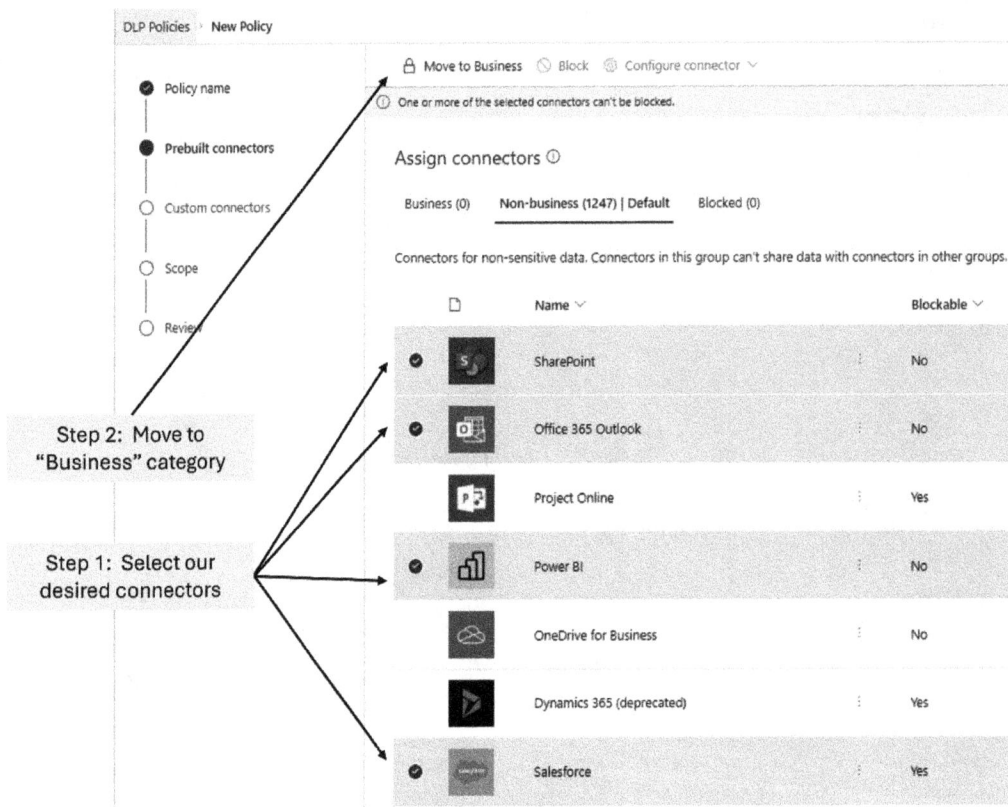

Figure 13.7 – Moving the selected connectors to the Business category

With that, we now have these connectors in the **Business** category, which means they're separate from all the other connectors. Looking at *Figure 13.8*, we now have four connectors that we can use for business purposes in the **Business** category, but these are now separate from LinkedIn, a social media site:

Business (4)	Non-business (1243)	Default		Business (4)	Non-business (1243)	Default
Connectors for sensitive data. Connectors in this grou				Connectors for non-sensitive data. Connectors in thi	

Connectors for sensitive data. Connectors in this grou

□	Name ∨
S	SharePoint
O	Office 365 Outlook
	Power BI
	Salesforce

Connectors for non-sensitive data. Connectors in thi

□	Name ∨
in	LinkedIn
in	LinkedIn V2

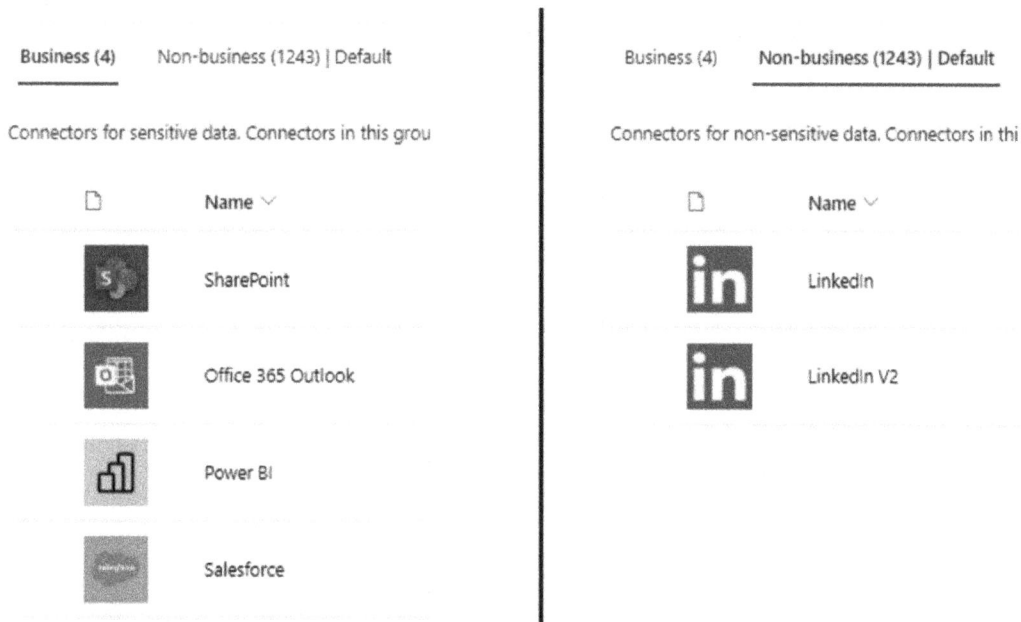

Figure 13.8 – Example of our four business connectors separate from a social media site

Step 5 – assign custom connectors

Like *Step 4*, this step allows you to categorize custom connectors, should you have any. Custom connectors are beyond the scope of this book, so we will skip this step for now. For more information on creating custom connectors, please visit Microsoft's site at `https://learn.microsoft.com/en-us/connectors/custom-connectors/`.

Step 6 – assign the environment scope of the policy

This step allows you to assign the overall scope of the policy. You have three options:

- **Add all environments**: This policy will apply to all environments in the tenant, including any new environments created in the future.

- **Add multiple environments**: This policy will apply to only those environments that you designate (which will add an additional step). Note that this policy will only apply to environments you select and will not apply to any new environments added in the future.

- **Exclude certain environments**: This policy will **not** apply to those environments that you designate (which will add an additional step). However, this policy will apply to any new environments created in the future.

For our purposes, we'll select the first option and have this policy apply to all environments. This is shown in *Figure 13.9*:

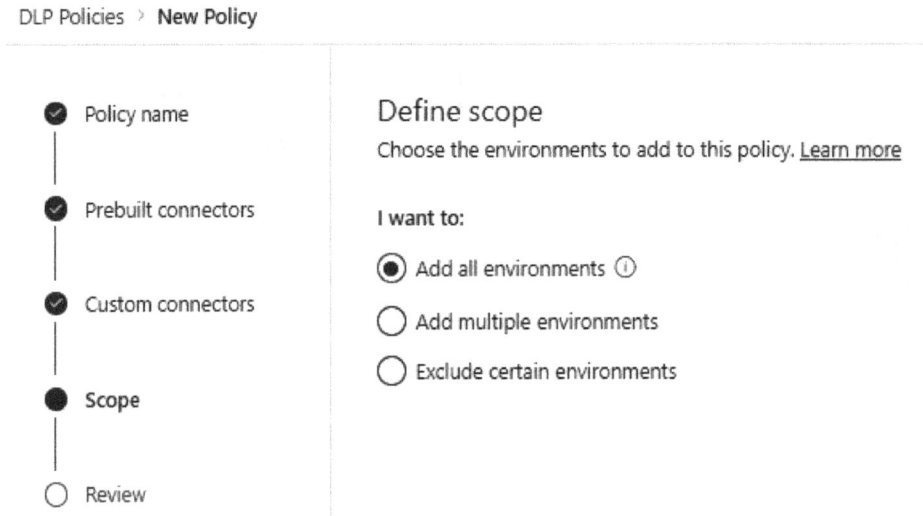

DLP Policies > **New Policy**

- ✓ Policy name
- ✓ Prebuilt connectors
- ✓ Custom connectors
- ● **Scope**
- ○ Review

Define scope
Choose the environments to add to this policy. Learn more

I want to:

- ◉ Add all environments ⓘ
- ○ Add multiple environments
- ○ Exclude certain environments

Figure 13.9 – Defining the environment scope of the data policy

Step 6A – specify selected/excluded environments (if applicable)

As noted in *Step 6*, unless the data policy is added to all environments, an additional step will be added to allow you to select which environments the policy applies or is excluded from.

Step 7 – do a final review

This is where you do a final review and then create the policy.

By setting up data policies, you can add guardrails to prevent people from exposing sensitive data. Now that we've covered these topics, let's learn how they can be applied in ALM.

Best practice tip

Implement a data policy as early as possible in the default environment. This environment is the most challenging to monitor/manage and delaying a data policy implementation can lead to complications. If users have already developed applications/flows with connectors you want to restrict, you will have to contact them, assess their needs, and potentially disrupt productivity. To avoid this, establish a restrictive data policy from the start to maintain control and security.

Understanding ALM concepts

ALM plays a critical role in managing applications across these environments. It's a process that encompasses the processes of developing, testing, deploying, and maintaining applications, providing a framework for **continuous integration (CI)**, delivery, and governance. Implementing ALM practices helps teams manage changes systematically, track versions, and ensure consistency across different environments. Scenarios where ALM becomes indispensable include large-scale projects with multiple developers, frequent updates, and complex integration requirements.

The following diagram provides a general outline of common ALM steps:

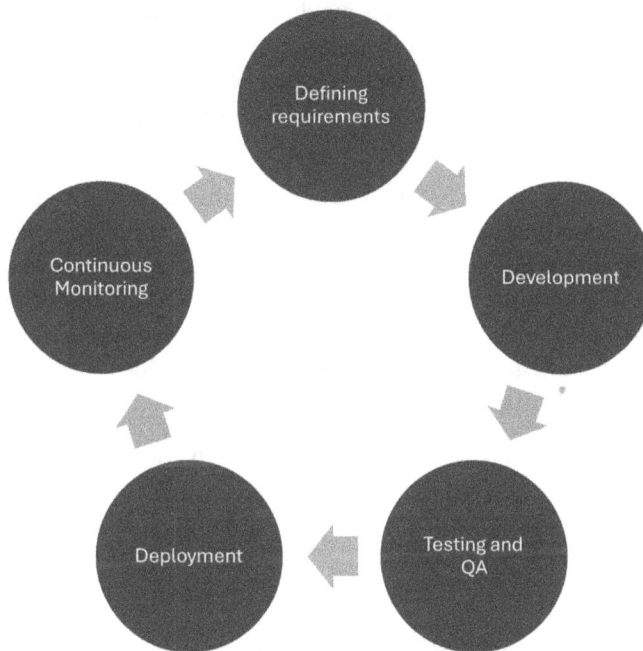

Figure 13.10 – Diagram of ALM steps

Let's provide an overview of each step:

- **Defining requirements**: This step involves gathering, analyzing, and documenting the needs and expectations of stakeholders, including business users, developers, and IT administrators. Clearly defined requirements ensure that all parties have a shared understanding of the project's scope, objectives, and constraints. We covered much of this in *Chapter 1*.

- **Development**: This step is where the actual development of the application takes place. During this phase, developers translate the requirements defined in the earlier stages into a functional product. The development stage involves designing the architecture, writing code, creating databases, and integrating various components to build the application.

- **Testing and quality assurance (QA)**: This step involves ensuring the application meets the defined requirements and functions correctly under various conditions. This involves systematically evaluating the application to identify and fix defects, verify performance, and validate that it meets both functional and non-functional requirements. Depending on the size of the project, there could be several iterations of testing, including initial testing, but also testing such as UAT, system testing, and/or integration testing. The overall objective here is to ensure the application works as expected.

- **Deployment**: This step is where the completed and tested application is moved to the production area for use by end users. We discussed developing an environment strategy earlier in this chapter, and this stage generally involves moving the application from a development (or testing) environment to a production environment. The overall objective is to make the live version accessible to the intended audience. Deployment can vary in complexity, depending on the application's architecture, the environments involved, and the specific deployment strategies employed.

- **Continuous monitoring**: This step focuses on ensuring the ongoing performance and security of the application after it has been deployed to the end users. This stage may involve using monitoring tools and techniques to continuously track the application's behavior, identify potential issues, and ensure it meets performance standards. Additionally, user experience monitoring helps capture real-time feedback from users, identifying areas where the application can be improved. Furthermore, as fixes or improvements are identified, these may result in new requirements, which starts the ALM cycle over again.

It is important to note that this section is not meant to cover all aspects of ALM. And, similar to project management, which we covered in *Chapter 1*, there are different variations of ALM. Our intent, however, is to provide overall concepts to the approach.

When to use ALM

ALM is a comprehensive framework that facilitates the management of an application's life cycle from its inception through development, testing, deployment, and maintenance. Implementing ALM is essential in scenarios where systematic and organized management of applications is crucial to ensuring consistency, quality, and efficiency across development stages.

One key scenario where ALM is indispensable is in large-scale projects involving multiple developers. In such projects, ALM tools and practices help coordinate efforts, manage code versions, and integrate changes seamlessly. By using version control systems such as Git, development teams can track changes, manage branches, and resolve conflicts effectively, ensuring that the application evolves in a controlled and predictable manner.

Frequent updates and feature enhancements are another scenario where ALM proves valuable. In dynamic environments where applications need regular updates, ALM provides a structured process for managing releases. CI and **continuous deployment (CD)** pipelines automate the testing and deployment of new features, reducing the risk of errors and accelerating the delivery of updates to users. This continuous delivery approach ensures that applications remain up-to-date and aligned with business requirements.

Complex integration requirements also necessitate the use of ALM. When applications need to integrate with various external systems, ALM practices help manage dependencies, configurations, and compatibility issues. Automated testing and deployment pipelines ensure that integrations are tested rigorously before deployment, minimizing the risk of integration failures in the production environment.

In addition to these scenarios, ALM is crucial for maintaining high standards of governance and compliance. Organizations often need to adhere to regulatory requirements and internal policies regarding data security and application management. ALM provides the tools and processes to enforce these standards consistently across all environments. By implementing ALM, organizations can ensure that applications comply with regulatory requirements, maintain data integrity, and minimize risks associated with non-compliance.

In summary, ALM is essential when you're managing large-scale projects, frequent updates, complex integrations, and strict governance requirements. By leveraging ALM tools and practices, organizations can achieve greater control, consistency, and efficiency in their application development and deployment processes.

Now that we've covered ALM in general, let's discuss key principles that can be applied in Power Apps.

Applying ALM in Power Platform

Applying ALM principles to Power Platform is essential for ensuring the structured, efficient, and secure development, deployment, and maintenance of business applications. As organizations increasingly rely on Power Apps, Power Automate, and other components of Power Platform to drive digital transformation, a well-defined ALM strategy becomes critical.

Here are some key areas that can help support your ALM strategy:

- **Implement a robust Power Platform environment strategy**: We covered this at the beginning of this chapter. Setting up different environments for development, testing, and production is important in supporting ALM.

- **Utilize Power Platform solutions**: We covered creating and using Power Platform solutions in *Chapter 2*. Solutions are vital for organizing, managing, and deploying your applications, flows, and customizations. Solutions also streamline the development process and ensure changes can be efficiently moved across your environments. There are also native deployment pipelines available in the Power Platform. These pipelines provide a streamlined, built-in solution for moving solutions across environments, making it easier to manage and automate deployments.

- **Implement governance and compliance**: We covered implementing data policies to help support your overall DLP strategy previously. The Power Platform Admin Center plays an important role in this and helps support ALM.

- **Change management**: Change management is a critical aspect of ALM within Power Platform, ensuring that updates, enhancements, and modifications to applications are handled systematically and with minimal disruption to users. Effective change management begins with a well-defined process for submitting, reviewing, and approving change requests.

- **Training and documentation**: Training and documentation are integral to the successful implementation of ALM. Training should cover not only the technical aspects of the platform, such as building and deploying applications, but also ALM best practices, governance policies, and compliance requirements. This ensures that everyone involved in the development and maintenance of applications understands their role in the ALM process and can contribute effectively. Training sessions can be conducted through workshops, online courses, or hands-on labs, and should be tailored to different roles within the organization, such as developers, administrators, and business users. Comprehensive documentation is also a cornerstone of a successful ALM. Documentation serves as a reference for current team members, a training tool for new hires, and a record of decisions and processes that can be used for troubleshooting and future development.

- **Ensure strong collaboration and communication**: Collaboration and communication are fundamental to the success of ALM. As Power Platform projects often involve cross-functional teams, including developers, IT professionals, business analysts, and end users, establishing clear and effective collaboration and communication channels is essential for ensuring that everyone is aligned and working toward the same goals. Establishing efficient feedback loops is essential for capturing user input and making necessary adjustments during the ALM process. This can be achieved by regularly soliciting feedback through surveys, user testing, and direct communication with end users.

There is much more that can be covered regarding ALM and Power Platform. Microsoft offers a significant number of resources and tools at `https://learn.microsoft.com/en-us/power-platform/alm/`.

Summary

In this chapter, we covered important topics around application governance, security, and ALM. First, we covered the importance of establishing a strategy for creating and using Power Platform environments. We provided different examples and some key questions to consider in establishing environments. We also covered having a naming convention strategy. Next, we discussed using service accounts versus individual accounts. We covered what a service account is and how it differs from an individual account while providing some potential use cases. Then, we discussed the concept of DLP in Power Platform environments and how data policies can be used to support this. Finally, we concluded by showing how these concepts can be applied to support an ALM strategy.

In the next chapter, we will consider the important topic of error handling.

14

Error Handling

Error handling is a crucial aspect of building robust and user-friendly applications in Power Apps. In any application, errors can occur due to various factors such as incorrect user input, development issues, connectivity issues, or unexpected conditions within the app's logic. Effective error handling ensures that these issues are managed gracefully, preventing crashes or unexpected behavior and providing users with clear guidance on how to resolve the problem.

In Power Apps, error handling is not just about preventing failures; it's also about improving the overall user experience. When errors are handled properly, users are less likely to encounter confusing or frustrating situations, which can lead to higher satisfaction and trust in the application. Moreover, well-implemented error handling can help developers identify and resolve issues more efficiently, leading to more stable and reliable apps.

In this chapter, we're going to cover the following main topics:

- Overview of error handling
- Using built-in error functions in Power Apps
- Using the `OnError` app property
- Creating custom error messages

Technical requirements

To successfully engage with the materials in this chapter, you'll need to have access to Power Apps for creating and managing your applications

Overview of error handling

In this section, we'll explore various types of errors that can occur in Power Apps, including validation errors, runtime errors, delegation errors, and connectivity issues. It will also cover the built-in functions and tools available in Power Apps to handle these errors effectively. By understanding and implementing proper error-handling techniques, developers can create Power Apps that are not only functional but also resilient and user-friendly.

Types of errors in Power Apps

Understanding the different types of errors that can occur in Power Apps is fundamental to implementing effective error-handling strategies. In this section, we will explore common categories of errors that might be encountered, including runtime errors, validation errors, delegation errors, and network or connectivity errors:

- **Runtime errors**: Runtime errors are issues that arise during the execution of an app. These errors can occur for a variety of reasons, such as dividing by zero, attempting to access a null value, or performing operations on data that does not exist. Unlike validation errors, which are often predictable, runtime errors can be more challenging to anticipate. Power Apps provides functions such as `IfError()` and `IsError()` to help developers catch and manage these errors dynamically, allowing the app to continue running smoothly even when unexpected conditions arise.

- **Validation errors**: Validation errors occur when the data entered by the user does not meet the predefined criteria or constraints. For example, if a form field requires a numeric input and the user enters text, this would trigger a validation error. These errors are essential to catch early to ensure data integrity and prevent incorrect data from being submitted to data sources.

- **Network and connectivity errors**: Network and connectivity errors occur when the app is unable to establish a connection to external data sources or services. These errors can result from various issues, such as network outages, server downtime, or incorrect configuration of data connections. Handling network errors is essential for maintaining the app's functionality, especially in scenarios wherein real-time data access is crucial. Developers can use the `IfError()` function to provide alternative actions or retry mechanisms when network issues are detected, ensuring that the app remains resilient in the face of connectivity challenges.

By understanding these types of errors, developers can better anticipate potential issues in their Power Apps and implement strategies to handle them effectively. The following sections will delve deeper into the specific functions and techniques available in Power Apps to manage these errors, providing a comprehensive approach to error handling.

Let's start by discussing some built-in error functions in Power Apps.

Using built-in error functions in Power Apps

Power Apps provides a range of built-in functions designed to help developers manage and respond to errors effectively. These functions allow for the detection, handling, and communication of errors within the app, ensuring that unexpected issues can be managed without disrupting the user experience. In this section, we will explore key error-handling functions available in Power Apps:

- `IfError()`
- `IsError()`
- `Notify()`
- `IsBlankOrError()`
- `Errors()`

IfError() function

The `IfError()` function allows you to handle potential errors by defining alternative actions or fallback values. It works similarly to an `if-else` statement, but specifically for error conditions. When an expression is evaluated, and an error occurs, `IfError()` can catch the error and provide a different outcome, ensuring that the app continues to function smoothly. It uses the following expression:

```
IfError(Value, Fallback, …, [Default Result])
```

- `Value`: This is the value that is being tested for an error
- `Fallback`: This is the alternate result to be provided if the value results in an error
- …: This allows you to add additional value-fallback pairs
- `Default Result`: This is optional, but allows you to provide a result if the value does not generate an error

Let's provide a specific example.

In *Figure 14.1*, we display a form with two sets of input fields. **Number_1** and **Number_2** are shown with both having **5** as their values. The results are then shown to the right using a simple formula that multiples the **.Text** values of each, thus showing **25**.

However, when **5** is replaced with **Five**, this generates an error as a number cannot be multiplied by text. Having this error could result in the app not working properly.

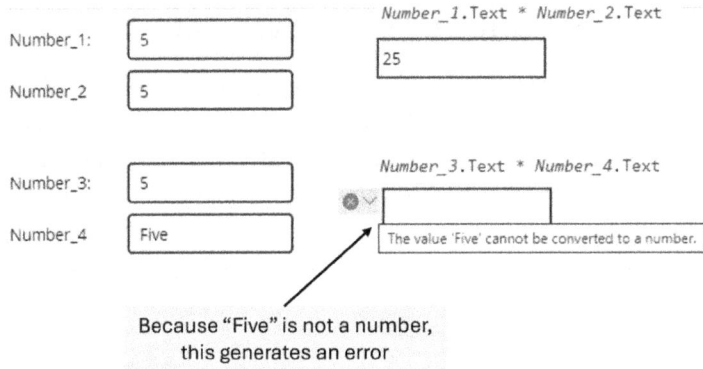

Figure 14.1 – Displaying a form that shows an error resulting from a formula

However, if we were to use an IfError() function, we could display a response for the user as shown in *Figure 14.2*.

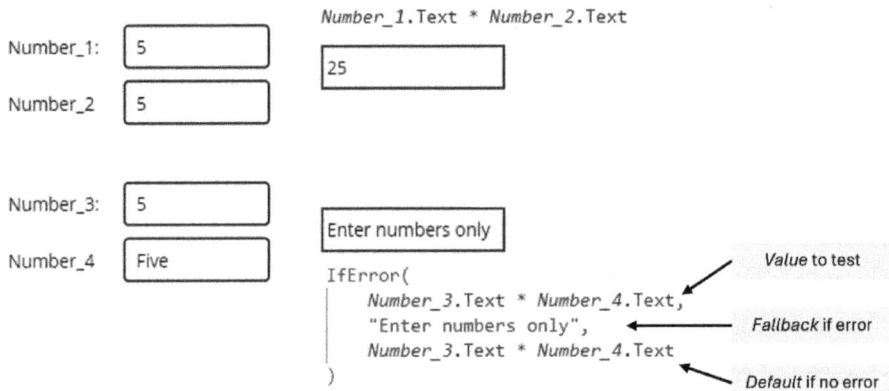

Figure 14.2 – Using the IfError() function to provide a notification

By using the IfError() function, we can not only prevent the app from malfunctioning but also display a notification for the user.

IsError() function

The IsError() function is used to test whether a particular value or expression results in an error. This function is especially useful when developers need to conditionally handle errors based on specific conditions within the app. It returns a Boolean true or false value. It uses the expression that follows:

```
IsError(Value)
```

`Value` is the value that is being tested for an error.

Let's provide a specific example. In *Figure 14.3*, we have a form with two numbers. The **IsError?** field shows whether the multiplication of these two numbers will result in an error.

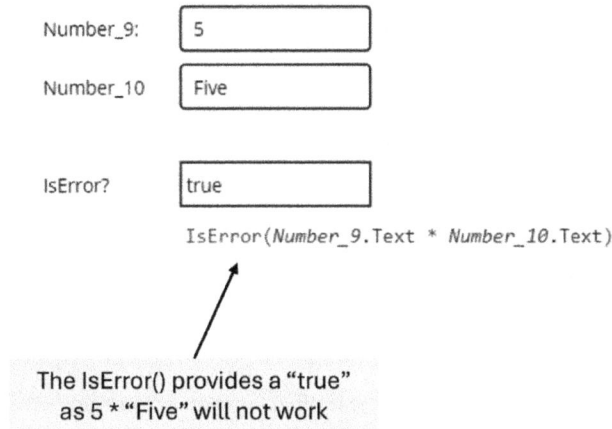

Figure 14.3 – Example of using the IsError() function

Let's take this a step further and also include the `Notify()` function.

Notify() function

The `Notify()` function is used to display messages to users, providing feedback when errors occur or specific conditions are met. Notifications can be used to inform users of errors, warnings, or successes, helping to guide them through the app's workflow. It uses the expression that follows:

```
Notify(Message, NotificationType, Timeout)
```

- `Message`: This is the message to be displayed.
- `NotificationType`: This determines how the message is displayed. Options include `.Error`, `.Information`, `.Success`, and `.Warnings`. It is an optional parameter.
- `Timeout`: This is a number that represents, in milliseconds, how long to wait before automatically closing the message. It is an optional parameter.

Let's expand our example in *Figure 14.3* and add a notification upon submission of the form. This is shown in *Figure 14.4*.

Number_9:   5

Number_10   Five

IsError?   true

Multiply Numbers

Button "OnSelect"
Property

```
If(
    IsError(Number_9.Text * Number_10.Text),
    Notify("There was an Error.  Please ensure that only numbers are entered.",
        NotificationType.Error, 5000),
    Navigate(HomeScreen)
```

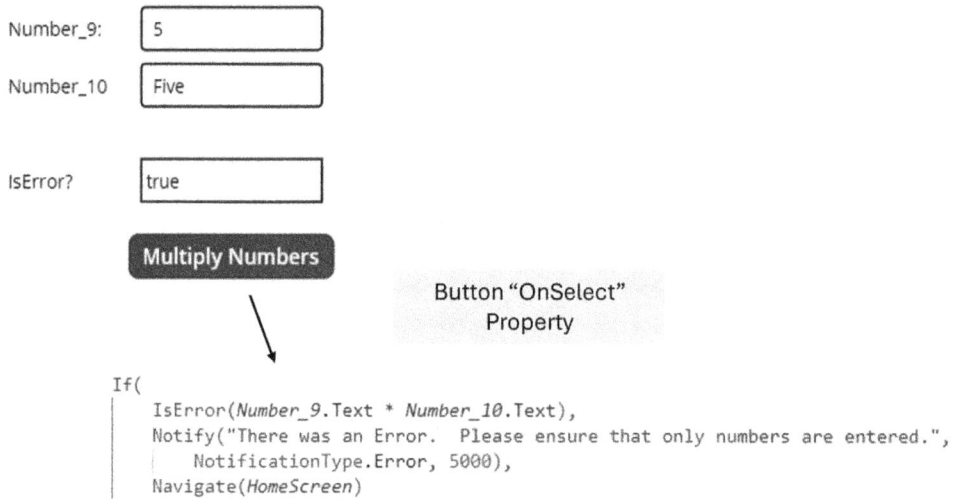

Figure 14.4 – Applying the Notify() function to a button

In this example, we've added an If() statement to test whether there is an error in the multiplication of these two values.

First, we use IsError() to check whether an error results from two fields being multiplied.

If an error results, this will generate the Notify() function, which will be displayed to the user. We also include a Timeout value of 5,000 milliseconds so the message will wait for five seconds until disappearing. *Figure 14.5* provides an example of the error that the user will see.

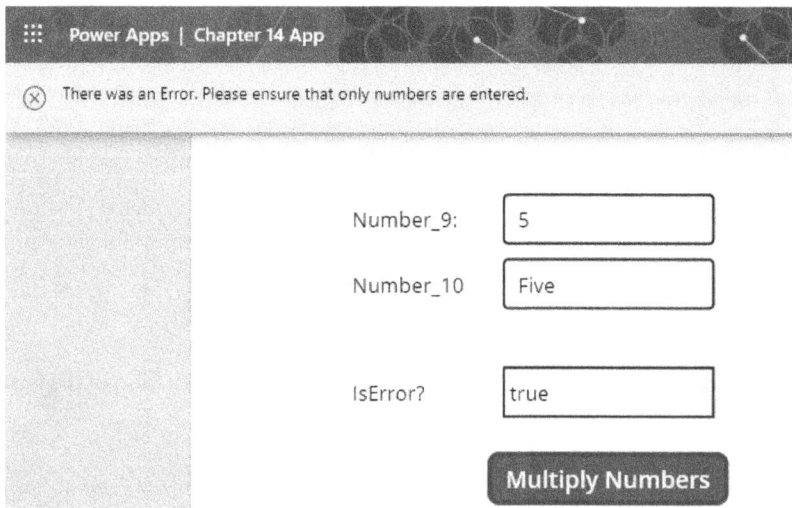

⠿ **Power Apps | Chapter 14 App**

ⓧ   There was an Error. Please ensure that only numbers are entered.

Number_9:   5

Number_10   Five

IsError?   true

Multiply Numbers

Figure 14.5 – Example of the error message being displayed

If no error results, the user will be then sent to a screen called HomeScreen.

IsBlankOrError() function

We previously covered the `IsError()` function. There is a similar `IsBlankOrError()` function. This function will test for both error and blank values. It also returns a Boolean `true` or `false` value and uses the following expression:

```
IsBlankOrError(Value)
```

`Value` is the value that is being tested for an error.

Errors() function

The `Errors()` function returns a table of error records related to a specific data source. This function is particularly useful when working with external data sources, as it provides detailed information about any errors that occur during data operations, such as adding, updating, or deleting records.

The table returned will contain the following columns:

- **Record**: This column contains the record from the data source that encountered the error. If the error occurred when creating a new record, this will be blank.

- **Column**: This column indicates the name of the specific column that caused the error, if applicable. If the error cannot be traced to a specific column, this field will be blank.

- **Message**: This will be a description of the error. This message can be displayed to the end user but just note that the information can be lengthy.

- **Error**: This will be an error code that provides insights into resolving the issue. The possible codes include the following:

Error code	Description
ErrorKind.Conflict	This indicates a conflict because another change was made to the same record. Use the `Refresh()` function to reload the record and retry the change.
ErrorKind.ConstraintViolation	One or more constraints were violated.
ErrorKind.CreatePermission	The user attempted to create a record but does not have the necessary permissions.
ErrorKind.DeletePermission	The user attempted to delete a record but does not have the necessary permissions.
ErrorKind.EditPermission	The user attempted to edit a record but does not have the necessary permissions.

Error code	Description
ErrorKind.GeneratedValue	The user attempted to modify a column that was automatically generated by the data source.
ErrorKind.MissingRequired	A required column value is missing.
ErrorKind.None	No error occurred.
ErrorKind.NotFound	The user attempted to edit or delete a record, but it could not be found.
ErrorKind.ReadOnlyValue	The user attempted to edit or delete a read-only column.
ErrorKind.Sync	The data source reported an error. Refer to the Message column for more details.
ErrorKind.Unknown	An unspecified error occurred.
ErrorKind.Validation	A general validation error was detected that does not fit into any of the other categories.

Table 14.1 – List of error codes and their description

Let's provide a basic example of using the Errors() function. In *Figure 14.6*, we have a simple screen that allows one to enter a person's name and age.

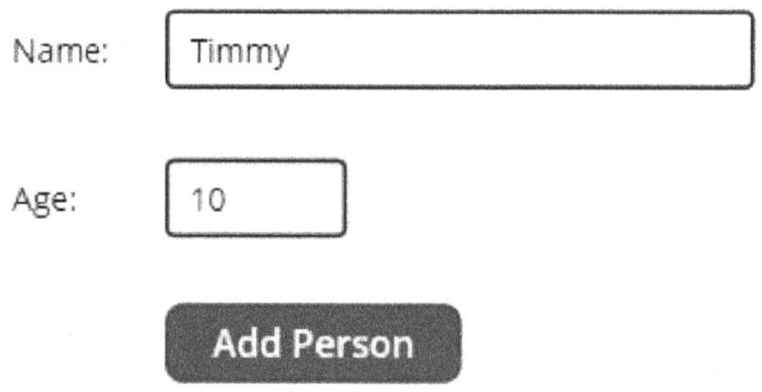

Figure 14.6 – Example of a form to add a person's name and age

The data source for this is a SharePoint list where the SharePoint column is a number format but has a setting to only allow numbers between the values of 1 and 110. This is shown in *Figure 14.7*:

Figure 14.7 – Showing the SharePoint list minimum and maximum values

We have created our app so that once the button is submitted to add a new person and their age, the user is sent to a success screen. This is shown in *Figure 14.8*.

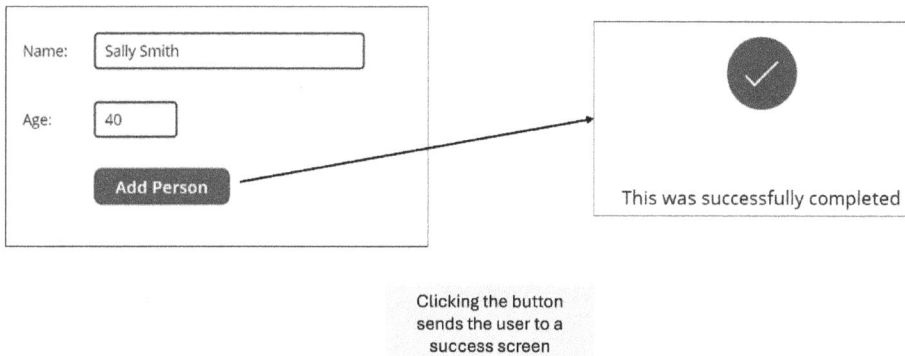

Figure 14.8 – Example of sending a user to a success screen once the Add Person button is selected

Let's also examine the code that we are using for the OnSelect property of the button:

```
Patch('Names', Defaults('Names'), {Title: TextInput1.Text, Age:
Value(TextInput1_2.Text)});
Navigate(SuccessScreen)
```

In the code, we are using the Patch() function to add a new record to a SharePoint list called Names and populating columns called Title and Age from two input controls on the app. Then, the user is sent to a screen called SuccessScreen.

However, what if the user entered a value that exceeds the maximum amount that we have set (which in our case is 110)? For example, let's say they entered 400 by accident. This will generate an error; however, the user is still sent to the success screen. This is shown in *Figure 14.9* with the error displayed at the top of the screen.

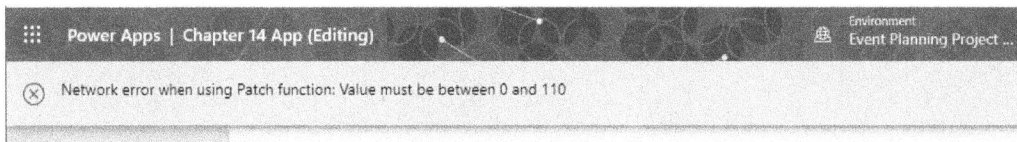

Figure 14.9 – An error is displayed on the success screen

One approach we could use is to modify our code for the button to test for any errors with the Errors() function. If there are any errors, a label is displayed showing the error.

Let's revise our button's OnSelect code to be as follows:

```
Patch('Names', Defaults('Names'), {'Title': TextInput1.Text, Age:
Value(TextInput1_2.Text)});
If(
     IsEmpty(Errors('Names')),
     Navigate(SuccessScreen),
     Set(
          varErrorMessage,
          Concatenate(
          First(Errors('Names')).Message,
          " located in column ",
          First(Errors('Names')).Column
          )
     )
)
```

This updated code applies the Patch() function to add the information to our SharePoint list called Names. However, the If() statement is used as follows:

1. It first checks to see whether there are any errors that result in the Names list. This is done by wrapping IsErrors('Names') around an IsEmpty() function. Remember, the IsErrors() function returns a table of any errors found. If this table is empty, as determined by the IsEmpty() function, then the If statement will return True.

2. If the outcome of *Step 1* is true then it will send the user to a screen called SuccessScreen via Navigate('SuccessScreen').

3. However, if the outcome of *Step 1* is `false`, meaning that an error did arise in `Names`, then it updates a variable called `varErrorMessage`. This variable will be updated with both the `Message` and `Column` names from the `Errors()` table. The result of this can be shown in *Figure 14.10*.

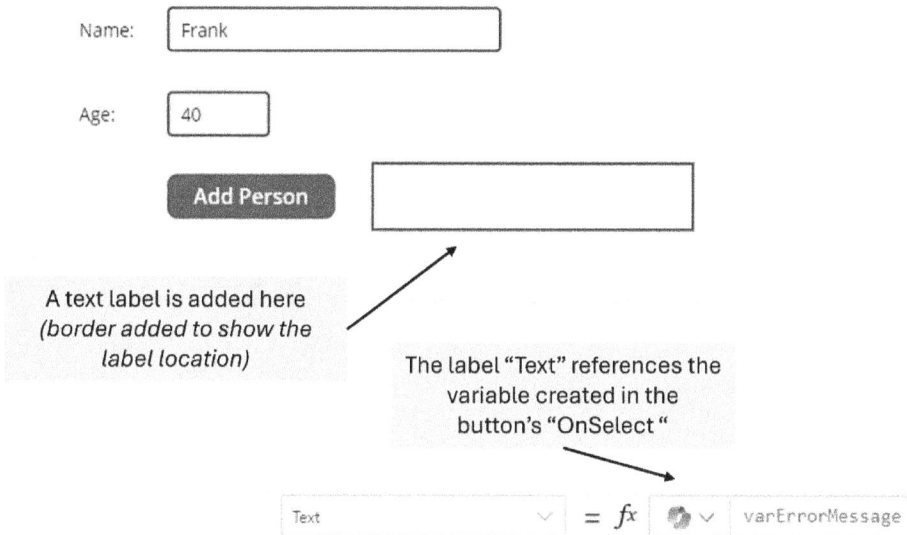

Name: Frank

Age: 40

Add Person

A text label is added here
(border added to show the label location)

The label "Text" references the variable created in the button's "OnSelect "

Text $= fx$ varErrorMessage

Figure 14.10 – Showing the label location and Text property

Now, when a value that exceeds the column limit is entered, the label displays the updated error message that is stored in the `varErrorMessage` variable. This is shown in *Figure 14.11*.

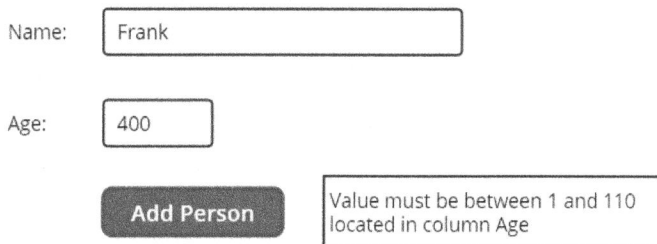

Name: Frank

Age: 400

Add Person   Value must be between 1 and 110 located in column Age

Figure 14.11 – The error message is displayed

Now that we have covered examples of key built-in error functions, let's look at another area that can be used to address errors. That involves using the `OnError` property for the overall app.

Using the OnError property

Another useful approach in handling errors is using the `OnError` property for the app itself. This property can be found as shown in *Figure 14.12*.

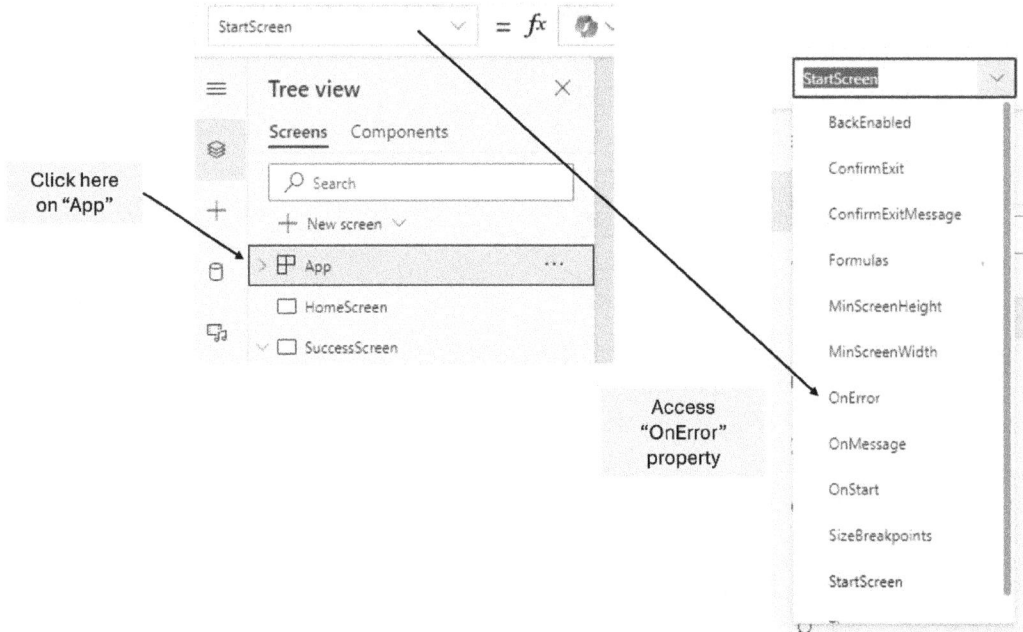

Figure 14.12 – Accessing the OnError app property

The `OnError` property is used to define how the app should respond to unexpected errors that come about during the execution of the app. This property is designed to handle errors that are not addressed by other error-handling methods, such as the built-in error functions that we reviewed previously (`IfError()`, `IsError()`, etc.). These types of errors are referred to as **unhandled** errors as they have not been handled by the error handling functions.

Examples of use cases for OnError

When an unhandled error occurs in a Power App, the `OnError` property is triggered. This allows developers to define a specific action or response when such errors arise. The typical use cases include the following:

- **Logging errors**: Developers can log error details for further analysis. This is particularly useful for debugging and improving app stability.

- **User notifications**: Custom error messages can be displayed to users, enhancing the user experience by providing clear feedback rather than generic error messages.

- **Suppressing errors**: In some scenarios, developers may choose to suppress error notifications to prevent user confusion, especially if the error does not significantly impact the app's functionality.

Let's provide a simple use case.

Creating a custom error notification

In this example, we want to provide additional information about what the error is. We'll take advantage of a record called `FirstError`, which includes the following fields:

Field	Description
FirstError.Kind	The category of the error
FirstError.Message	The message about the error
FirstError.Source	The location where the error originated, such as the control that caused the error
FirstError.Observed	The location where the error is shown to the user
FirstError.Details	This will be details about the error; it will only be provided for network errors

Table 14.2 – List of FirstError fields

With this information, let's create a custom message that includes the error message, where it originated, and where it was observed by the user:

```
Notify(
    Concatenate(
        "Message: ",
        FirstError.Message,
        ", Source: ",
        FirstError.Source,
        ", Observed: ",
        FirstError.Observed
    ),
    NotificationType.Error
)
```

This is also shown in *Figure 14.13*.

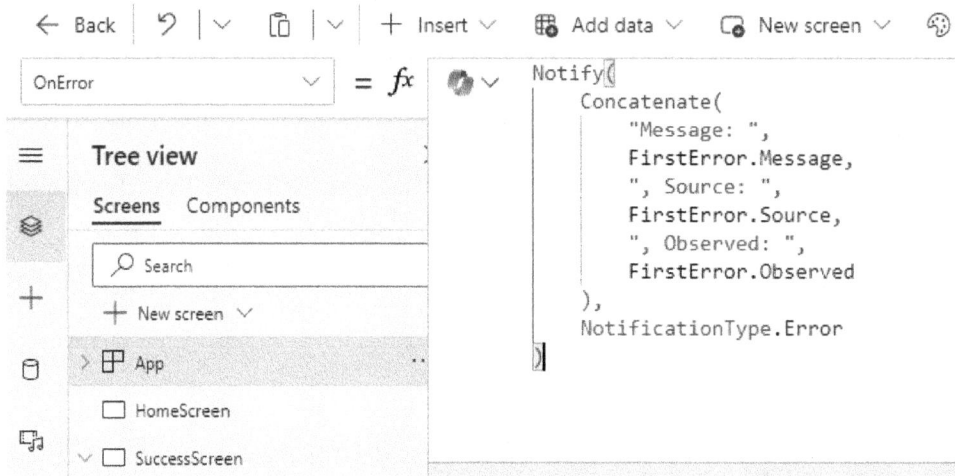

Figure 14.13 – Added a custom error messaging to the OnError App property

To see this in action, we'll revisit our simple app that was previously used to add a person's name and age. In this example, we'll enter an age number that exceeds the maximum allowed value: 125. In *Figure 14.14*, our custom message created in the OnError property provides additional information.

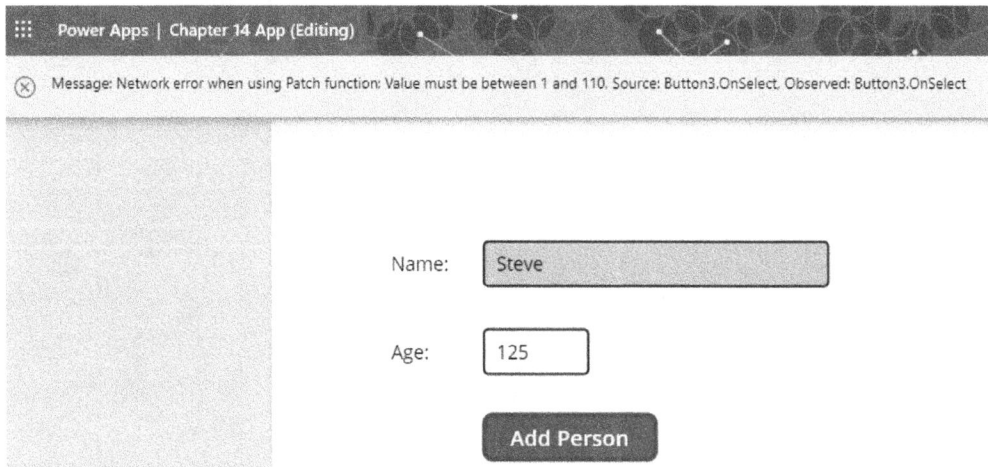

Figure 14.14 – Viewing the custom error message created in the OnError property

This is just one example of how the OnError property can be used. However, there are additional approaches. Let's walk through two examples, including logging the error to a separate data source, as well as sending an automated email to the development team.

Logging the error to a separate data source

Our example will log the error and related information to a separate data source such as a SharePoint list. The steps are covered in the following sections.

Step 1 – create a SharePoint list to log all errors

For our example, we'll call the list `Error Log`. This list will contain just a few columns, including the following:

- **Title**: We will just use this column to log the `Error Received` text
- **Date logged**: The date when the error occurred
- **User**: The user who encountered the error
- **Error message**: The message from the error; for this, we can make use of the `FirstError.Message` function that we just covered

An example of our SharePoint list is shown in *Figure 14.15*.

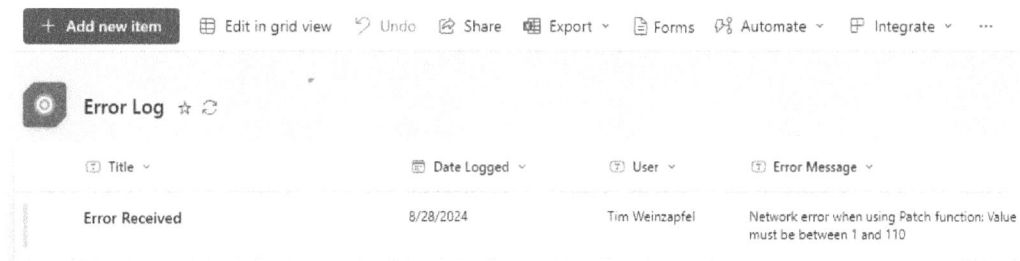

Figure 14.15 – Example of our Error Log SharePoint list

Step 2 – use the Patch() function in the OnError property

We'll return to the app and go to the `OnError` property. We will add the following `Patch()` function:

```
Patch('Error Log', Defaults('Error Log'),
    {
    Title: "Error Received",
    'Date Logged': Today(),
    User: User().FullName,
    'Error Message': FirstError.Message,
}
)
```

When an error occurs, this Patch() function will update the SharePoint list error log, make a note of the date when it occurred, the user who experienced the error, and then the error message. By doing this, you can develop a log of errors when they happen.

Now that we have done this, let's also take advantage of Power Automate and alert the development team that this has occurred.

Triggering a Power Automate flow

Let's walk through a few simple steps to set up the notification flow.

Step 1 – create the Power Automate flow directly from Power Apps

To do this, select the Power Automate flow icon on the left side of the navigation area. This is shown in *Figure 14.16*.

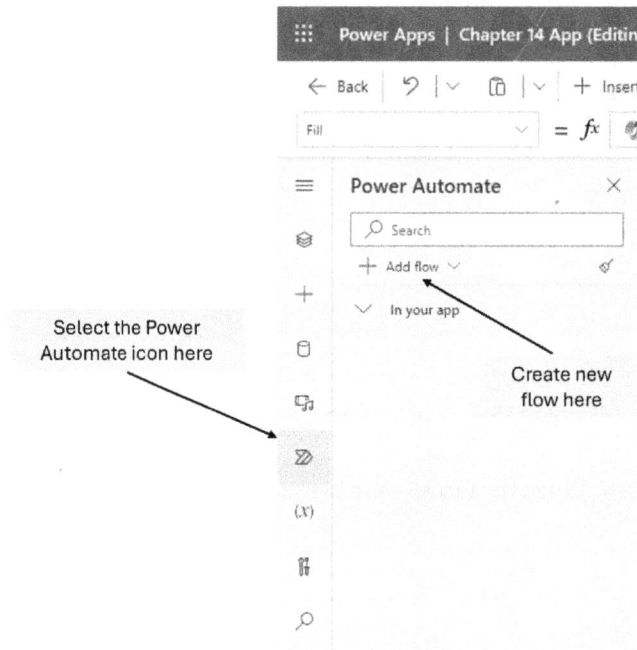

Figure 14.16 – Selecting the Power Automate icon in Power Apps

From there, select **Add flow**, which is also shown in *Figure 14.16*. This will open the Power Automate interface directly in Power Apps.

Step 2 – add inputs to the Power Automate trigger

In Power Automate, we need to add the input boxes that we want to include in our email message. For ease of use, we'll just send the following three items to our flow:

- **UserName**: This will be the user who experienced the error

- **DateOccured**: The date the error occurred

- **ErrorMessage**: The message of the error

We'll create input boxes for these three items as shown in *Figure 14.17*.

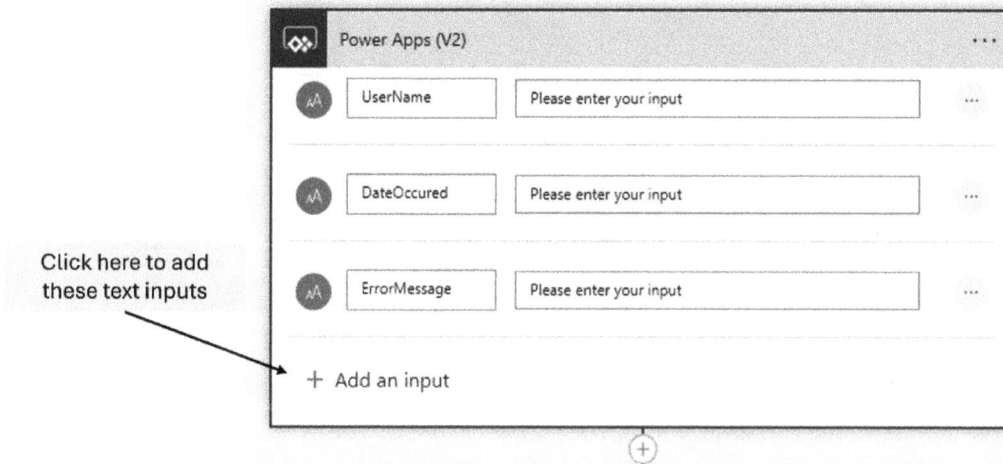

Figure 14.17 – Adding the three input boxes for our flow

We'll then add one more step to send an email to the development team. For this, we'll use the **Send an email (v2)** action from the Outlook connector. You may need a different connector depending on your email application. *Figure 14.18* provides a view of this step.

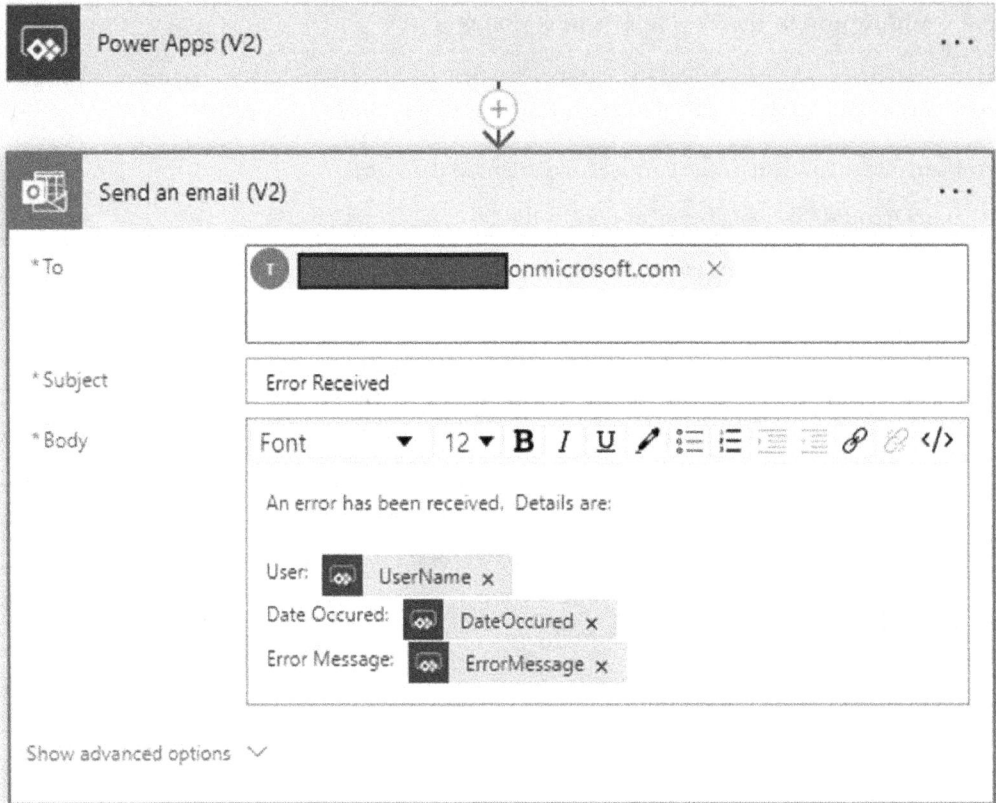

Figure 14.18 – Adding the Send an email step to our flow

We then just need to name our flow and save it. For our purposes, we'll save this as Chapter 14 Error Flow.

Step 3 – execute the Power Automate flow in the OnError property

We now return to our Power App, specifically the OnError property. We just reference the flow and add the required parameters. Let's provide the code; we will then walk through it:

```
Chapter14ErrorFlow.Run(User().FullName, Today(), FirstError.Message)
```

The first part of Chapter14ErrorFlow.Run is how we execute the flow. The flow name is used (removing any spaces), followed by .Run, to execute it. We then need to add any applicable parameters that our flow requires. In our case, it requires three parameters: UserName, DateOccurred, and ErrorMessage. Remember – these were the three input parameters that we added in our Power Automate flow. Refer to *Figure 14.17*.

In our case, we use the following fields to correspond with those parameters:

Power Automate field	Power App value
UserName	`User().FullName`
DateOccurred	`Today()`
ErrorMessage	`FirstError.Message`

Table 14.3 – Mapping our Power Automate fields to Power App values

That is it. When an error occurs, we will update an error log in SharePoint. Also, our development team can be advised that an error occurred. Let's now move on to creating custom error messages.

Creating custom error messages

Creating custom error messages in Power Apps is a good practice for enhancing the user experience. By providing clear and actionable feedback when errors occur, custom error messages help users understand what went wrong and how they can correct the issue. In this section, we'll explore how to design and implement custom error messages using context variables and the Notify function, as well as displaying error messages on the screen using conditional logic.

Designing user-friendly error messages

The key to effective error handling is making sure that the messages users see are both informative and easy to understand. When crafting custom error messages, consider the following best practices:

- **Be specific**: Clearly describe what went wrong and why. Instead of saying **Error**, use messages such as **Please enter numerical values only**.

- **Provide guidance**: Offer suggestions on how the user can fix the error. For example, **Your password must be at least 8 characters long and include at least one symbol**.

- **Avoid technical jargon**: Use language that non-technical users can easily understand. Instead of **Null reference exception**, say **This field cannot be left blank**.

Implementing custom error messages with variables

Global or context variables in Power Apps allow you to store temporary information that can be used to control the display of error messages. We covered these variables in *Chapter 6*. By setting and resetting global or context variables based on user actions or app conditions, you can manage when and how error messages are shown.

Let's look at an example of using a variable to store information caused by an error. In *Figure 14.19*, we revisit a simple app that captures the name and age of a person. This is the same app that was used in *Figure 14.11* and as a reminder, **Age** requires a number entry between 1 and 110. If a user enters a value in excess of 110, an error will be generated. Information on this error will be maintained in a variable called varErrorMessage.

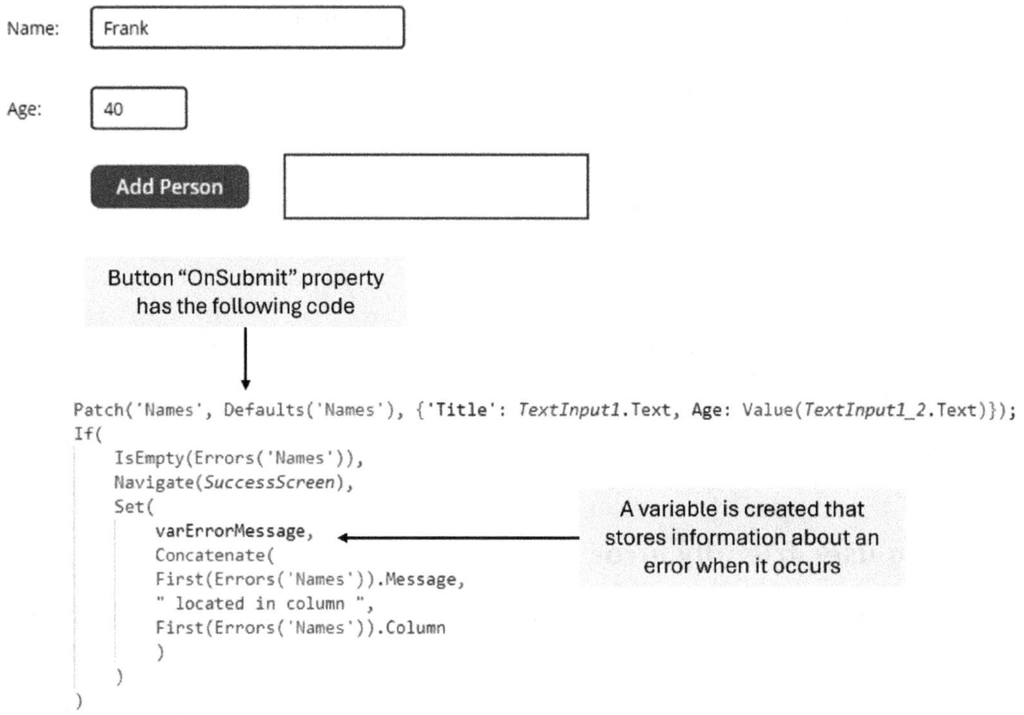

Name: Frank

Age: 40

[Add Person]

Button "OnSubmit" property
has the following code

```
Patch('Names', Defaults('Names'), {'Title': TextInput1.Text, Age: Value(TextInput1_2.Text)});
If(
    IsEmpty(Errors('Names')),
    Navigate(SuccessScreen),
    Set(
        varErrorMessage,                    A variable is created that
        Concatenate(                        stores information about an
        First(Errors('Names')).Message,     error when it occurs
        " located in column ",
        First(Errors('Names')).Column
        )
    )
)
```

Figure 14.19 – Displaying the button's OnSubmit code that will store error information in a variable

This variable can then be displayed on the form when an error arises, thus providing immediate information to the user. This is shown in *Figure 14.20*.

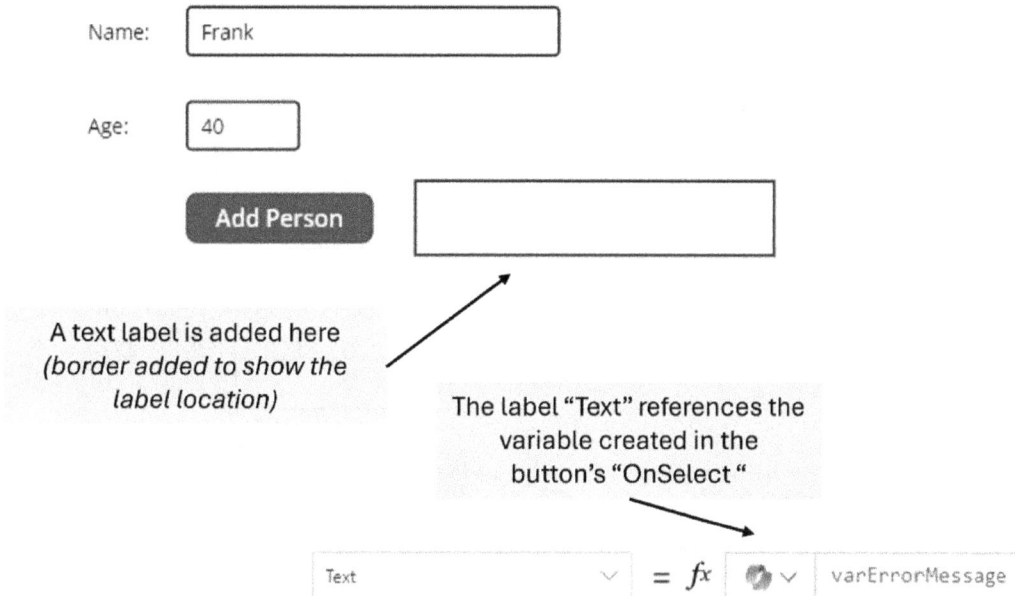

Name: | Frank |

Age: | 40 |

Add Person | |

A text label is added here
*(border added to show the
label location)*

The label "Text" references the
variable created in the
button's "OnSelect "

| Text | ∨ | = *fx* | 🎨 ∨ | varErrorMessage |

Figure 14.20 – Displaying the error message stored in the varErrorMessage variable

This is just one example of storing error information in a variable. Let's move on to using the Notify() function.

Using the Notify() function for dynamic error messaging

The Notify() function is a versatile tool for showing messages to users. We covered this function earlier in this chapter. It can be combined with conditional logic to dynamically display different error messages based on specific circumstances. Let's revisit a previous example that we used when explaining the Notify() function. We displayed an example in *Figure 14.4* but we will bring that example back up in *Figure 14.21* for ease of understanding.

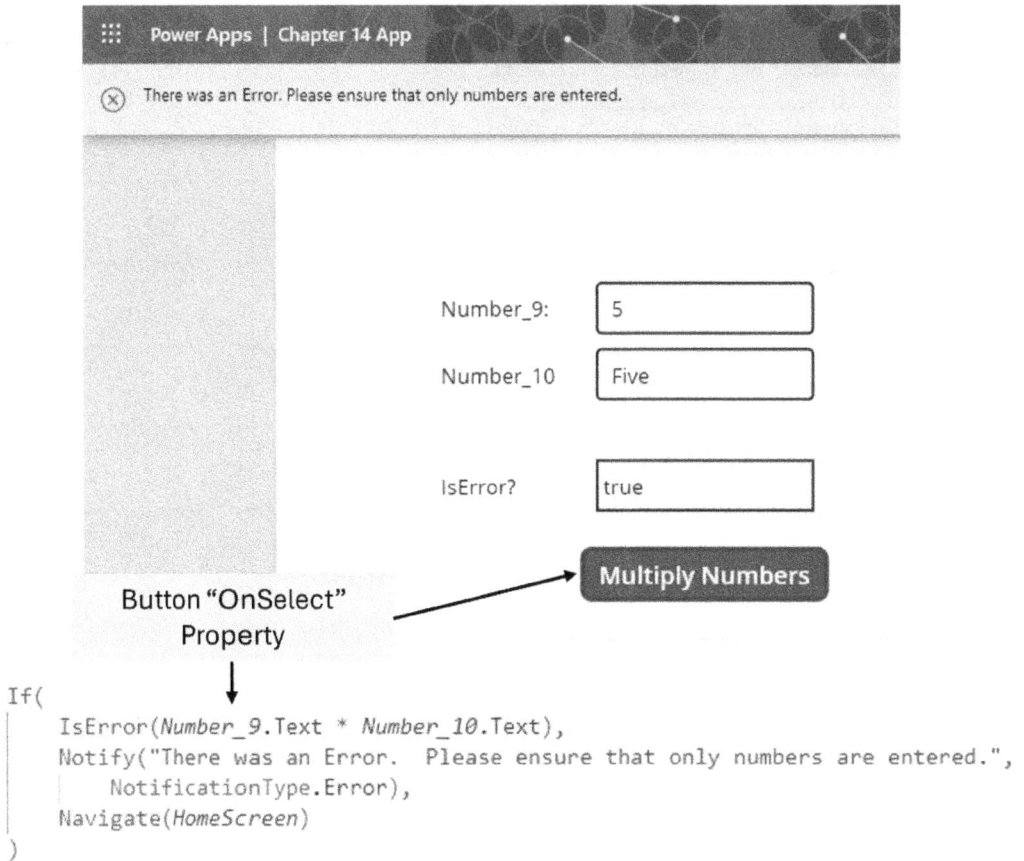

Figure 14.21 – Example of using the Notify() function to display a user-friendly message

As you can see, we used a user-friendly message via the `Notify()` function: **There was an Error. Please ensure that only numbers are entered**.

Displaying error messages on the screen

In addition to using notifications, you can display custom error messages directly on the screen, providing users with immediate visual feedback. This approach can be especially useful in forms where multiple fields require validation.

Figure 14.21 provides an example of having a simple `Text` label control that displays a message when an error occurs.

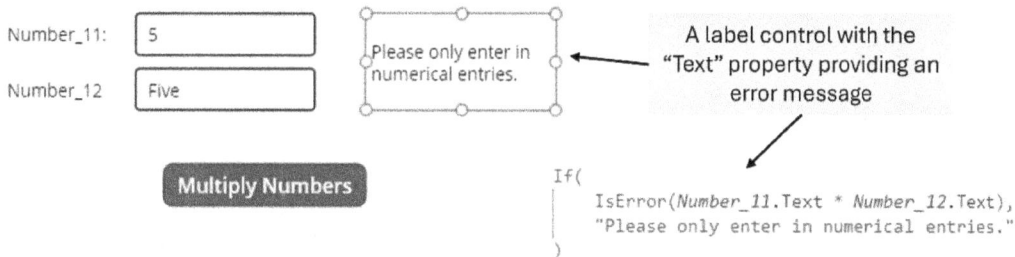

Figure 14.22 – Example of displaying an error in a Text label

In this example, a simple `If()` statement is used. It checks to see whether the values of the two input controls will generate an error via the `IsError()` function. If so, then the text is displayed.

These are just some examples of separate ways to alert users that an error has occurred.

> **Best practice**
>
> It is important to prioritize implementing data validation first to prevent errors at the source. Only after establishing strong validation should you focus on handling error techniques. This approach ensures that most issues are caught early, reducing the need for extensive error handling later.

Final note on error handling

Throughout this chapter, we have provided different functions and approaches to handling errors. It is important to remember that there are always different ways to approach validating data and handling errors. We believe that the key factor is understanding what functions are available and how they can be used for your given situation.

Summary

In this chapter, we covered the important topic of error handling. First, we provided an overview of various errors that may occur within your app. We then discussed key built-in error functions. These functions can be used to provide real-time feedback to users. We also covered using the `OnApp` property for the overall app that can be used to address unhandled errors. Lastly, we provided techniques for creating custom error messages. Effective error handling is important to develop user-friendly applications.

In the next chapter, we will conclude the book by discussing how to register an app in Azure.

15

Registering a Power App in Azure

In this chapter, we will explore the process of registering a Power App in Azure. This crucial step ensures secure communication between your Power App and Azure services, allowing for a more integrated and robust application experience.

You will learn the step-by-step process of registering a Power App in Azure, including creating an app registration, obtaining the necessary client ID and secret ID, and configuring these within your Power App. This knowledge is essential for developers looking to enhance the security and functionality of their Power Apps by leveraging Azure's capabilities.

By the end of this chapter, you will be able to register a Power App in Azure, understand when to add the client ID and secret ID within Power Apps, and grasp the importance of this registration process for ensuring secure and efficient app performance.

In this chapter, we're going to cover the following main topics:

- The importance of registering an app in Azure
- Registering a Power App in Azure
- When to add the Azure app ID, client ID, and secret ID within Power Apps

Technical requirements

In this section, we'll outline the technical requirements for this chapter. Ensure you have the following technologies and installations ready:

- **Azure subscription**: An active Azure subscription is required. You can start with a free account that includes $200 in credit for 30 days. A **pay-as-you-go** subscription is typically sufficient, with minimal costs for basic app registration and development.

- **Azure Active Directory (AAD)**: Your subscription must include access to AAD. Ensure you have the necessary permissions, such as AAD administrator or appropriate **role-based access control (RBAC)** roles, to create and manage applications.

- **Access to Power Apps**: `https://make.powerapps.com/`.

- Basic knowledge of Azure portal navigation.

The importance of registering an app in Azure

Registering an app in Azure is a crucial step in the development and deployment of Power Apps, especially when integrating with other Azure services. This process provides several key benefits that enhance the functionality, security, and management of your applications. Let's look at some of the primary reasons why registering an app in Azure is important.

Enhanced security

For applications requiring advanced security features, such as authentication through AAD or OAuth 2.0 protocols, registration is a must:

- **Authentication and authorization**: Registering your app in Azure allows you to use AAD for authentication and authorization. This means you can leverage AAD's robust security features to control who has access to your app and what they can do within it.

- **OAuth 2.0 protocol**: By registering your app, you can use the OAuth 2.0 protocol to securely obtain access tokens, which are necessary for accessing Azure resources. This helps in protecting user credentials and ensuring secure communication between your app and Azure services.

- **Scenario: Secure employee portal**: Imagine you're developing an internal employee portal that stores sensitive HR data, such as personal information and payroll details. For this portal, advanced security features are crucial to protect sensitive data and restrict access. By registering the app in Azure, you can leverage AAD for authentication and authorization, ensuring that only authorized employees can access the portal. This scenario highlights the importance of app registration when enhanced security is required.

Seamless integration with Azure services

If your Power App needs to interact with other Azure services (e.g., Microsoft Graph, Azure Storage), registration is required to enable secure and seamless connections:

- **Access to Azure APIs**: Once registered, your app can easily access various Azure APIs, such as those for Microsoft Graph, Azure Storage, and other services. This enables you to create more powerful and feature-rich applications by leveraging the full suite of Azure offerings.

- **Service connections**: Registering the app allows you to set up service connections, making it easier to integrate with other Azure services and automate workflows using tools such as Power Automate.

- **Scenario: Automated document processing**: Consider a scenario where you need to create a Power App that automates the processing and storage of documents in Azure Storage. By registering the app in Azure, you enable secure and seamless connections between the Power App and Azure Storage. This registration allows you to integrate with other Azure services, streamlining document management and improving overall productivity.

Improved application management

Here are some points to improve application management:

- **Centralized management**: Azure provides a centralized portal where you can manage all your registered applications. This includes configuring settings, monitoring usage, and updating permissions. It simplifies the management of multiple apps and ensures consistency in their configuration.

- **Scalability and maintenance**: With your app registered in Azure, you can take advantage of Azure's scalability features. This ensures that your app can handle increased load and performance demands. Additionally, it facilitates easier maintenance and updates.

- **Scenario: Managing multiple business apps**: Suppose your company has multiple Power Apps deployed across different departments, each with specific configurations and permissions. Registering these apps in Azure allows you to manage them centrally through the Azure portal. This centralized management simplifies operations, ensures consistency in app configurations, and makes it easier to monitor and update the apps as needed, making it ideal for organizations with multiple applications.

Compliance and governance

If your app operates in a regulated environment or requires strict compliance and governance controls, registering it in Azure ensures you can enforce these policies and manage permissions effectively:

- **Policy enforcement**: Registering your app in Azure allows you to enforce compliance policies and governance controls. You can ensure that your app adheres to organizational and regulatory standards, which is particularly important for industries with strict compliance requirements.

- **Audit trails**: Azure provides detailed logging and audit trails for registered apps. This helps in tracking user activities and changes, which is essential for security audits and forensic investigations.

- **Scenario: Healthcare application with HIPAA compliance**: Imagine you're developing a healthcare application that must comply with HIPAA regulations. Registering the app in Azure allows you to enforce compliance policies and manage permissions effectively, ensuring that the app adheres to strict regulatory requirements. This scenario emphasizes the need for app registration in regulated environments where compliance and governance are critical.

Custom branding and user experience

Here are some points to consider while looking at branding and user experience:

- **Brand identity**: Through app registration, you can customize the branding of your app's login and consent screens. This provides a consistent user experience that aligns with your organization's brand identity.

- **User consent**: When users log into your app, they are presented with a consent screen that informs them about the permissions your app is requesting. This transparency builds trust with your users and ensures they are aware of how their data will be used.

- **Scenario: Corporate intranet portal**: Suppose you're developing a corporate intranet portal using Power Apps. By registering the app in Azure, you can customize the branding on login and consent screens to align with your organization's identity. This enhances the user experience by providing a consistent brand presence every time employees log in, demonstrating how app registration can be used to reinforce brand identity.

Support for multi-tenant scenarios

Let's learn about multi-tenant scenarios and a related scenario:

- **Multi-tenancy**: Registering your app in Azure allows it to support multi-tenant scenarios, where a single instance of your app can serve multiple organizations. This is particularly useful for **software-as-a-service** (**SaaS**) applications, enabling you to build scalable and flexible solutions that cater to multiple clients.

- **Scenario: SaaS product for multiple clients**: Imagine your company is developing a SaaS product using Power Apps to serve multiple clients. Registering an app in Azure enables support for multi-tenancy, allowing each client to have their own isolated instance of the app. This scenario highlights the flexibility and scalability of app registration for SaaS products, where serving multiple organizations from a single app instance is necessary.

Registering your app in Azure is a foundational step that brings numerous benefits, from enhanced security and seamless integration to improved management and compliance. By leveraging Azure's capabilities, you can build more secure, efficient, and scalable Power Apps that meet the needs of your organization and its users. Whether you are developing a simple application or a complex enterprise solution, the importance of registering your app in Azure cannot be overstated.

Integrating a Power App in Azure

In this section, you will learn how to register your Power App in Azure. We will walk through creating an app registration in the Azure portal, which will provide you with the app ID, client ID, and secret ID necessary for integrating your Power App with Azure services.

Here are the step-by-step instructions for doing so:

1. **Sign in to the Azure portal**:

   - Open your web browser and go to `https://portal.azure.com/`.

   - Sign in using your Azure account credentials.

2. **Navigate to Microsoft Entra ID**:

   - In the left-hand navigation pane, click on **Microsoft Entra ID**.

What's changed with AAD?

Microsoft AAD has been rebranded as **Microsoft Entra ID** as part of the broader Microsoft Entra suite. This change reflects Microsoft's expanded focus on identity and access management across various platforms. While the underlying functionality and services remain the same, the rebranding aligns with Microsoft's broader strategy for comprehensive identity solutions beyond Azure.

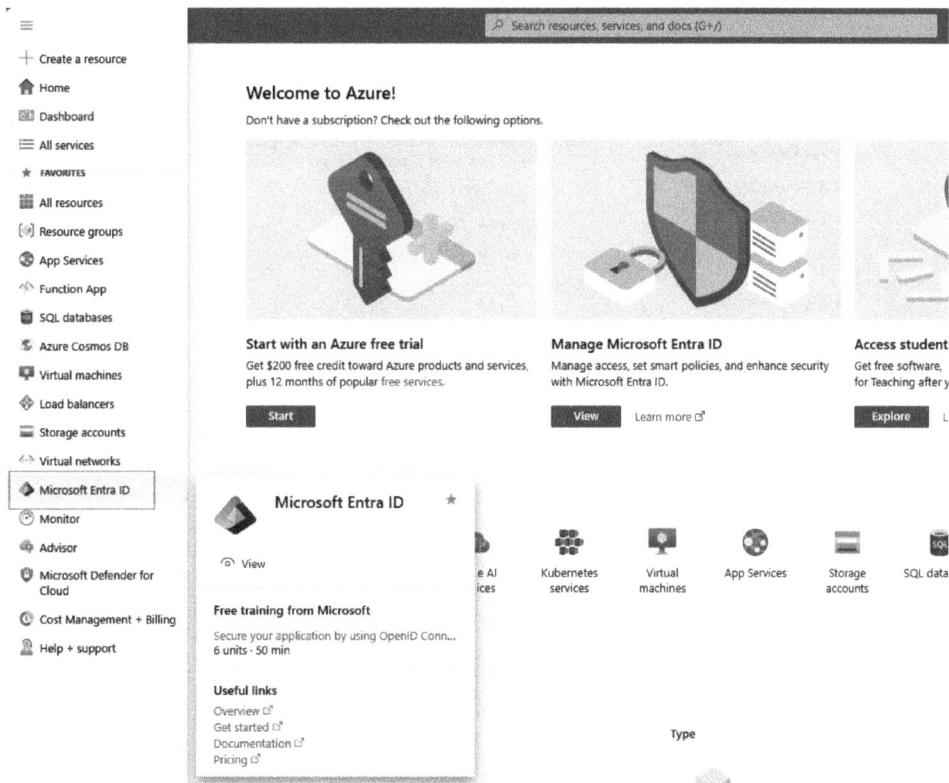

Figure 15.1 – Microsoft Entra ID in the left navigation panel

3. **Create a new app registration**: In the **Microsoft Entra ID** menu, there are two different ways to create a new app registration:

- **Option 1**: Click on **App registrations** from the **Manage** drop-down menu on the left-hand side.

- **Option 2**: Alternatively, click on **+ Add** in the top ribbon of the **App registrations** page and select **App registration**.

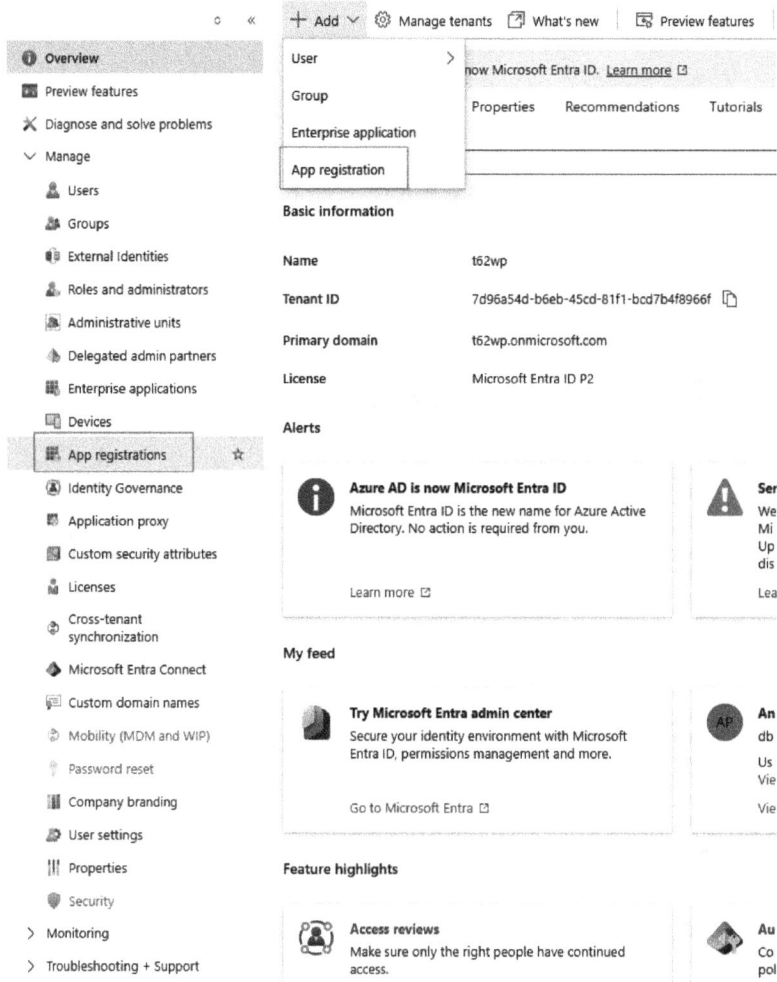

Figure 15.2 – App registrations in Azure

- If *option 1* is chosen, click on **+ New registration** at the top of the **App registrations** page.

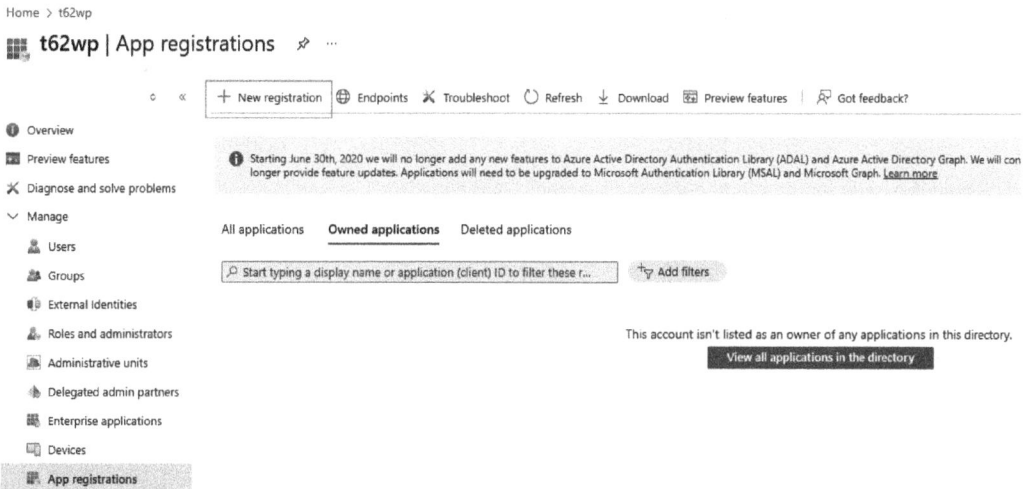

Figure 15.3 – New registration in Azure

4. **Fill in the app registration details**:

- **Name**: Enter a name for your application (e.g., Event Planning App).

- **Supported account types**: Choose who can use this application (e.g., accounts in this organizational directory only).

Supported account types meaning

Accounts in this organizational directory only (single tenant): This restricts access to users within your organization only.

Accounts in any organizational directory (multitenant): This allows users from any Microsoft Entra ID tenant (different organizations) to access the app.

Accounts in any organizational directory (multitenant) and personal Microsoft accounts: This includes users from any Microsoft Entra ID tenant and personal Microsoft accounts (e.g., Skype, Xbox).

Personal Microsoft accounts only: This limits access to personal Microsoft accounts only, excluding organizational accounts.

- **Redirect URI (optional)**: If you have a specific redirect URI, enter it here. This URI is where Azure will send the authentication response. If unsure, leave it blank for now.

5. Click **Register** to create the app registration.

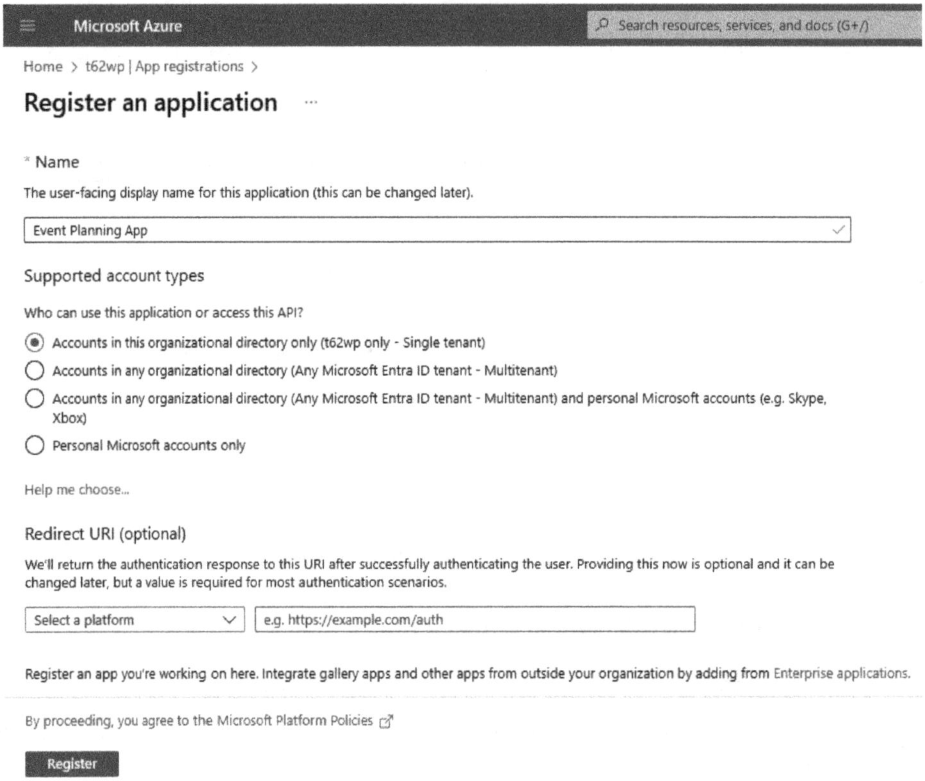

Figure 15.4 – The Register an application form

6. **Record the application (client) ID:**

   - After the registration is complete, you will be directed to the app's overview page.

   - Copy the **Application (client) ID** information as you will need it later.

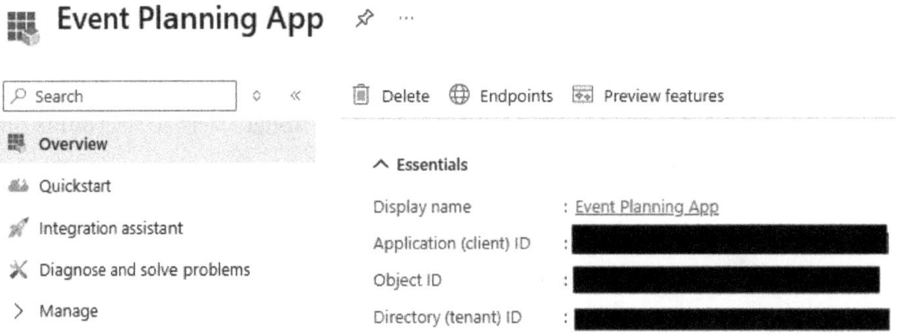

Figure 15.5 – Overview of Event Planning App in Azure

7. **Generate a client secret:**

   I.   In the left-hand menu, under **Manage**, select **Certificates & secrets**.

   II. Under **Client secrets**, click on + **New client secret**.

   III. Add a description for the client secret and choose an expiration period:

   - **Description**: Event Planning Client Secret

   - **Expires**: Recommended: 180 days (6 months)

   IV. Click **Add**.

   V.   Copy the **Value** of the client secret immediately as it will be hidden once you leave the page.

8. **Configure API permissions:**

   I.   In the left-hand menu, click on **API permissions**.

   II. Click on + **Add a permission**.

   III. Choose **APIs my organization uses**.

   IV. Search and select **PowerApps Service** to configure permissions for interacting with Power Apps services.

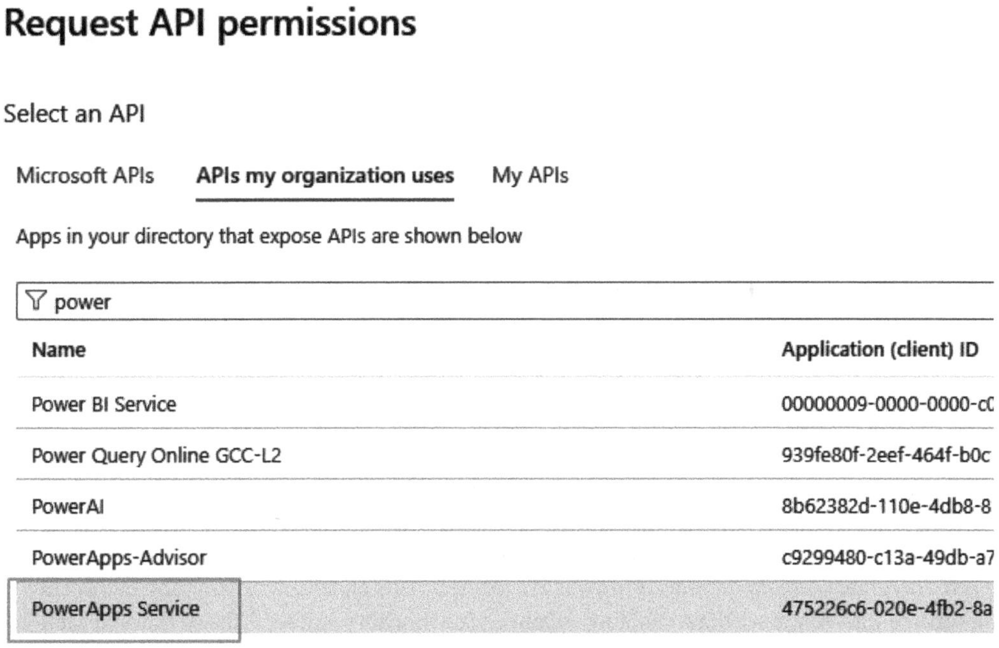

Request API permissions

Select an API

Microsoft APIs   **APIs my organization uses**   My APIs

Apps in your directory that expose APIs are shown below

⟁ power	
Name	**Application (client) ID**
Power BI Service	00000009-0000-0000-c0
Power Query Online GCC-L2	939fe80f-2eef-464f-b0c
PowerAI	8b62382d-110e-4db8-8
PowerApps-Advisor	c9299480-c13a-49db-a7
PowerApps Service	475226c6-020e-4fb2-8a

Figure 15.6 – PowerApps Service

V.   Choose the appropriate permissions (e.g., **Delegated** or **Application**) based on how your app will interact with these services:

i.   Select **Delegated | User**.

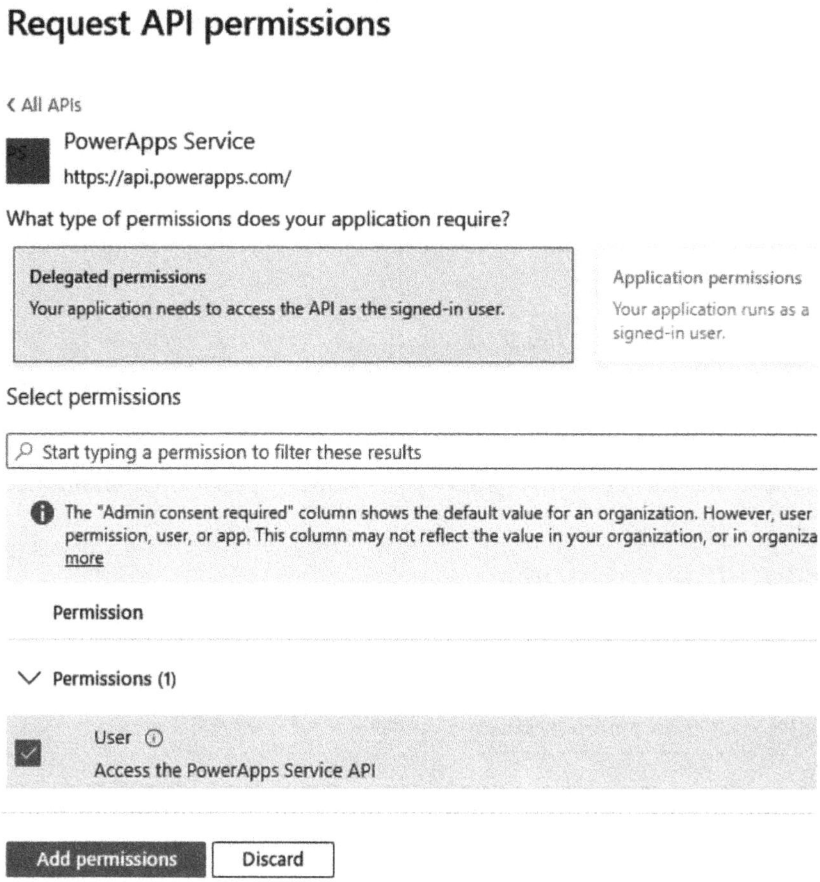

Request API permissions

< All APIs

PowerApps Service
https://api.powerapps.com/

What type of permissions does your application require?

Delegated permissions	Application permissions
Your application needs to access the API as the signed-in user.	Your application runs as a signed-in user.

Select permissions

🔍 Start typing a permission to filter these results

ⓘ The "Admin consent required" column shows the default value for an organization. However, user permission, user, or app. This column may not reflect the value in your organization, or in organiza
more

Permission

∨ Permissions (1)

☑   User ⓘ
Access the PowerApps Service API

Add permissions | Discard |

Figure 15.7 – Request API permissions

ii.   Click **Add permissions**.

• If necessary, click **Grant admin consent for** *[Your Organization]* to approve the permissions.

With your Power App registered and the necessary permissions configured in Azure, the next step is to integrate these credentials into your Power App. This ensures that your app can securely communicate with Azure services, enabling enhanced functionality and security. In the next section, we will demonstrate when and how to add the application ID within Power Apps.

When to add the Azure app ID, client ID, and secret ID within Power Apps?

Within Power Apps, the app ID (also known as the application ID or client ID), client secret, and secret ID play a crucial role in enabling secure communication between your app and Azure services or external APIs. These credentials are particularly important when your app needs to authenticate and interact with other systems, whether within the Microsoft ecosystem or with third-party services.

Understanding the role of Power Apps credentials

Let's quickly look at the credentials:

- **App ID (client ID)**: This is a unique identifier for your registered application in Azure. It tells the service which application is making the request.
- **Client secret (secret ID)**: This acts as a password that your app uses in combination with the client ID to authenticate with Azure and access protected resources.

For beginners, think of the app ID as the username and the client secret as the password that your app uses to log in to another system or service. Together, these credentials allow your Power App to perform secure actions, such as accessing data, automating processes, or integrating with other services.

Scenarios requiring app ID, client ID, and client secret in the event planning application

Let's consider our event planning application as a practical example to illustrate where and how you might need to use these credentials:

1. **Custom connectors**:

    - **Use case**: When creating a custom connector to interface with an external API for event management, you may need to provide an app ID, client ID, and client secret to authenticate requests.

    - **Example**: Suppose your event planning application needs to integrate with an external venue booking service that provides an API. By configuring a custom connector with these credentials, your app can securely interact with the service to book venues or retrieve availability data.

2. **Azure service integrations**:

   - **Use case**: Integrating with Azure services for event-related features might require credentials for secure access.

   - **Example**: If your event planning app uses Azure Cognitive Services to analyze feedback from event surveys, you would use the client ID and secret to authenticate the service calls. This ensures that only authorized applications can use these services to process and analyze data.

3. **Custom applications**:

   - **Use case**: If your event planning solution includes a custom-built application that requires authentication against AAD, you will use these credentials.

   - **Example**: Imagine developing a custom dashboard for managing event registrations that need to authenticate users via AAD. You would configure the dashboard with the app ID, client ID, and client secret to enable secure sign-in and access.

4. **Power Automate flows**:

   - **Use case**: Automating processes within your event planning application using Power Automate may require OAuth 2.0 authentication for API interactions.

   - **Example**: If you set up a Power Automate flow to automatically send confirmation emails to event attendees, and the email service requires API authentication, you would configure the flow with the app ID and client secret to authorize these email requests.

5. **Third-party integrations**:

   - **Use case**: Integrating with third-party services for event management might need the app ID and client secret for secure API access.

   - **Example**: If your event planning app integrates with a third-party service for ticketing or event promotion, you would use these credentials to authenticate API requests. This ensures secure and authorized access to third-party resources.

6. **API access**:

   - **Use case**: For accessing specific APIs that require OAuth 2.0 authentication, the app ID, client ID, and secret ID are necessary.

   - **Example**: If your event planning app needs to pull data from an external calendar API to sync event dates, you would use these credentials to authenticate the API requests, ensuring secure data retrieval.

Integrating the application ID (client ID) and client secret into your event planning Power App allows it to securely communicate with various external services and APIs. Whether connecting through custom connectors, Azure services, third-party applications, or API access, these credentials ensure that your app can interact effectively and securely with these resources.

Summary

In this chapter, we covered the critical process of registering a Power App in Azure. We walked through the steps to create an app registration in Azure, generate the necessary client ID and secret ID, and know when to configure these within Power Apps. Understanding this process enhances the security and integration capabilities of your Power Apps, ensuring they can leverage Azure's powerful services.

Congratulations on completing this comprehensive journey through the best practices and essential techniques required to build robust, scalable, and efficient Power Apps. Here's a recap of what skills we learned throughout this book:

- In *Chapter 1, Understanding Requirements and Project Planning*, we learned the importance of determining project scope, understanding client requirements, and building a solid project foundation

- In *Chapter 2, Working with Solutions*, we learned how to create a solution and publisher, develop components within a solution, and when to create a solution

- In *Chapter 3, Power Platform Environments*, we learned about the differences between development, testing, and production environments, deploying solutions, and understanding data across environments

- In *Chapter 4, Choosing the Right Tool – Navigating Canvas Apps, Power Pages, and Model-Driven Apps*, we learned how to choose the appropriate app type based on project requirements and understanding the strengths of each app type

- In *Chapter 5, Data Connections*, we learned how to connect data from SharePoint to Power Apps, use dataflows, and understand Dataverse

- In *Chapter 6, Variables, Collections, and Data Filtering*, we learned about filtering data, working with large datasets, and using variables and collections for dynamic content and logic

- In *Chapter 7, Canvas App Formulas*, we learned about using various formulas to bring apps to life, navigating between screens, and connecting forms to Dataverse

- In *Chapter 8, Conditional Formatting and URL Deep Linking*, we learned how to apply conditional formatting, display data dynamically, and utilize deep linking with URL parameters

- In *Chapter 9, Integration with Power Automate/Teams/Outlook*, we learned about integrating Power Apps with Teams and Outlook using Power Automate, sending emails, calendar invites, and Teams notifications

- In *Chapter 10, Integrating with Power BI*, we learned about embedding Power BI reports in Power Apps and Power Apps in Power BI reports, and advanced techniques with embedded apps

- In *Chapter 11, Integrating Power Apps with SharePoint*, we learned about embedding Power Apps in SharePoint sites, integrating with document libraries, and data access and security

- In *Chapter 12, Integration with Power Virtual Agent/Copilot*, we explored Copilot Studio, connected to data sources, and created custom topics

- In *Chapter 13, Governance, Security, and Application Life Cycle Management*, we learned about ensuring robust security and proper user access management in Power Apps

- In *Chapter 14, Error Handling*, we learned about upgrading Power Apps, incorporating error handling, and the importance of technical documentation

- In *Chapter 15, Registering a Power App in Azure*, we learned about registering a Power App in Azure, adding the client ID and secret ID within Power Apps, and the importance of app registration in Azure

Final thoughts

By now, you should have a comprehensive understanding of how to create, manage, and deploy Power Apps effectively. The skills and knowledge you have gained will help you build applications that are not only functional but also scalable and secure. Whether you are just starting or looking to enhance your existing apps, these best practices will serve as a valuable resource. As you continue your Power Apps journey, remember to keep exploring, learning, and implementing new techniques to stay ahead in the ever-evolving world of app development. Thank you for embarking on this journey with us, and we wish you success in all your future Power Apps endeavors!

Index

‹packt›

Other Books You May Enjoy

If you enjoyed this book, you may be interested in these other books by Packt:

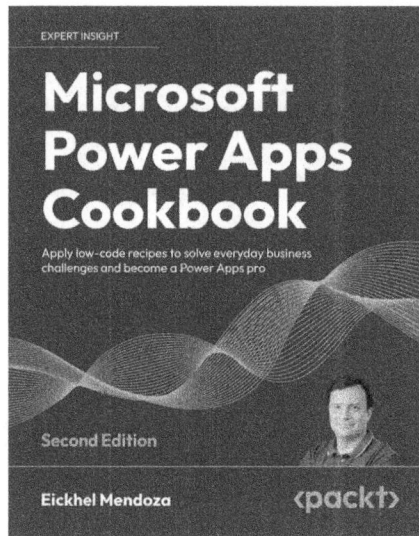

Microsoft Power Apps Cookbook, 2nd Edition

Eickhel Mendoza

ISBN: 978-1-80323-802-9

- Learn to integrate and test canvas apps
- Design model-driven solutions using various features of Microsoft Dataverse
- Automate business processes such as triggered events, status change notifications, and approval systems with Power Automate
- Implement RPA technologies with Power Automate
- Extend your platform using maps and mixed reality
- Implement AI Builder s intelligent capabilities in your solutions
- Extend your business applications capabilities using Power Apps Component Framework
- Create website experiences for users beyond the organization with Microsoft Power Pages

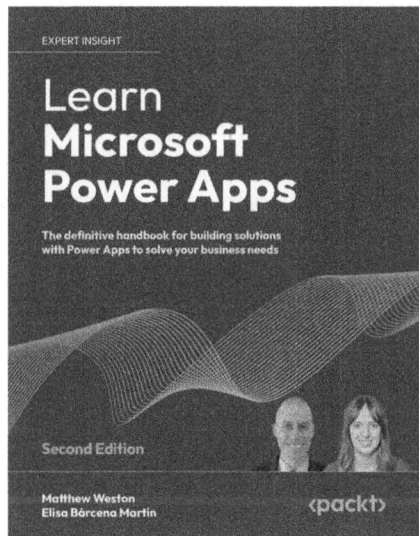

Learn Microsoft Power Apps

Matthew Weston, Elisa Bárcena Martín

ISBN: 978-1-80107-064-5

- Understand the Power Apps ecosystem and licensing
- Take your first steps building canvas apps
- Develop apps using intermediate techniques such as the barcode scanner and GPS controls
- Explore new connectors to integrate tools across the Power Platform
- Store data in Dataverse using model-driven apps
- Discover the best practices for building apps cleanly and effectively
- Use AI for app development with AI Builder and Copilot

Packt is searching for authors like you

If you're interested in becoming an author for Packt, please visit `authors.packtpub.com` and apply today. We have worked with thousands of developers and tech professionals, just like you, to help them share their insight with the global tech community. You can make a general application, apply for a specific hot topic that we are recruiting an author for, or submit your own idea.

Share Your Thoughts

Now you've finished *Power Apps Tips, Tricks, and Best Practices*, we'd love to hear your thoughts! Scan the QR code below to go straight to the Amazon review page for this book and share your feedback or leave a review on the site that you purchased it from.

`https://packt.link/r/1-835-08007-3`

Your review is important to us and the tech community and will help us make sure we're delivering excellent quality content.

Download a free PDF copy of this book

Thanks for purchasing this book!

Do you like to read on the go but are unable to carry your print books everywhere?

Is your eBook purchase not compatible with the device of your choice?

Don't worry, now with every Packt book you get a DRM-free PDF version of that book at no cost.

Read anywhere, any place, on any device. Search, copy, and paste code from your favorite technical books directly into your application.

The perks don't stop there, you can get exclusive access to discounts, newsletters, and great free content in your inbox daily

Follow these simple steps to get the benefits:

1. Scan the QR code or visit the link below

https://packt.link/free-ebook/B21327

2. Submit your proof of purchase
3. That's it! We'll send your free PDF and other benefits to your email directly

www.ingramcontent.com/pod-product-compliance
Lightning Source LLC
Chambersburg PA
CBHW081038220326
41598CB00038B/6914